Prowling an Uncharted Desert

by Stephen Jerik

I dedicate this book,
as all things in my life,
to the Great Creator.

Jonathan shouldered his backpack and headed down the unblemished dirt track in the direction of Armijo Canyon. His face and hands were deeply tanned despite the fact that he was rarely without his wide-brimmed hat and leather gloves, especially during the middle part of the day. His long pants and long-sleeved shirt protected his arms and legs from the sun, as well as from the stiff branches and sharp thorns that were everywhere. His meager beard had once matched the sand and dirt of the desert, but lately it had acquired the frost of advancing years. His gear and clothing were all various shades of brown and green causing him to blend in quite nicely with the landscape, much in the manner of snakes and lizards and coyotes. He had been walking the backcountry for a long time now, ever since he first arrived in New Mexico some 34 years ago. At times he wandered about aimlessly, but often he searched for signs of the past: abandoned mines, old camps scattered with the rusted carcasses of vintage automobiles and rare wood stoves, traces of the Spanish roads left behind centuries ago by the **carretas** of gold prospectors, the crumbling ruins of rock houses once inhabited by primitive sheep herders. Today he was searching for something different, something he'd never thought of trying to find before. Today he was searching for Euclidian geometry. Jonathan had read a lot of books in his lifetime, and recently he'd read a book written by a physicist who lamented the quandary in which quantum mechanics now found itself, but took great comfort in the fact that Euclidean geometry was still a solid part of his world. If that were true, Jonathan thought, then he should be able find it, because he was good at finding things, especially out in the desert.

Jonathan surveyed the ragged outline of the distant mountains, now colored by the early morning sunrise, and alternately watched his shoes scuff the loose gravel as he strode across the rolling hills, walking at a good clip, energized by the cold morning air and several cups of hot coffee. His stalwart hiking companion was Emma. Emma was young and strong and beautiful. Her well-developed muscles were sharply defined, much as those of an accomplished body-builder, giving her a voluptuous figure. Jonathan was in love with her, perpetually gazing at her in adoration. Emma's boundless energy, her amazing agility and gracefulness, inspired him to push onward, following her as she ran across the softly undulating hills, playfully dodging the sage and leaping over the dense stands of grasses.

Emma had been paying no attention to Jonathan all morning, but suddenly she stopped and stared at him. The intensity in her brown eyes was captivating. She had a broad smile on her face and he could see that she was very happy. He nodded his head in approval. Emma immediately swiveled and ran at full speed down the road until she was just a small figure off in the distance. She turned around and looked at him once again, but this time she was too far away to signal anything. It didn't matter. Emma knew that he approved. Her wild and enthusiastic behavior was one of the main reasons they were both out here, to give her an opportunity to exercise her muscles and maintain her fitness. Jonathan once jogged up and down these backroads himself, long before Emma was born, but now that he'd turned 65, he limited himself to walking. Climbing up and down the hills all day with a pack full of water and gear was plenty of exercise for a man of his age.

Jonathan turned his attention away from Emma and focused on the surrounding landscape, leaving Emma to her own recognizance. What he enjoyed most about this open country was the fact that he could see for a long way in every direction, and wherever he looked, he found many interesting geological formations waiting for him to inspect them. Canyon walls meticulously sculptured by the forces of erosion, hills and ridges speckled with endless creosote and sagebrush, the magnificent mountains always dominating the horizon. Up ahead the ridges became networked into complex patterns and interlaced with deeply cut arroyos, offering him numerous routes to follow. He could identify some of the rocks that he unearthed along the way, but he had dropped out of geology class in college and his knowledge of rocks was rather poor. He was much better at identifying plants, yet in all this countless multitude of objects, both living and inanimate, there wasn't a single geometrical figure of any kind. If Euclidean geometry was a genuine part of his world, as the physicist had clearly claimed in his book, then why couldn't he find it out here?

Jonathan looked down and examined the rubble surrounding his feet, studying the shapes of the fractured and highly irregular rocks, but he saw no triangles or squares—not even crude approximations of them. He could easily draw these figures on a piece of paper, but he could not find them anywhere in the desert. Most people saw the desert as being empty and barren, so perhaps that was the problem.

Emma knew better. She knew that the desert wasn't barren at all. The whole place was full of small creatures, all of whom lived underground for protection from the extremes of temperature. Emma could smell them and stopped regularly to excavate promising burrows. She'd learned from experience just how difficult it was to actually meet the sources of these olfactory sensations, but as Jonathan watched her exploring the ground, her head suddenly disappeared from view as she thrust it into the opening of a newly discovered tunnel. He knew that she wasn't going to find the animal that was hiding down there—even if there was one—and it was equally clear that she wasn't going to find any geometry either.

The sun was rapidly climbing in the sky and even though the air temperature was still rather cool, the rays of the sun were getting quite warm. Emma had a short coat and in spite of the fact that it was nearly white, she didn't tolerate the sun very well. Jonathan had been disappointed when he first discovered this because he had selected her primarily for her color. She had been the only white pup in a litter of nine—five females and four males —all the others decidedly yellow. She had also been the smallest pup in the litter, which was good, because labs could get quite large and he had wanted a somewhat smaller dog. He'd since learned that the smallest puppy did not grow up to be the smallest adult. Emma had turned into a substantial beast, brawny and barrel chested, and she had developed a personality to match her physique. Emma loved to pick on Jonathan and play games with him. She knew that when he was closing up the truck and getting ready to hit the trail, he would always have his leather gloves in his back pocket. She had spotted the gloves again this morning and snuck up behind him, clamping down on the exposed corner of one of the gloves. Jonathan protested as she knew he would, but she didn't care. She yanked the glove out of his pocket and pranced about with it in her mouth, bucking like a little pony, wildly excited by the acquisition of this highly cherished prize. Jonathan just stood there and watched her with a smile on his face. Emma turned and faced Jonathan so that she could taunt him. Keeping her eyes fixed on him, she violently shook the glove, the loose end flapping loudly against the sides of her face, then galloped around battering the limp glove mercilessly in furious eruptions of zeal, stopping frequently to focus all of her attention on the task. Finished with her ostentatious display of prowess, she disdainfully cast the glove aside, flinging it into the tall grass alongside the trail, and happily ran away. Jonathan grabbed his pack and locked up the truck, then went over to look for his glove. Being dirty and worn, now a dingy and faded shade of yellow, it blended in well with the desert. He finally spotted the crumpled, lifeless piece of leather lying pitifully next to a bush and stooped to retrieve it as he headed out in pursuit of Emma.

Jonathan finally caught sight of Emma wandering between the bushes and noticed that she was panting heavily. He dropped his gloves to the ground, about a foot apart, bent his legs and dropped his knees on top of the gloves. The rough gravel combined with Emma's frequent waterings caused the knees of his pants to wear out prematurely, long before the other parts had gotten much use. Jonathan sometimes sewed patches over the knees after the material had begun to disintegrate, but this was a laborious and time-consuming practice and he sought to minimize the effort and delay the chore as long as possible. The gloves helped, but they did not completely solve the problem. Jonathan also darned his socks. He found that this doubled or tripled their lifespans. He also kept a stockpile of insoles handy for his shoes. Whenever his feet started hurting on the trail, he was always surprised to find that the insoles had worn completely through.

Jonathan always purchased low-cut, leather hiking shoes which allowed his ankles a full range of motion. He had tried to wear boots because the clerks at the shoe stores always insisted that he needed the ankle support, but the restrictions on his ankles only put more stress on his knees and hips and these joints invariably gave out on him. Boots also felt like shackles and they made him walk around stiffly like a cow. Boots were heavy and cumbersome and he felt stupid lugging around these dead weights all day. He wanted to be quick and nimble on his feet. Apparently a lot of modern people had unusually weak ankles, because no matter how many miles he walked, his ankles never got sore. Native Americans had never worn boots and Jonathan didn't recall ever reading about them having chronic ankle problems.

Emma spotted Jonathan kneeling in front of his pack. She'd been focusing on hunting and tracking and hadn't been thinking about water, but when she saw Jonathan pulling items out of his pack, she realized that she was thirsty. She ran over to him and laid down next to him as he filled her dish, then hastily lapped up the refreshing water. As soon as she had guzzled down the usual amount, she leapt to her feet and ran away, leaving Jonathan kneeling in front of his pack. He poured the leftover water back into her slobber bottle, secured the lid, and packed everything up.

He needed to track down some shade so that Emma could cool off. He also needed to find some geometry. Perhaps geometry was out here somewhere, but its presence was not so obvious. Maybe he just needed to use his imagination a bit more. He glanced up at the sun and thought about the triangle it formed with him and the horizon. An aphorism among physicists said that "if you can measure something, then it must be real." If he could measure the altitude of the sun, then he'd know that it was real and he'd have located the mathematics he was looking for.

Jonathan found the perfect place to relax. He sat down on a massive block of limestone hidden beneath the wide umbrella of an alligator juniper and rested for a bit. Emma saw him and came over to similarly take a break. She laid down beside him and took full advantage of the coolness of the stone. Of course, he couldn't just settle on the number. He needed to first construct an instrument, a mathematical object that he would then use to determine the angle. He was going to need a sight stick, a plumb line, and a protractor. He left his pack in the shade and set out looking for materials. There were all sorts of things laying around in the desert, but there were no protractors. A protractor was, by its very nature, the embodiment of mathematical ideas. It had to be constructed using geometry as a guide and that was why no matter how long and hard he searched, he would not find one. There were no protractors out in the desert because nature couldn't use the theorems of mathematics to make things. He could make a protractor only because he understood the ideas behind it. He had been taught these ideas many years ago while he was a student in grade school.

Making an accurate protractor out of the crude materials available to him in the desert was not an easy task, all alone in the middle of nowhere, save for his dog who was of no help to him in this endeavor. Other instruments would facilitate the construction, perhaps a ruler and a compass. He tied two twigs together at one end with a blade of grass and made a compass, but the ruler was a particular problem because nothing in the desert was perfectly straight. Jonathan finally located a flat, smooth rock and scratched a line across it. He used his crude compass to determine another line that was at a right angle to the first. He was still a long way from having a protractor. Dividing a quadrant into ninety equal arcs seemed an impossible task under the circumstances. He found it strange that he had so quickly and unexpectedly arrived at an impasse. A mathematical object, whether it be a protractor or an imaginary triangle in the sky, could only be made using other mathematical objects, that is, objects that were deliberately designed and fabricated in accordance with mathematical principles. Thus he couldn't create the mathematics he was seeking without first having mathematical objects in his possession, and he couldn't get his hands on any of these because nothing out in the desert was mathematical in nature. He was stuck in a quandary. He had no way to get started on his assignment.

The fact that mathematical objects were required to generate other mathematical objects demonstrated that mathematics was self-contained and therefore self-referential. Everyone had to start with mathematics and finish with mathematics since it was not based on anything else. But mathematics wasn't just independent and self-sufficient, it was closed. No one could enter it from the outside, and once established, no one could exit from it, that is, move on to something else. Mathematics was not only a world unto itself, but a world set apart. Mathematics simply had to be assumed, taken for granted, imposed without justification or cause. Nothing other than mathematics led to mathematics and mathematics could only lead to more mathematics. Mathematics was a trap from which there was no escape. Once people had taken the grand leap and fallen into the lair of mathematical thinking, there was no hope for them and they would never find anything other than more mathematics. However, anyone who stood their ground and resisted the temptation to begin imagining mathematical forms everywhere would remain free and unbiased, leaving open the possibility of at least seeing things as they truly were.

Jonathan realized that there were other problems with his project as well. If he happened to be out at sea, the horizon would be well-defined and a suitable reference point for making this measurement, but here in the desert, the horizon was masked by the uneven terrain and obscured by the distant mountains. If he dropped a plumb line, however, the string would be perpendicular to the line between him and the horizon. The right angle was one of the key concepts in Euclidean geometry.

Beyond these technical difficulties lay a much greater issue. Jonathan's inability to produce a protractor had merely shown that something like the altitude of an object in the sky was a very queer idea, an idea that would never have occurred to anyone under any circumstances—that is, unless they had already been taught geometry in school. Jonathan surveyed the various places he was considering. The sun was very far away on the other side of the sky, the horizon was a long way off, several days' hike at least, and he was currently standing on a ledge, his further path blocked by a large arroyo. He never would have guessed that these diverse places were all somehow connected. A normal person might wonder: what could they possibly have in common? The answer was that they were all vertices in a polygon with invisible edges. Despite the mathematician's protestations and claims to the contrary, the answer was not obvious. The altitude was part of a system of highly abstract ideas, a way of thinking and not an aspect of the world. People had to be expressly taught to formulate these ideas in their heads simply because they were so bizarre, so unnatural, so devious. Still in the end everyone had to admit that there was no triangle in the sky and the vain imagination was just make-believe. Jonathan glanced up briefly at the sun. The day was getting warm. He took a step back and turned around, deciding to go retrieve his pack.

Once Jonathan started measuring angles by imagining nonexistent lines, there was no end to what he could do. Angles were everywhere because he could put lines wherever he wanted them to be. He could sight lines to clouds or prominent landmarks and calculate the angles between them, but what was the point of doing this? The altitude of the sun was different because the motion of the earth gave the sun a somewhat regular motion in the sky and this gave the altitude a special meaning. The sun could now be treated as if it were a mathematical object and this allowed the angle to be placed within a system of other mathematical ideas. These relationships could enable further calculations and permit predictions to be made. Planetary motions were key to the origins of modern physics because these seemingly insignificant events provided ample material for mathematical formulations. In all other respects, the measurements of celestial positions were as pointless as the measurements of cloud positions or the locations of landmarks. Physicists elevated what were in truth minor phenomena to the highest possible status simply because they facilitated their program of mathematizing the universe.

Jonathan could see that if someone wanted mathematics in the desert then they would have to create it themselves, because the mathematics wasn't out here. One way this had been done in New Mexico was with the U.S. Geographical Survey. Many years ago steel posts had been driven into the ground to serve as benchmarks for making determinations regarding all locations: circular metal discs engraved with range and township numbers and little pictures showing the corners of these imaginary squares. These posts represented the fictional points of a nonexistent grid. Rock cairns were often assembled around the posts to make them more visible.

Jonathan found these posts in the craziest of places because they took no account of the surrounding terrain. They were not derived from the earth, but rather came from another place altogether, the realm of mathematical ideas. A surveyor couldn't just put them wherever it might be natural to put them, say at the base of a hill or along the rim of an arroyo, but instead he had to put them halfway up an impossible incline or at the summit of an unscalable pile of rocks, because their existence was dictated by something that was not of this world.

Jonathan sometimes tried to picture to himself the hardy individuals who had been engaged in this monumental task, people who had trekked countless miles across rugged country leading pack horses loaded down with survey equipment. The Herculean effort that was required to populate a vast wilderness with such a large army of miniature sentinels was truly amazing. After many months of intense labor, Jonathan imagined one of the guys saying to the other: "Hey, Joe! Come on, I'm really tired. We've been working at this for a long time now. Let's just put the post here and call it a day. We're out in the middle of nowhere. No one is ever going to find this post, and even if they did, how would they ever know if it was in the right place or not? These posts don't relate to anything around here."

The posts referred to a way of thinking rather than the world, and only a select group of people ever engaged in such thinking. They understood the mathematics and knew that this was what gave the locations significance, meanings that were not at all self-evident and which could only be unveiled through a highly unusual perspective and a complicated series of measurements and calculations. A casual observer would first have to be indoctrinated into a set of beliefs and adopt a peculiar way of looking at things, and only then could this person picture the grid in his mind and thus make a ruling on the legitimacy of this particular post.

The survey crew had taken something that didn't exist and they had made it real. By virtue of these steel posts, numbers were now assigned to every location in the desert. Anyone could walk into the County Clerk's office and hand them a scrap of paper with the appropriate numbers written on it and the employees stationed behind the counter could identify the parcel in their plat books. Thanks to the efforts not only of the surveyors but of a larger society of like-mined individuals, the mathematics had become institutionalized, much in the same manner as the institutionalization of religious doctrines through the building of churches. The kingdom of God was invisible, but by erecting magnificent structures, true believers had made it not only tangible, but beautiful as well. The architectural masterpieces were not built for the benefit of the skeptics, who weren't convinced by such things anyway, but for the faithful, who demanded some sort of palpable support for their beliefs. The survey grid was also a temple to an invisible kingdom, but unfortunately the people who engaged in these types of projects typically forgot one thing: the building of churches did not prove the existence of God, and the fabrication of mathematical objects did not prove the existence of a mathematical universe.

As strange as these survey markers were, Jonathan found fence lines to be even greater anomalies. These structures were entirely out of place—perfectly straight lines in a land where nothing was straight, irrationally plunging into the most precipitous ravines and reaching the pinnacles of the most formidable ridges, often scaling dramatic heights at the oddest places. The fences could have been placed in harmony with the existing landscape and still have demarcated ranges with predetermined areas, but instead they slashed across the desert in laughably unswerving formations, imposing geometry on a chaotic region where there had been no geometry before. The insistence of building fences in straight lines seemed absolutely ludicrous. The mathematics of straight lines looked ridiculous out in the desert and didn't fit in with the natural flow of things. The strange forms bulldozed across the landscape and forced strange concepts on an unwilling recipient. The results weren't pretty. The jarring and inappropriate constructions destroyed the inherent beauty of the earth and disrupted the continuity of geological features which had a logic of their own.

Emma suddenly appeared out of nowhere and ran directly in front of Jonathan, causing him to stumble and miss a step. She had been chasing geckoes all morning and had now lathered herself up into a frenzy. He stood for a moment and watched her at work. She darted and zigzagged, kicking up dust, mimicking the behavior of her prey as if that would increase her chances of capturing it. The gecko plunged into the safety of a group of bushes and disappeared from view. Emma pulled up short of the barrier and leapt straight up into the air, pausing briefly several feet above the ground, frantically scanning the area behind the bushes. She reminded Jonathan of a professional basketball player making a jump shot, rising above the outstretched arms of a defender trying to make a play, defying gravity by remaining suspended in midair for an impossible length of time. Her legs hung uselessly in front of her as she pivoted her head back and forth looking for signs of her elusive target. Jonathan searched the ground along with her. They both realized that the small lizard had somehow escaped. Gravity finally took over and Emma landed gracefully back on the earth, becoming a dog once more. She hit the ground running and immediately resumed the hunt for the others.

Emma never knew where the next gecko would come from so she concentrated all her attention on the desert floor. The ground became her entire world. For reasons that were understood only by Emma, she suddenly looked up and saw a group of antelope standing on a low rise up ahead, their faces all turned toward her, their large eyes watching her intently. She immediately ran toward them forcing the antelope to make a dash through the sage and creosote. She gained ground on the antelope but did not chase them. Instead she turned and raced alongside of them, keeping pace only a short distance away. From this position she gradually merged into their midst and happily joined the group, sharing in the fun. The antelope were graceful and from Emma's point of view they were capable of impressively long strides. They were so inspiring that she began to emulate their actions, leaping wildly across the open desert in exaggerated arcs, vaulting over the shrubbery as if she too were an antelope. The antelope did not seem to mind. They were not threatened by the fact that Emma had joined them and payed no attention to her. None of them suddenly changed direction or attempted to evade Emma's proximity. They all ran together as one, reveling in nature as kindred spirits, wild animals endowed with four legs and the innate urge to bound across the landscape.

Emma began to tire and started to lose ground to the antelope. They continued on without her, steadily climbing the extended slope of a low hill. Suddenly realizing that they had lost their companion, they stopped and waited for her to catch up, but Emma had already turned around and abandoned them. She spotted Jonathan standing in the sage and made a beeline for his position. He noted the excitement in her eyes as she approached. Today was a very special day. Today she had gotten to run with the antelope. Emma galloped in a wide circle showing off her new skills for Jonathan, then raced up and down still jumping in the manner of an antelope. Eventually the excitement wore off and she returned her attention back to hunting geckoes, but now she had an additional spring in her step, something she had learned from her brief encounter with a species of animals that could run with the wind.

Jonathan likewise returned to what he'd been doing, thinking about mathematics and the nature of the universe. When Jonathan was still a young man and very enthusiastic about acquiring new skills, he'd learned to silkscreen his own t-shirts. He would lay a sheet of clear acetate over a photograph or a painting—or sometimes a decorative ceramic tile or a dinner plate—and trace out a pattern, choosing which features of the original to copy and which to neglect, sometimes altering certain aspects to achieve certain desired effects, but in general remaining faithful to the original image. He would then paint a screen in a darkened room with a light-sensitive coating, lay the acetate sheet over it, and turn on a light for a specified length of time. The coating hardened where it had been exposed to the light, but it could easily be washed away in the places where the light hadn't reached it, behind the lines he had drawn on the acetate. This process resulted in a stencil which was a good facsimile of the pattern he had drawn on the acetate. Back then Jonathan thought that this was what mathematics did, copy the

9

patterns of the world and reproduce them in another medium, but since then he'd learned that this wasn't the case at all. The mathematics of physics created patterns of its own, patterns not suggested by the world in any way.

Jonathan paused and tried to recall what he remembered about geometry. He asked himself: "What was geometry's most basic tenet?" The answer came instinctively: every pair of points determined a line. He could start there. Perhaps he could find an example of this out in the desert. Jonathan searched for something he could consider to be points. The leaves of the creosote bushes were quite small so he could use them. He selected a leaf on a bush nearby and another leaf a few hundred yards away. There were so many leaves that it was hard to keep track of them. He would need to paint the two leaves white with a fine point artist's brush and then mark the bushes with vertical poles stuck in the ground. Geometry said that there was precisely one straight line passing through both of these leaves. Jonathan walked along in the direction of the second bush, first mounting a small hill and then dropping down into a shallow arroyo, and thought to himself: how could he possibly locate this line? How could he ever get his hands on it? What could he do to identify it and make this line come to life? He knew the line must exist, therefore it must be here somewhere, floating directly in front of his eyes, but he could not see it. Suspending a rod or stretching a string or wire that far was impractical. If he had a laser, he could hold the laser next to one leaf and point it at the other leaf and then he could use the beam of light to signify the line, but there was no way he could see the other leaf that far away. Strangely physicists felt no need to substantiate the line. They didn't find it necessary to corroborate its existence, nor did they see any reason to provide a realistic representation of it. The line was merely a convenient fiction. It was not supported by the universe in any way, but if physicists made a calculation based on the existence of this line and then came up with a correct result, that would make the line a reality for them. Jonathan found the reasoning confusing. Ideas were supposed to be based on realities and not the other way around.

Jonathan closed his eyes for a moment and saw himself sitting in class, an eager freshman at a state university wanting nothing more than to soak up the most arcane knowledge and learn everything there was to know about the universe. His physics professor started drawing straight lines and marking off distances on a blackboard, adding various angles and arcs. The professor persisted and the drawing became quite elaborate, the picture of something Jonathan wasn't familiar with. Jonathan patiently observed this artistic composition unfold, then finally blurted out an objection without bothering to raise his hand or get the attention of his professor: "Wait a minute! Where did you get all of these lines? Nature didn't provide us with any lines. What do you think you're doing? You need to erase these lines right now and get them out of here. I didn't come here to study the figments of your imagination. I came here to study the universe. We've barely just begun and you've already turned this whole affair into a circus show. I'm not going to stand for it. I want my money back right now!" The professor stopped, turned around, and smiled at him. Jonathan felt that maybe he should just leave. Instead he continued his protest: "So you've got nothing else—it's these lines or nothing?" The professor stared blankly at Jonathan and shrugged his shoulders. Jonathan replied to the gesture by saying "That's alright. I've got better things to do." Jonathan brusquely packed up his books and notepads and headed out the door, the other students casting him vacant looks. The real world was waiting for him outside. He took a deep breath and looked around. He didn't see any lines, any complex linear configurations, any fanciful patterns. What in the world did the professor think he was doing? What was he talking about? His whole presentation was just a dream. The strange part was that everyone in the class believed him. They were all fine with the idea that the entire world was nothing more than a bunch of crazy geometrical shapes, but as far as Jonathan was concerned, the professor needed to prove that. The deduction was anything but obvious. A realization then came to Jonathan. The physicist made his entire living by seeing things that weren't there. He understood for the first time that the physicist was a visionary, a person who would not rest until everyone saw the same personal vision that he saw. The physicist turned every situation into a visualization until the entire universe was cast in mathematical imagery and everyone was bewildered by this surrealistic fantasy. When Einstein had said that "Imagination was more important than knowledge," he was referring to the notion that imagination was the basis for all scientific theories and thus it was the key to the further development of physics, but Jonathan wondered if Einstein ever realized that there was a more important issue at stake here. The fact that knowledge was based on imagination was a serious problem.

Jonathan pictured the other students still sitting in class and he began to regret his actions. He found himself standing at the entrance to the library. He hadn't planned on going to the library that day, but he'd followed the path out of pure habit. He paused briefly before turning around. He couldn't really blame the professor for what he had done earlier in class. Since physicists couldn't possibly see the ultimate reality of the universe, they were going to have to employ some sort of imagery. They were going to have to visualize this reality in their minds. In fact everyone was going to have to do this, but Jonathan recognized that there were many ways to go about it. Instead of envisioning mathematical forms, Jonathan wanted to imagine something else. He obviously didn't live in a mathematical universe so he needed to explore other options, find another view that fit his experiences, generate a perspective that more closely matched his observations.

Jonathan understood the impulse to draw make-believe pictures since it had been a long-standing tradition among cosmologists and metaphysical thinkers to do this, but he wondered: if someone truly wanted to envision the ultimate reality of the universe, why would they ever choose mathematics of all things? Where in the world did anyone ever get this insane idea? The professor was essentially telling his students that mathematics was the foundation of everything. When Jonathan examined the world around him, he could find no evidence to support this claim, think of no argument that would lead to this conclusion, see no indication that natural phenomena were in any way mathematical. Nothing seemed to suggest this view, but if the secret mathematics gave no outward signs of its presence and left no discernible clues, then how could physicists simply proceed with it anyway?

Jonathan actually knew why physicists were so comfortable drawing lines freely all over the place. They'd spent their entire lives watching mathematicians do this so they figured that it must be okay. Since they were going to apply the same mathematical thinking to the world, they simply took their cues from mathematicians, but mathematicians were not bound to anything other than the possibilities and limitations of their ideas and they had no qualms about making everything up in their heads. Physicists didn't realize that in attempting to capture the essence of physical reality they were facing an entirely different problem and they could no longer act like mathematicians, freely calculating designs that did not exist in the world. That kind of behavior was totally inappropriate and it would never lead them to the truth. Mathematicians had only mathematical truths, and they were nothing at all like the substantial truths of the real world.

Geometry had never been a necessary assumption. The world had gotten along fine without it. Jonathan wanted to know what was absolutely essential to the world, what was central to its character. He could easily throw geometry out the window because geometry didn't matter at all. Geometry didn't mean anything to nature. Jonathan asked himself: "Why did physicists turn all this around and make the least important thing in the whole world the most important thing in the whole world?" Jonathan knew right away that the answer to that question would shed light on the true nature of physics.

Long-time students of cultures and civilizations have typically characterized Western thinking as domineering. Westerners felt that the world was at their disposal and they could utilize its resources without questioning the wisdom of doing so or considering the consequences of their actions. Physicists were no different. They were free to enlist nature in their cause and take what they needed from their surroundings. They were at liberty to manipulate the objects of nature in any way they saw fit. When they looked around them, they didn't treat the world with reverence and respect, but with an eye for using these objects to their own advantage, as a means to an end. This attitude was so ingrained in the Western psyche that its validity was generally taken for granted and its premises remained unquestioned. Westerners didn't let the world guide them; instead they guided the world to a destination of their own choosing.

The physicist wanted his mathematical constructions to be real but he knew that the straight line he had drawn wasn't real because he had just made it up, so he transferred the existence of the leaves to the existence of the line. He was not imagining the leaves hanging on the branches and fluttering in the wind, but he was imagining the line between them, a structure suspended in the empty air, so he'd have the leaves make the line a reality for him, turn it into an object that was as real as the leaves themselves. In that way he could extend existence to cover the products of his imagination, enlarging the domain of "what was" far beyond its traditional limits. He could start drawing lines all over the place—axes and tangents, centerlines and boundaries, the lines that formed angles and established distances—and feel confident that he was still studying reality, although he was no longer thinking of the leaves as leaves, but as mathematical points. With this subtle move he was no longer in the world of nature, but inside the mind of the mathematician.

The physicist's view of the world then became mass confusion: entities masquerading as real began intermingling with entities that were in fact real, things that were nothing more than definitions became cloaked in the guise of physical objects, mirages started parading around as perceptions, strings of deductions acted as if they were chains of causality. After a while Jonathan found that it was hard to tell who was who, reminding him of the sci-fi arc where the hero had to discriminate between the aliens who had taken over the bodies of humans and the humans who had not yet succumbed to the invasion. He could no longer distinguish between the constructions and the realities because they all looked alike. They all got mixed together in a hodgepodge of disparate elements that then became the body of a scientific theory.

The line not only didn't exist, there was no sense in which the line was real. The line didn't correspond to anything. The line couldn't correspond to the leaves because the existence of the leaves did not imply the existence of the line. Leaves could never produce lines, so where did the line come from? If only there had been a rational explanation for the appearance of the line Jonathan would have been fine with it, yet the physicist, without evidence, shamelessly defying all logic and reason, put the line there then acted as if this were the most natural thing in the whole world. The line was obvious, not only the correct way to analyze and understand the objects and events of nature, but a perfectly normal extension of what existed. He felt this way only because geometry had been drilled into his brain from a very young age. The physicist was now so familiar with these constructions that they flowed effortlessly out of his head to the point where they seemed to already be out in the world as integral parts of nature and basic facts of life, but in truth the axioms of geometry were anything but self-evident. The term 'self-evident' referred to the interplay of ideas within geometry and not to the harsh realities of physical existence. The comparisons of geometrical forms to the elusive irregularities of an indeterminate world were highly suspect.

Leaves were not put on creosote bushes so that lines could exist. The leaves were there for other reasons. The leaves enabled the plant to exist and the plant enabled the leaves to exist. The physicist employed leaves in his service by investing them with a purpose for which they were never intended. He was doing something totally unnatural with them. A natural order did exist in the world, a harmony in nature, and the leaves had a place in it, a role to play, but creating lines was not one of the functions of a leaf.

Whenever Jonathan read a textbook on physics, he got the impression that mathematical relationships were supposed to reveal the existence of the underlying connections in the world. If two leaves could be connected by a straight line, then they were somehow connected in a real way, that is, by something other than the line. The idea that when Jonathan was thinking about mathematical relationships he was picturing in his mind the way things really were was simply not true. Physicists saw the concepts as accurate portrayals of independent truths and therefore felt that their thoughts captured the realities of the world, but beyond the line what else was there? The geometrical construction was based on what? The leaves? The line was based on the idea that leaves were not leaves, but rather impossible entities in a crazy dream.

Jonathan's professors had led him to believe that the axioms of geometry were formulated specifically to capture the crucial aspects of space. Mathematicians, being clever people, looked at the world around them and summarized its essential features, then expressed these truths in the concise language of symbols. With the theorems of geometry, mathematicians had put spatial relationships in a nutshell. Geometry was everything anyone ever needed to know in order to have a complete understanding of space. His teachers then launched into long-winded speeches talking about things that didn't exist. They needed to get out a pencil and paper in order to show Jonathan what they meant because they were fabricating designs in their heads, copying down their own thoughts on paper. This wasn't the real world. All anyone had to do was look around them. Geometry was about fitting ideas together, seeing what mathematicians could do with definitions and concepts, playing with them as children played with toys. Physicists then fashioned corresponding shapes out of metal and glass and plastic and measured the mathematical relationships of these material objects, then said that this proved the validity of the ideas, but in fact these objects were simply the ideas themselves now cast in the tangible bodies and concrete substrates of man-made artifacts. The physical world could take on all kinds of shapes and forms—including mathematical ones—but what did that prove? Nothing, because in a similar fashion Jonathan could construct all sorts of non-mathematical forms and then by using the same logic he could show everyone that the world wasn't mathematical at all.

Still Jonathan could say that the straight line he had imagined in his head was comparable to a stretched wire, and therefore the reality of the line must be comparable to the reality of the wire. What was the problem with this reasoning? First of all, if there actually were a wire stretched between the bushes, Jonathan would have to deal with it, that is, acknowledge its existence by stepping around it. The reality of a stretched wire was very different from the reality of no wire at all. Second, in a simple case such as this, Jonathan could consider the line as if it were real and proceed on his way, but then as he fabricated more complex mathematical constructions, he carried this error forward and because he had granted the line a reality it didn't possess, he could now say that everything that followed was also real and a minor discrepancy turned into a major blunder in thinking. Instead of following a strict course, he had taken a trail that diverged imperceptibly and for a long time the two trails paralleled each other. He could readily cross back and forth between them, but eventually the two courses struck out in different directions and the second trail led him to a completely different place, cut off from his original destination by an impassable barrier. The line was never real in the first place, even after he had put it there in his mind, and everything that followed shared that characteristic.

The hardest thing for the physicist to accept was that the straight line distance between two material bodies wasn't a reality of the universe. The line was a definition in geometry and a definition in geometry could never be a reality of the universe. Definitions were the products of peculiar ways of thinking and this made the straight line a mental deduction. The line was not observable and the universe provided no evidence for it. Since the universe wasn't thinking in terms of straight lines, these absurd fictions were totally irrelevant. The universe wasn't contemplating distances when it created phenomena, which was to say that the universe wasn't built around these ideas. Physicists could still go ahead and draw straight lines, but they were all meaningless, essentially pictures of nothing. The distances they created were easily calculated and they became the basis for all sorts of further geometrical constructions, allowing physicists to delude themselves into thinking that the truths of these mathematical relationships were also the truths of the universe. The failure of the straight line to represent any kind of reality was obvious whenever huge numbers of material bodies were gathered together and placed in intricate relationships such as in the desert landscape, but out in space nothing stood in the way of these whimsical fantasies, nothing directly contradicted them, so they seemed perfectly fine.

Jonathan stopped himself at this point. What right did he have to talk about the "reality" of the world? Someone could easily come along and challenge his use of the word. Some people might say that he couldn't talk about "reality" because no one could agree on what that was and it might just be an illusion, but for Jonathan, reality was just another term for brute existence. He was merely referring to whatever was out there without specifying what that was. All philosophical and scientific inquiries assumed that something was out there, even if they then decided to disregard it or label it as unknowable. If the world was, as the mystic claimed, merely a dream, Jonathan could easily disprove all the theories of physics by simply asserting that the world didn't exist. He could sweep everything aside and his job was done. The mystic put consciousness first in a way that shamed even the mathematician. The mystic withdrew so far into his own consciousness and remained within his mind for such a long period of time that the physical world no longer seemed real. Such an extreme philosophical position was sufficiently far-fetched that Jonathan felt no need to address it, however, the mystic and the mathematician shared something important. Both believed in putting their own thoughts and their own consciousness first. The mystic did this to such an extent that the world no longer needed to be contemplated, while the mathematician only went so far as to say that the world was based entirely on

14

his own ideas. Plato was a stout champion of mathematical thinking and a bit of a mystic himself, perhaps showing the natural connection between the two.

Still the creosote leaves made the straight line seem like an indisputable fact. The clear-cut and unequivocal character of the definition combined with the fact that it was based on real objects made the line seem like an actuality. Jonathan pointed out that he could draw different lines, a sine wave for example, or a sawtooth wave or a square wave, between the two points. These obviously didn't exist or represent any kind of reality. They were just his own personal visions. He'd merely used mathematical concepts to define regular, uniform, linear patterns, but the problem was that the straight line was also just a definition. These other lines didn't relate to anything in the universe, so why was the straight line an exception, a definition that was more than a definition, a concept that had some kind of universal legitimacy? Jonathan could complicate his imaginary lines by modulating the amplitudes and frequencies of his waveforms, then start making the lines haphazard and irregular, forming jagged lightning bolts or smooth and flowing swirls, or perhaps an erratic combination of both. As he elaborated his ideas, these lines steadily became harder and harder to describe and at some point they ceased to be definable. This made them difficult for Jonathan to picture in his mind, but they still connected the two points, just as the straight line had done. Why couldn't he say that these lines were also the basic facts and true realities of the universe, in the same sense that physicists considered the straight line to be? Jonathan could find no reason to believe that the straight line was in any way privileged, that it had an intrinsic priority over these other lines. Did physicists see the straight line as being a reality simply because it was so easy to envision and thus they could readily believe that it was actually there? Or did their deep love for mathematics cause them to invest the line with a significance that it did not possess or warrant? Establishing straight line distances in geometry made perfect sense, but out in the real world, determining the length of a nonexistent line segment seemed rather pointless. Physicists merely assumed that since the straight line was so vital to geometry and so important to the further development of higher mathematics that it must be equally crucial for understanding the physical universe, a strange premise to say the least, given that mathematics and materiality were so unalike. The fact that no reality lay behind the straight line rendered it empty and hollow. A fictitious entity could never play a role in anything.

Jonathan had begun his quest for the truth by feeling the downy texture of a leaf between his fingertips and noting the tug of the woody branch that connected it to a haphazard assortment of other leaves—but that was where reality ended. No one could deny that he had started out with the objects of the material world, but then in a shot out of the blue, he introduced a purely imaginary concept and flipped over to pure reasoning, not fully understanding the impact of his actions, that in doing so he was leaving the real world behind and entering a world of free associations and made-up mental images.

The situation was analogous to a high-diver at the pool: the diver took two steps on the springboard of reality to build his momentum before he leapt into the air and careened into the pool of the imagination with a splash. The diver said: "This leaf is real and that leaf over there is real," and then, his voice trailing off as he flew into the emptiness, "The line between them is real." In the desert, everyone could see that geometrical constructions were absurd, yet they were so convincing in textbooks, coordinate systems attached to little drawings of objects similar to the flags flown on motor vehicles letting everyone know that the drivers were rebels, pirates, patriots or fans of some particular football team. But here in the desert, there were no coordinate systems, or any other types of geometrical constructions.

This might not seem important, but it upset the whole scientific method, which depended on the mathematics being derived from the world and not added to it. But even in geometry, two points did not give the mathematician a line. In fact, two points did not even suggest a line. The points merely created an opportunity for the mathematician to construct a line in a further independent step. The points made it possible for a line to be added over and above the points, but Jonathan could do other things with those points, put them on the circumference of a circle for example. He could also tie them together with non-mathematical concepts and talk about their harmony and friendship, or divide them apart and deny any connection at all.

Simply by using the same word "point" in each case, mathematicians were letting language make the connections for them. The term "two points" paved the way for the line to be introduced, easing them into that conclusion by expressing an underlying bond already assumed to exist between points. The fact that points were complete nonentities lacking the substance and character of real places allowed them to be connected freely without reservation. If instead they were bestowed with unique identities, connecting them might become a problem. A pebble in the arroyo and a wafer of bark on the nearby tree did not suggest any association and Jonathan wondered about putting a line between them, perhaps suspecting that it was inappropriate to do so, but in mathematics a connection between all points was already implied by the terminology. Thus drawing lines between disparate objects did not raise eyebrows and physicists could easily envision such connections. What people saw in their minds depended to a large degree on how they chose to talk about things. Mathematicians and physicists employed a quirky use of language to create an unusual and highly suspect vision of the world, very much at odds with how Jonathan spoke about the world and how he envisioned it. Points had a logic of their own, and it was not the logic of the natural world. If physicists were going to employ mathematical thinking everywhere, then they were going to have to justify such a move, because out here in the desert, mathematics was anything but obvious.

A common misunderstanding in mathematical thinking was that the axioms implied the ensuing mathematics. Axioms were only stepping stones, launching points, springboards. Later mathematical proofs carved out places for these axioms within a larger context of ideas, as parts of a greater whole, as essential pieces of a grand puzzle. The theorems were not contained within the axioms, but built on top of them. The question was always "What was the mathematician going to do with these axioms and how was he going to proceed?" But when the mathematician put the further ideas together in his head, he envisioned that this was the way the ideas actually fit together, so the ideas assembled themselves and the theorems automatically appeared as if nothing more had been added. The mathematician imagined that the axioms implied the theorems and therefore the theorems had been contained within the axioms all along. The further elaborations were necessary consequences of these seminal ideas, but he was assuming a way of thinking as also being self-evident, in addition to the axioms themselves. He could take mathematical thinking for granted because he considered such thinking to be universally true and not something that was open to debate. Given this assumption, the mathematician was left with just the axioms, and in that case the axioms necessitated everything that followed. A person who had not been previously indoctrinated into the standard methods of mathematical reasoning might not see where the axioms could possibly lead him, but the mathematician was convinced that once this person was shown the natural course of events and the indisputable veracity of the supporting logic, he would be forced to concur. At least, that was what happened in the Platonic dialogues. But if the constructions were all just imaginative nonsense, then the mathematician was stuck playing meaningless games and fiddling with logical relationships that had no basis in anything other than what he had imparted to them through pure flights of fancy.

Jonathan found himself standing on the crown of a ridge with a good view of entire area. He pushed the brim of his hat up with his gloved hand and raised his eyes. The broad expanse of creosote and sage was cut at regular intervals by spacious canyons. The immense collection of plants extended unabated to the horizon and beyond, continuing for many miles in every direction. The sheer number of leaves was unfathomable, yet every pair of them was connected by a straight line. He pictured an immense spider web of filaments entangling the vegetation, then realized that the leaves weren't connected by anything at all. Mathematical ideas connected things that weren't connected. That was because mathematical relationships weren't based on the realities of the world. The leaves were connected to each other only through small networks of crooked branches and that was it. By focusing on mathematical ideas, the physicist had become captivated by the vision of another world, a world very different from this world, and he had begun creating an extravagant work of fiction, a tale spun in a sense by the leaves themselves, but a fantastic story nevertheless.

Jonathan asked himself: "Why couldn't physics simply be a description of the universe?" Because in that case it became just an interesting story and as with all literary accounts, physics was only someone's point of view—or rather in this case the shared perspective of a small group of individuals. The written text of a mathematical description necessarily took its place alongside the varied styles of a diversity of authors. Mathematics might be a unique form of expression, an original and highly creative way of putting things, but it was nonetheless nothing more than another take on the world. If physics was merely descriptive, then physicists were only discussing the universe in the flowery prose of equations. All they had done was invent a clever way of forming sentences using a vocabulary that no one had ever used before. Jonathan knew that this was not what physicists wanted.

When an author created characters based on his past experiences with the people who had entered his life, he could only present the sides that they had revealed to him and certainly much about their temperaments and personalities remained hidden from view. The way an author portrayed these individuals in his writings might not reflect their true natures or accurately record the events surrounding their lives as they themselves perceived them. Physicists didn't see themselves doing anything like this. Physicists saw themselves as revealing the hidden truths of the universe, truths that were universal and objective, truths that had existed long before they uncovered them and brought them to light. A description, on the other hand, was only a summary of a person's observations, a personal rendition of events cloaking the subject matter in intimate anecdotes and idiosyncratic narratives. Physicists would only be telling stories and storytelling was what people did sitting around the campfire at night. Scientific theories would all be just talk, as was the case with every book. But according to physicists, physics transcended authorship and ultimately the material was written by none other than creation itself. Physicists attempted to factor themselves out of the equation in order to convince everyone that these were nature's words and not theirs. Their account of the universe was not their own personal version of the truth, but an absolute truth, something that could not be disputed or ignored by anyone.

When physicists said that objects and events were obeying mathematical laws, the mathematics was an integral part of the action. When Newton's Law served to explain what was observed in a laboratory, when mathematics became the reason why things happened, mathematics was not just someone's perspective. Physics wasn't just an interesting form of expression. The ideas had to be more than just words on a sheet of paper—they must actually exist in the world. Physics was reality and truth, the hidden structure of the universe, the foundation of all that existed, and as such, it ceased to be merely a description and became the actual mechanisms underlying the events. Physics might in fact be a tall tale, but that was certainly not what physicists were attempting to create.

Jonathan had recently read a curious story, the latest headline from the physics community plastered on the cover of a magazine: "Do we live in a holographic mirage from another dimension?" The statement referred to a dubious principle in string theory, but the idea that knowledge in physics was similar to a hologram was much more interesting. Constructing realities out of mathematical concepts generated images of things that weren't really there, yet physicists could still see them, but only as ghostly apparitions of the mind, unlike the substantial existences of physical objects.

Physicists saw nothing wrong with this. Their fascination with mere appearance was even carried over into the popular culture and occasionally portrayed in movies. In one such scenario, the science fiction hero fell in love with a hologram, a projection of light, an unreal phantasm that had been cast in the form of his highest desires, the beautiful girl of his dreams, the ideal companion possessing a perfectly compatible personality. This depiction of a romantic relationship was symptomatic of a deeper problem, the fact that physicists were hopelessly in love with illusions and fakery, the insubstantial forms of mathematics, the allure of mere appearances, as well as the promise of the unattainable. If they couldn't get their hands on reality at least they could make it appear as if they held something in their hands, even if it was only a paltry sketch of reality, a montage of bare forms and mere outlines, a mental image of something that could only be imagined. When they embraced their wonderful theories, they were putting their arms around a void, a mere representation of beauty, a body of propositions that didn't capture the essence of actual existence. They only wanted to be in love with their own creations because that was the highest form of love, the love of oneself. As long as they could make it appear that they had the truth in their possession, they didn't need to back it up with anything real, the genuine understanding that grasped the true nature of things.

Jonathan always seemed to run into the same old story, that "the scientist began with the world of experience and experiment." The slogan was practically universal in the scientific literature, an obligatory statement of creed, the standard starting point for every physicist who embarked on a discussion of the scientific method. But an experiment was already an elaborate mathematical construction and thoroughly impregnated with mathematical ideas. The devices and instrumentation that were required for an experiment, whether they were simply a ruler and a compass or the 14,000 ton behemoth of wires and silicon wafers located at the Large Hadron Collider in Switzerland, were designed and built with mathematical precision using the commonly agreed upon concepts of mathematics. This was what made the experience of the laboratory so different from ordinary experience. The physicist fabricated a new kind of experience by bringing mathematical ideas into the picture, a type of experience that had been completely unknown to indigenous people throughout the world prior to its invention by scientists, an experience involving a new kind of imagination—lines, angles, and numbers that weren't really there. The scientific method embraced a new way of thinking about things. Experimentation introduced everyone to a unique and unprecedented perspective on the world, conflicting with, and often incompatible with ordinary experience.

The afternoon wore on and Jonathan discovered that he needed to cross an extensive area of sand dunes in order to return to the truck. His feet sank into the soft sand and walking became slow and difficult, his path flanked on either side by yuccas and beargrass as he laboriously attempted to gain the crest of the nearest hill. Once on top, to his surprise, he found that the other side of the hill had been completely cut away by a large arroyo. A sagebrush was standing precariously on the edge of the shear cliff, the venerable plant about to be undermined by the rapidly disappearing substrate. He deposited his pack into its waiting arms and reclined on an adjoining pallet of flat compacted stones, peering down into the void left by the absent hillside. The spot overlooked a number of interlaced riverbeds threading their way toward a narrow band of trees, a green canopy several miles to the east that shrouded the blue waters of the Rio Grande. He knew that the waters were actually brownish-green in color, but from this perspective they appeared to be blue due to the reflection of the sky on the glassy surface. Jonathan removed his hat and gloves and carefully studied the panorama presented to him by his elevated vantage point, focusing on the details of the surrounding terrain, but what he saw were not points connected by lines. That vision was the product of a perverted mind. He had no reason to imagine the desert in that way.

He'd heard the story that scientists were just like everybody else, in that everyone observed the world around them, only scientists were much more adept at it, but from Jonathan's point of view, seeing geometrical figures everywhere was a disturbing hallucination rather than a heightened state of awareness. The unnatural urge to cast everything into geometrical shapes was an aberrant instinct not characteristic of a healthy mind. Why would anyone take the inherent beauty of this wild landscape and transform it into something so alien?

Many years ago Jonathan had pictured himself to be a physicist and a mathematician. Prior to that he had been a hopeless romantic, completely at odds with the cold rationality of abstract thought, but then in a roundabout way, he fell under the spell of logic and mathematics. During his high school years, he became captivated by a strange dream. He imagined that he was the chosen one, the special person who would do what no one else had been able to do: solve the mystery of the soul using mathematical principles. He decided that he would dedicate his life to this noble cause and sacrifice everything for an understanding of the meaning of consciousness, but first he needed to master the current state of knowledge, so he enlisted in the scientific community and studied physics. He embraced the perspectives of physics and strove to understand its methods. He did the work, but he never really had any intention of actually becoming a physicist. He saw himself instead as a romantic hero, an eccentric who labored tirelessly in his private study and custom-built workshop, a creative genius who didn't need anyone else. Of course, he never figured anything out and later abandoned the project, but he had never been able to completely dispel the physicist he had created within himself. The physicist became a permanent part of his psyche and as hard as he tried, he could not free himself from this alter ego. Although he had since adopted a very different way of looking at things, he continued to wrestle with this adversary. He talked to him and imagined that this devil's advocate talked back to him. Sometimes he carried on long conversations with this person as he walked alone with no one else around to hear his words, no one to challenge his statements, no one to question his beliefs, other than a ghost from the past who relentlessly haunted his consciousness.

Jonathan's thoughts gratefully flew away at this point and he became one with the desert. After relaxing silently in the warm sunshine for a while, he realized that there was no hope for his quest. Perhaps he might find geometry elsewhere, beyond the boundaries of this vast desert. He imagined himself traveling to other parts of the world and much as an archeologist unearthing ancient ruins in the jungle, hiking into some remote region and stumbling upon a small fragment of mathematics, a leftover skeletal outline of precise relationships neatly preserved in the hardened sediment, a tiny piece of a lost empire. The afternoon was getting late and he was tired of searching. One place where he was certain to find geometry was in the library, so he sat and wondered in which direction he would travel from this desolate location to reach the nearest library, and how many days it would take to get there. Emma looked at him and told him with her eyes that she was primed and ready for such a long and arduous journey, but he didn't think that he was, at the moment, equal to the task. Jonathan then realized his mistake. The author hadn't been talking about Jonathan's world. He'd been talking about his own world, the world of physics, and to a lesser extent the modern world, a prodigious domain teeming with endless tributes to Euclid's vision and ingenuity.

Jonathan collected his belongings and retraced his steps down the hill. Once again standing on the desert floor, he circumnavigated the hill and transected the series of dry washes he'd meticulously examined from above, then struck a diagonal up the opposite side of the broad river valley, mounting the ridge where he was treated to another grand vista. He maintained his elevation and stuck to the backbone of the ridge for a while, enjoying the spectacle before him, as the evening shadows lengthened, emphasizing the intricate folds and contours of the hills and valleys. During the middle part of the day the sun shone on every part of the desert equally causing the landscape to appear flat and washed out, but as the sun declined, the shadows brought everything into stark relief, highlighting each rock and outlining every detail of the surface. The interplay of light and darkness transformed the desert into a special place and suddenly Jonathan saw it for what it was. He dropped down into the shadows of the next arroyo and wound his way along the parched riverbed staring up at the channel sandstones, the passage being enclosed on both sides by these sculptured terraces. Emma was running wildly a hundred yards ahead of him. Twilight was her favorite time of the day because it offered her so many hunting opportunities, numerous chances to surprise nocturnal creatures as they awakened and ventured out of their dens and burrows.

The desert was completely hushed and languid until a coyote unleashed a series of hysterical cries nearby, the explosion of yips and screams drowning out the gentle and very pleasant sound of Jonathan's shoes rhythmically crunching the brittle grass and mashing the loose mixture of clay and gravel. Emma stood still and didn't make a sound. She really didn't have to because the coyote was making enough noise for the both of them. Emma moved in to investigate further and disappeared into the foliage ahead. The coyote continued its yapping and yelling. Jonathan watched the two emerge from the bushes, first the coyote running away, followed by Emma playfully running after it. He noticed that Emma was galloping in a strange way. She was moving forward by holding her legs stiffly and raising her front and back alternately, what a child's rocking horse would look like if it abruptly took off running across the landscape. Emma had no further interest in the coyote so she quickly abandoned the chase and playfully loped around in a circle, maintaining the seesaw motion of the rocking horse as she headed back toward the bottom of the arroyo. She looked like she was having great fun, but the coyote abruptly halted its flight and watched her from a low hill, then bolted toward her. Apparently the coyote had thoughts of confronting Emma and perhaps attacking her. Jonathan didn't hesitate. He swung into action shouting curses in a deep voice as he ran toward them, doing his best impression of the voice of God commanding all of creation to obey, showing the coyote in no uncertain terms that he was a formidable opponent and he was coming to join in Emma's defense.

The coyote was not intimidated by Jonathan's boisterous display and brazenly intensified its pursuit of Emma, vigorously plunging into the bushes at the same spot where Emma had just disappeared. As he reached the area, Jonathan's sustained tirade finally had an effect. The coyote reluctantly emerged from the bushes and slowly moved away, then hesitated for a moment, took a few steps back toward the bushes, turned around once again and retreated a bit further, then cautiously approached Jonathan, staring directly at him. The coyote appeared to be sizing him up, trying to assess the danger he posed to it. Jonathan reckoned that the animal was a female with a den of pups tucked into the bank of the arroyo and she was seeking to protect her litter from the invasion of the predators. Since the coyote was somewhat larger than Emma, it might have been the male fulfilling his paternal duties in place of the mother, but Jonathan wasn't sure if male coyotes actually participated in the defense of the den.

Emma had no interest in either stealing the pups or fighting the parents so she ran toward Jonathan and departed the scene the minute he called her. She was excited by the antics of the coyote and was glad that she had gotten the chance to meet it, but the encounter held nothing more to offer so she was willing to move on to the next adventure further down the trail.

Jonathan focused his attention on the last remaining stretch of the journey, anxious to get back to the truck. The sun had already set and the desert was rapidly growing dark. He whistled for Emma. She came bounding up onto the flat shelf of land where he was standing surveying the river bottom below, studying the course he'd just taken for the past mile or so. He picked up a stone and held it in his hand. Physicists said that there were more atoms in this one stone than there were stars in all the galaxies of the universe. They claimed that the stone was alive with activity, electrons spinning at fantastic speeds through elaborate, layered complexes of orbits, electrical forces attracting and repulsing these particles in intricate patterns. Jonathan tried to picture this in his mind, but he couldn't. He felt that he should be able to hear the incessant flurry or perceive some sort of hum, perhaps detect a faint buzz or feel a slight vibration, but there was nothing. The stone was inert, lifeless, a dead weight in his hand. He let it go and it fell to the ground with a dull thud. A mild breeze brushed against his face, caressing his hair. He tried to picture the immense swarms of molecules being swept along, the phenomenal number of high-speed collisions occurring right next to his head, but it was no use. The wind was an entity unto itself, a spirit, and he felt its invisible presence. The strange apparition quietly passed him by with a faint whisper. Scientific explanations were of no use to him in the desert and he wanted to be told different stories. All the machines that purported to see atoms were hundreds of miles away at universities and large institutions, basking under bright florescent lights in concrete rooms, expensive palaces erected expressly for their benefit. As darkness fell around him, the atomic world seemed very insignificant and far away.

The stars came out as he reached the truck. He hadn't noticed the vehicle until he was almost right on top of it and nearly walked into it, an indeterminate black shape lost in an endless expanse of indeterminate black shapes, a blackness that was no different from the overall blackness of the desert. Only a few traces of the sunset's afterglow reflecting here and there on the metallic surfaces finally gave it away. He circled around to the driver's side where the truck was now silhouetted against the last glimmers of the fading sunset. From this angle he could clearly distinguish the outlines of the side-view mirrors and the roof-top carriers.

Twenty-five years ago, long before they were married, Jonathan and Sarah had camped in the boot heel of New Mexico to begin an auto-tour along the border in Jonathan's 1968 Nova. He didn't know exactly what happened that first night—perhaps it was just a spectacularly clear night—but the stars came out in fantastic numbers and they shone with an unreal brilliance. He left their lonely camp on foot and slowly walked down the road in complete silence, lost in the maze of stars hovering above his head. He had never seen anything like it before. The stars were not etched onto the dome of the heavens as they usually were, but suspended inside of it, hanging down into the empty vault of utter blackness. He felt as if he were no longer on the earth, but out in space with them, floating in the void, completely surrounded by an impossible number of sparkling lights. He let go of his ordinary existence and succumbed to the dizzying spectacle. The earth was as black as the night sky and it had mysteriously disappeared. He saw only stars everywhere. He'd only caught a glimpse of this phenomenon on a few occasions since then. He had no idea what caused it.

Jonathan learned the constellations when he was in high school. Since then he had tried to point out the constellations to other people on numerous occasions, waving his arms about and tracing imaginary lines in the sky, but there were lots of stars up there and he never knew if the other people saw the same patterns that he saw. As everyone knew, the stars in a constellation were not related to one another. They did not dwell in close proximity to each other, something they appeared to do when viewed by someone standing on the earth. A constellation was not an object that actually existed somewhere out in space, yet he could easily identify the familiar pattern in the sky whenever he was out walking in the desert at night. The constellation had a configuration that could be specified through the exact measurements of distances and angles, an arrangement that could be laid out in unequivocal fashion using the universally sanctioned techniques of mathematics, yet the constellation was really just an illusion, a matter of perspective, an apparition caused by a handful of stars that were widely scattered throughout the universe. By employing a long-standing trick in physics, Jonathan could simply use the measurements to define an object and then give this object a name. A fictitious entity now had a life of its own and it could play a role in his further thoughts. He could make statements about the constellation Orion and perhaps tell everybody that Orion was his patron saint, a mighty personage who watched over him and guided him home at night. Orion became a part of Jonathan's world in spite of the fact that Orion ultimately traced back to nothing and essentially vanished whenever he tried to pin it down. Every object in physics, every arrangement of force and energy, every field and particle, was just an observable mathematical pattern, and perhaps shared Orion's fate of being merely an empty reference.

The idea that an ultimate reality lay hidden behind a veil of appearances was an ancient one. Jonathan could understand that the senses did not give anyone an accurate picture of this reality and that no one could see it with their eyes, but what made anyone believe they could think it with their minds. Why weren't people's thoughts considered to be as deceptive and unreliable as their senses? Part of the answer might be that the senses were well-defined and limited in scope, but thoughts were infinitely malleable and therefore capable of being linked together in inexhaustible ways. Somewhere in this boundless range of possibilities, a group of people called physicists believed they could discover those special ideas that constituted the real world. All they had to do was keep searching through the infinite combinations. Or perhaps the effort was as futile as searching for geometry in the desert.

— 2 —

Jonathan was still in grade school when his class performed an updated version of Galileo's famous experiment using an inclined plane. Along with the other students, he was given a small cart the size of a child's toy, little more than a wooden block with metal wheels attached to it. The wheels fit into a straight wooden track. He also received a buzzer which had a special inked point at the end of the armature. He hooked the coil of the buzzer up to a battery. The magnetic field pulled the armature down, breaking the circuit. A spring then pulled the armature back up, reconnecting the circuit. The endless cycle of making and breaking the circuit caused the buzzer to vibrate. He fastened a ribbon of paper onto the back of the cart, of the type commonly used back then in adding machines and cash registers. He fed the paper through the buzzer so the inked point would print a dot each time the coil pulled the armature down.

He positioned the cart at the top of the inclined track and let it go, the cart dragging the ribbon of paper behind it. He measured the speed of the cart by the spacing of the dots that were printed on the paper by the buzzer. As the cart accelerated, the spacing between the dots increased. Counting the number of dots per inch along the length of the ribbon gave Jonathan a crude idea of the cart's motion.

Everybody in the class got different results. The experiment was a complete disaster and Jonathan was heartily dismayed. Obviously, the demonstration hadn't worked at all. The teacher had kept the carts and tracks and buzzers crammed into a large cardboard box that sat in the back of a closet under a pile of other stuff, dragging them out only once a year for the benefit of each new group of students. Some of the wheels did not turn freely or were too loose and wobbly to fit snuggly in the tracks, and most of the buzzers were temperamental, sticking and intermittently changing the rate of vibration. The paper would get snagged and not feed cleanly through the small gap between the coil and the armature.

After witnessing the erratic behaviors of the carts, Jonathan knew that Galileo had been wrong and instead he wanted to write his own physics. Still the moral of the class had not been wasted on him, because he learned a valuable lesson that day, something he'd carried with him ever since. In his mind, only one part of the class had been missing: a summation. The teacher should've stood before the students at the end of the period and made a declaration: "I hope you were all paying close attention today. Today we learned the three great Laws of the Universe: (1) every location was different from all the others, (2) every moment was a unique experience, and (3) every action of any significance was utterly irreproducible. As hard as physicists tried to suppress these fundamental truths, the existential nature of the universe always reared its ugly head at the most inopportune moment. Looking at the results today, I'd say that the experiment was a complete success."

23

So why weren't the inconsistent behaviors of the carts the real law of nature? Jonathan was present that morning and he saw for himself what had happened. He could easily blame the wide variation in the outcomes on the poor condition of the equipment, or even more on the overall design of the experiment. Getting that kind of crude mechanical setup to work right would be hard for anyone to do. But using more sophisticated equipment did not eliminate the problem. As an experiment became more complex and the apparatuses became more intricate, the larger scale of the enterprise introduced many more variables, and little things that didn't seem to matter could still influence the final result. Nonetheless, physicists believed that by going to great lengths and exerting large amounts of energy they could successfully counter the natural tendency of the world to act in strange and unpredictable ways.

The fact that the class project had not revealed the intended law of physics was a serious problem for Jonathan. The experiment was supposed to demonstrate the invariant and universal relationship between the force of gravity and the acceleration of an object, but all he could do at that point was take the principle on faith and he was extremely unhappy with that. A law of nature was unyielding and unapproachable. It was utterly reliable and the polar opposite of what he had seen in class. Natural law stated that certain actions had unavoidable consequences—nothing at all like the erratic results produced by his classmates, which was pretty much everything under the sun.

Even Emma knew that certain events followed other events. When she and Jonathan returned home from one of their walks, she knew that he would start fiddling around in his pocket looking for the keys, then fumble with them in his hands, often dropping them onto the steps. Eventually he would insert the proper key into the lock. He would turn the knob, push the door in, and they would go inside. That was the way the world worked. One thing followed another in a prescribed order.

The truth was that the world didn't work that way at all. Only the man-made world of mathematical objects worked that way, but these days that was everyone's world. People had forgotten about the natural world and they assumed that the entire universe mirrored their little world of geometrical fabrications where everything behaved just as the door did, but as soon as Jonathan donned his pack and set off across the desert, he realized that this perspective was utterly wrong. The universe was not like the door on his house. People had turned the universe into something strange, a series of clearcut alternatives, a plain menu of options, a place where one thing followed another in a definite order. Either the door was locked or it was unlocked. Either it was open or it was closed. No one seemed to understand how peculiar that was.

Jonathan compared the door to a lightning storm he had observed last summer while driving back from the mountains. Darkness had fallen so he wasn't able to discern the shapes and textures of the clouds, but that only served to emphasize the specific nature of the lightning. At first the storm cell sent vertical columns of searing white-hot streaks straight down to earth in the immediate vicinity, unnervingly close to where he had pulled off the road, blazing pillars of light with almost unbelievable breadth and diameter, the entities appearing to be standing upright on the ground. The immensity and ferocity of each electrical discharge was scary, even for someone who had spent his entire life repeatedly exposed to lightning storms. Witnessing the shear power of these colossal blasts made Jonathan tremble in fear. If lightning strikes had been granted a scale of magnitude similar to the one that had been invented for tornados, these bolts would have certainly been F-5's or Fingers of God. Then without warning the storm changed its character. Patches of the sky lit up, the lights darting and dancing across the heavens, the various clouds flashing brightly, on and off in sequences and patterns, but never showing any signs of bolts or streaks. After a long succession of such luminous displays, once again the storm suddenly changed its behavior. Jagged lightning bolts, now thin and yellow rather than thick and white, twisted and turned up in the sky, tracing out complicated patterns. The huge number of very short segments were all facing different directions yet curving and arcing as one, at times completely turning around and heading the other way, always striking out in new directions yet sometimes backtracking and filling up the empty spaces they had left behind. This was an entirely new phenomenon quite unlike the ones that had gone before. Not only did the storm leap into new patterns, but the series of examples in each case were very different from one another. Jonathan tried to predict what would happen next when the unexpected happened. The storm suddenly fell silent and he was lost in the ensuing darkness.

Jonathan didn't really care much about the door. It was what it was. But he never considered postulating the unlocking and opening of the door as an example of how the world worked. Physicists, however, realized that they could capitalize on the unique behavior of mathematical objects. In the same manner as the door, the objects in a laboratory produced definite outcomes in definite patterns and physicists could make something out of the ensuing order and regularity. This was a great opportunity for them, a chance to do something that had never been done before, namely, assemble a universe of precise relationships into a mathematical system. The door was similar to the apparatus of an experiment. The door could be opened to various degrees and the extent to which it was opened could be expressed in terms of the angle it formed with the frame. In performing experiments, physicists were operating on the premise that they could study the physical world by studying the behavior of mathematical objects, however, Jonathan's experiment with the door—determining whether he could find a way to let himself and Emma into the house—didn't tell him anything about physical reality, but only about the ideas that had been embedded in the door, the mathematical forms that had been created during the course of its manufacture.

The use of mathematical objects to create simplicities in the laboratory produced nothing but falsehoods. If there was one thing that could be said about the world, it wasn't simple. Physicists tried to understand the world in terms of its opposite, spinning the illusion that underneath all the chaos and confusion there was a precise mathematical system. Nothing could be further from the truth. Underlying the chaos and disorder was only more chaos and disorder. Digging deeper did not unearth a neat and precise conceptual foundation, yet the physicist could postulate an invisible mathematical realm and no one could contradict him, for after all, it was invisible. The real problem wasn't finding out how the order and structure of the laboratory dissolved into the chaos in the natural world, but how the chaos and disorder of the natural world pulled itself together and managed to do something, and not just anything, but an amazing number of wildly different feats.

The cart in Jonathan's grade-school class could have been simple or complicated, plain or ornate, ugly or beautiful, inspirational or disgusting, because nothing about the cart mattered. The cart could have been a replica of Louis Chevrolet's race car, the one he had used to tear up the brickyard at Indianapolis and shatter the windshield of his finely-tuned machine, driving a shard of glass into his eye, or it could have been a replica of Bruce Wayne's Batmobile, the one he had used to foil the Joker. If the class had been given a wide range of vehicles to choose from, a student might not want to trade his space shuttle for a locomotive and become confused when the teacher told him that in terms of the experiment they were all one and the same. He had not yet learned to turn a blind eye to the reality of the world. He needed to understand that the experiment wasn't about the reality of the world. The cart was merely a marker, a placeholder, and it had the blankness of a symbol. The experiment was solely about the mathematics of the setup, the geometry of the layout. The success of the experiment required only that the mathematics of the phenomena follow the mathematics of the apparatus and in order for that to happen, no obstacles could stand in the way, no mitigating circumstances, no confounding influences, no complications. The path had to be cleared of all debris, the way opened for the initial mathematics to express itself. The track the cart followed had to be absolutely straight and uniform, the wheels perfectly round and smooth, the axles completely free-wheeling and frictionless, because only in that way could the mathematics of the experiment dominate the phenomena, exercise absolute control over the ensuing events, be in complete charge of the situation. Only in that way could a mathematical law be established in the pristine environment of the laboratory where the mathematical forms were impeccable, unblemished and flawless, and they were able to determine the outcomes of events without interference or corruption.

The problem was not unique to the laboratory, but plagued everyone who cherished mathematics over everything else. The beautiful lines of a new car, if not passionately protected, quickly devolved into an ugly mess. Models that were worshipped when they were fresh off the assembly lines became unsightly wrecks after many years at the mercy of the elements: peeling paint and cracked windshields, major dents and minor scratches, canted bumpers and loose moldings, hoods that wouldn't close properly, corrosion eating away at the fenders, taillights missing chunks of plastic allowing the white light of the bare bulbs to shine through. Once the mathematical perfection of the pure forms were corrupted, the vehicle wasn't the same anymore. Instead of this fantastic otherworldly creation, it was now a member of the earth, raggedy and disheveled, worn out and beaten down, lop-sided and askew. The only way to prevent this natural progression was to keep the car locked up in a secure place, away from the ravaging forces of nature, the dust and dirt of the world, the harsh sunlight of the outdoors, the humidity and heat of summer and the freezing cold of winter, the minor mishaps in parking lots, the dings and scratches of ordinary usage, yet the materials still degraded, the rubber tires slowly stiffened and cracked, the plastic parts became brittle, the fabric seats lost their resiliency. The chrome pitted, the lenses clouded over, mildew took up residence in the carpets and cushions, and the mathematical purity was compromised.

Jonathan and Sarah were neglectful and thus very hard on their vehicles. They explicitly did not worship machines. They did not lust after exotic and expensive models and they knew how vain and ridiculous it was to try to impress others. They had a saying between them: "If the car got you to Walmart and back, what more was there? If the tires rolled, the doors closed and the steering wheel was still connected to the tie rod ends, who cared what it looked like? It was just a car." Mathematical objects didn't appeal to them so they didn't devote a lot of time to cleaning and polishing their vehicles or routinely repairing cosmetic defects, trying to keep the original mathematical purity intact. It was a losing battle anyway, so why postpone the inevitable? Why not just let nature take its course? The struggle in the laboratory to maintain the purity of the mathematical forms, not just to impose these mathematical forms but then hold onto them, was also a daunting project that required a great deal of attention and entailed a sustained and relentless effort to ensure that everything conformed to the ideas of the experiment and never wavered.

All mathematical forms in nature rapidly disintegrated and quickly perished. They were stamped out, swamped by all sorts of irregularities, obliterated by hosts of crooked forms all successfully competing with it, buried under mountains of intricate complications. A mathematical form could never survive out in the wild for very long. It had to be nurtured and protected in the safety of an artificial environment, cultivated under unnatural circumstances, preserved in a museum where everything was kept under glass cases. The mathematics was easily corrupted, its clearness became clouded, its perfection became polluted. The abstract ideas and airy concepts readily became lost within the maze of conflicting influences that was the real world, dissolving into the swirling mixtures of non-mathematical forms that were flowing everywhere. The laboratory was the one place where that didn't happen—well, in principle at least.

The students could have rolled all sorts of model vehicles down the ramps. The models were concrete objects but they didn't have engines and gears and were only representations of actual vehicles, things that were based on ideas. However, the experiment wasn't about the cart as a concrete object, but rather it was about the cart as a mathematical object, a blank with a certain form. Only the form mattered. The cart was an 'x' with wheels. Thus the experiment wasn't about physical reality. The cart had a specific role to play in the experiment and in order to fulfill that role it had to have certain symmetries and geometries. The experiment did not consist of rolling stones and pine cones and bird eggs down irregular slopes littered with twigs and rivulets formed by natural runoff, something a bored child stuck out in the desert might do for entertainment. Natural objects produced no mathematics because they had no mathematics of their own.

Jonathan no longer played games with toys as he once did in his grade-school physics class. He and Sarah now concerned themselves with the practical objects of daily life. These included three widely distributed properties and three different vehicles. For camping, Jonathan had an Airstream trailer, a pop-up, slide-in camper for his full-size pickup truck, and three tents of widely different sizes. They had keyed gates, garages, tool cabinets, bike locks, roof carriers and storage sheds. Jonathan had lots of keys and he couldn't possibly carry the whole collection around with him all the time. Rather than put them on one large key ring that wouldn't fit into his pocket anyway, he divided the keys between a number of smaller rings and typically carried three of these at any one time, depending on where he was and which vehicle he was taking that day. He found that regardless of the specific key he required, that particular key would invariably appear on the last ring he pulled out of his pocket. The order in which he had originally put the key rings into his pocket didn't matter and it didn't matter which key he happened to need at the moment. He would always pull the other two rings out first. He found this phenomenon to be highly annoying. When he and Emma got back to the house, Emma always waited patiently as he performed this hackneyed ritual. She sighed deeply and cast a disapproving look at him, treating him as if he were an idiot. He often looked down at her and said: "It's not my fault, Emma. It's a law of nature".

Well, it wasn't actually a law of nature, because every once in a while, by some miracle, he would pull the right key out of his pocket on the first try. A law of nature had to be ironclad and never allow an exception to the rule. Although the lock on the door was a mechanism built with mathematical precision and based on simple physics—a precise arrangement of regular forms: springs and cylinders, circles and straight lines, torques and pressures—it was not a law of nature that inserting the key into the lock and turning it would open the door because on rare occasions a tumbler pin would stick, either from an accumulation of dirt or from a coating of ice in the wintertime, and the lock would fail to disengage the bolt as intended.

But the results of a scientific experiment were even less reliable than the lock on Jonathan's door or his ability to find the right key in his pocket. A law of nature was an abstraction, an obscure pattern in a bunch of numbers decipherable only to the trained eye of a specialist, a ripple in the titanic sea of digits and numerals that were cranked out by cluttered machines with tangled wires and long equations with Greek letters in them. Physicists acted as if it were more than that, a real entity occupying a prominent place in the universe, the foundation of everything that existed. They had elevated mathematical law to this illustrious throne forgetting its humble origins. A law of nature was really just a calculation, a relationship between quantities, numbers that referred to empty concepts, things that didn't exist in the world. Jonathan asked himself why something as important as a law of nature should depend on someone's crazy ideas or be the product of a completely oddball way of thinking.

Ordinary events could never lead to anything as stupendous as a law of nature because a law of nature was something that could only be built using mathematical objects that were arranged in mathematical patterns. Events had to follow each other in consistent and reproducible patterns, even though erratic changes occurred throughout nature, irregular cycles of more rapid changes superimposed over irregular cycles of more gradual changes. The idea that the laws of physics were eternal and universal was nothing more than a giant leap of faith. The early scientists had appropriated the idea from medieval scholasticism where God and heaven were set in opposition to the corruptible realm of becoming, perhaps as Nietzsche had suggested, as a resentment of the inescapable burden of birth and death. The only argument in favor of the everlasting and all-embracing dominion of physical law was that it made sense, but the notion only made sense to the rational mind laboring under the spell of logical ideas. The ideas of logic and mathematics, once they became accepted, were from that point on forever fixed and incontrovertible, and thus they formed a suitable replacement for the eternal God, but without an independently existing deity, the model for physical law rested on the ideas themselves. The laws of physics were eternal in the same sense as the mathematics that had been used to express them.

If the results of an experiment were the same regardless of where or when the experiment was performed, in the minds of physicists this fact showed that the structure of the universe was everywhere the same, but the observed uniformity was actually the result of the mathematical structure of the experiment always being the same. The universe could not override the physicists' imposition of forms and passively submitted to the physicists' rather preposterous ideas and harebrained schemes. Physicists could force physical reality to perform all sorts of stunts and to perform these same stunts over and over again, but as soon as physicists allowed nature to act on its own, anything could happen and the results were always unpredictable. As Jonathan traveled across the desert and saw each day unfold, the universe was always presenting him with unique situations and the possibilities for the future were, if not truly unlimited, at least far-ranging enough as to refute the idea of natural law. If the physicist similarly let go of the phenomena by removing the constraints and allowed nature to decide the outcomes of events rather than dictating them himself, then the results were always different. Physicists guided phenomena into restrictive boxes so that they could pin down the phenomena, thereby creating the observed uniformities that supposedly tied the universe together. The structure of the universe was really only the structured existence of the laboratory, ultimately the result of the physicist's way of thinking.

The physicist had decided ahead of time that the cart in the experiment would strictly follow the track, but the world had other ideas. The world took every opportunity to create something new out of the circumstances that were handed to it. The carts in Jonathan's physics class did all sorts of things. They twisted and rode part of the way down on two wheels. They skidded and scraped the sides of the tracks hindering the rotation of the wheels. They stuttered and stopped. They pulled the ribbon of paper out. They did flips in the air. The world loved to invent new patterns. But if the physicist was able to successfully stifle the artistic impulses of nature, then the cart could do only one thing: exactly follow the physicist's commands. This compliance was what the physicist meant when he said that the world obeyed the laws of physics. The physicist dictated the course of events based on what he personally wanted to happen. When events successfully unfolded according to what he had previously envisioned in his head, he claimed that the law of physics was revealed. But nature was miffed. Nature had tried to contradict the physicist's ideas and did its best to prove the physicist wrong. This was not at all what nature wanted to do with the carts. Consequently the experiment did not depict the character of the natural world, but only represented a set of ideas, and in fact an extremely bizarre set of ideas.

The very notion of the universe having a structure made no sense at all. Structure referred to a conceptual universe where mathematical forms stood in definite relationships to one another. These relationships could be mimicked in a laboratory, but once out in the real world, a universe that was concrete, non-conceptual and non-mathematical, nothing stood in definite relationships to anything else and thus there was no structure to speak of. Physicists had to create a structure for the universe because it did not possess one of its own. They did this principally by drawing lines and fabricating mathematical objects. They envisioned geometrical constructions in their heads then related these geometries to the geometrical outlines of their devices. These fanciful patterns and designs were then taken to be the structure of the universe. Physicists had made the amazing discovery that all mathematical objects stood in definite relationships to one another, but the best part was yet to come—get this—that these relationships proved that everyone lived in a structured mathematical universe, in spite of the fact that prior to the arrival of mathematicians, the universe was composed entirely of non-mathematical objects.

The idea of the universe being everywhere the same was a direct consequence of the demand for reproducibility that had inspired the concept of the experiment in the first place. The fabrications of mathematical arrangements in the laboratory were deliberate acts expressly designed to establish reproducibility in the phenomena. In truth, the universe was connected only through these phony arrangements. Physicists had this crazy idea that they were allowing nature to act freely in the laboratory, giving it options after they had expressly forbidden it to have options. The mathematical forms of the experiment were fixed and set, so new behaviors would have had to appear out of nowhere and for no reason making the phenomena highly unpredictable and erratic—the way that objects and events were out in nature—however, objects and events in the laboratory were expressly prohibited from doing this by the constraints that had been placed on the variables. The fact that the observed phenomena never suddenly cast off their shackles didn't prove the universality of physical law, but only proved the universality of the means by which objects and events could be manipulated according to preconceived ideas. The physicist claimed that he hadn't determined the outcomes of events because he hadn't foreseen these outcomes ahead of time, but this ignorance did not relieve him of the responsibility for having created them anyway. He freely created all sorts of geometries, oblivious to the fact that the world could never be understood through geometry.

The sun was already up when Jonathan finally found a place to park the truck. As he opened the door, a cold breeze entered the truck and greeted him. The frosty morning air was not unusual for this time of year, the low desert in early spring. The freezing temperatures of another lonely, star-filled night would linger well into the afternoon until the warm sunshine finally mustered enough strength to loosen winter's grip and the day became quite pleasant. Jonathan braved the chilly morning in order to take full advantage of the mild afternoon, hiking at a slightly higher elevation than usual.

He let Emma out of the cab and circled around to the back of the truck. He lowered the tailgate then collected various items and put them into his pack, finally zipping up all the compartments and locking the doors. He'd barely set out across the desert when he noticed that he'd already lost Emma. He surmised that she probably picked up a scent somewhere and taken off searching for the culprit. He needed to make sure that she knew which way he was going so he waited for her to show up, carefully studying the vegetation for signs of her whereabouts. He stood in place shifting the weight of his pack on his shoulders while scanning the terrain. After a while he spotted a cloud of dust erupt from behind a bush a couple hundred yards away and watched it slowly drift along the desert floor. A few moments later a second cloud emerged from the same spot. Then a third cloud appeared. The series of clouds all floated above the ground driven by a gentle breeze, each one following the others until finally one by one they disappeared from view, dropping down to earth just as Newton had predicted they would, only not in the manner of an apple. If Newton had been out here with Jonathan today patiently listening to his ramblings, the great personage would not have been able to calculate the trajectories of all those swirling dust particles or find the delicate balances between buoyancy, wind and gravity that gave rise to their forms and motions, or put all of these factors together into a simple, elegant formula, but of course Newton would have had better things to do. The important point was that Emma was sending Jonathan valuable information encoded in the dust clouds, a form of communication similar to the smoke signals employed by Native Americans. The message they conveyed was clear: she was hard at work excavating a rabbit warren.

The rabbits shared their domiciles with rock squirrels, pocket mice and kangaroo rats. Jonathan had never seen a kangaroo rat other than in books because they never came out during the day. For a long time he never even knew they existed, but he had always known that rabbits existed. Jonathan started walking toward the dust clouds. If Emma had a rabbit cornered somewhere inside the large mound of dirt with its numerous openings and networks of tunnels and chambers, then he'd have to forcibly drag her away. Twice she had managed to pull a rabbit out from one of these hiding places so she knew that she could do it and wouldn't quit until she had succeeded. He'd watched her routine on many occasions. She ran up to the mound, poked her head into one of the openings and smelled the rabbit. She thought "Rabbit here! Dig here!" She furiously worked her front paws causing the dirt to fly, spraying out behind her in a long arc. The rabbit retreated to another part of the warren so the smell diminished. Emma ran around to the other side and checked one of the other openings. She smelled the rabbit. She thought "Rabbit here! Dig here!" Again her paws started churning, sending up another cloud of dust. The pattern repeated itself until the entire warren lay in ruins, an ugly mess of open craters, caved-in tunnels, and large piles of fresh dirt.

Jonathan marched toward the place where Emma was busy at work. As he reached the site, he found that the front half of Emma's body was completely submerged and only her tail and rear end were sticking up where he could see them. He tapped her on the rump. She began to back out of the hole wherein he immediately grabbed her collar. He turned her around to face the desert and started walking away. After a few steps, he let go of her collar. He only needed to break the cycle momentarily, forcibly interrupt the reflex arc, and the lock on Emma's behavior was broken. Emma reminded Jonathan of a phonograph record that had gotten stuck and started skipping, playing the same sequence of notes over and over again. All he had to do was lift the needle over to the next groove and the record started playing normally again. Emma suddenly remembered that rabbits were everywhere. She excitedly began searching for them, now thoroughly engrossed in an entirely new pattern of behavior, caught up in a different cycle of perceptions and actions that no longer had anything to do with the rabbit that was still hiding back there in the warren.

Jonathan raised his eyes and surveyed the terrain up ahead. A cluster of large, towering hills, each several hundred feet high—what were often called mountains here in the desert—were the only landmarks on a comparatively level plain. He arrived at the base of the closest hill and struck a diagonal up the lower skirt to avoid the highest parts, which were much rougher than he had expected. From a distance the hill looked smooth and the gentle contour invited him to scale it, but upon closer inspection the hill was just a pile of rocks with limitless opportunities to stumble and fall and perhaps twist an ankle. Since the summit promised no benefit other than a slightly better view of the plain than he already had, there was no advantage in taking the considerable risk involved in reaching it.

Jonathan's diversion away from the pointless conquest took him to the backside of the hill where he found the mouth of a deep canyon. A large arroyo had somehow managed to cut right through the middle of the hills and split them in half. The cliffs on either side of the canyon were quite impressive. He entered the opening and disappeared into the sheltered enclosure. The low angle of the sun in winter had prevented its radiant heat from reaching this shadowy realm and that had allowed a hefty amount of snow to accumulate on the canyon floor over the winter. Even though the last vestiges of snow had long disappeared elsewhere, this isolated canyon was still an icy microcosm. Emma loved the snow and ran jubilantly, cavorting across the thick blanket of powder, enjoying the feel of soft snowflakes on her paws and constantly kicking up ice crystals with her nose.

Each winter at least one major storm system managed to drive an arctic airmass down through the middle of the state. The cold air settled into the basin of the lower Rio Grande valley where it remained for a few days—or sometimes for a week or more—throwing the ordinarily mild climate into a bit of a deep freeze and giving the inhabitants a good taste of winter, a generally welcome relief from the long, hot summers. This year everyone had been treated to an impressive event, a major blizzard, even by the standards of residents in the northern latitudes.

Jonathan remembered carefully following the weather forecasts that week, but the storm hadn't materialized as the weathermen had promised, so he figured that it would simply pass to the north and never affect this part of the country. Convinced that the prediction was just another gross miscalculation, he set out to conduct his usual afternoon walk, swinging the truck onto the interstate and speeding toward the faraway hills with

Emma by his side. The day was unsettled and windy but otherwise mostly sunny. He decided to leisurely tour one of his traditional hiking destinations, however as he approached the hills with the range of mountains in the background, he could see neither of them. They were both enclosed by a peculiar-looking cloud, a large circle of haze and smoke. He thought to himself that a forest fire couldn't possibly be burning at this time of year. He really had no idea what was going on so he just shrugged his shoulders and continued driving, following his usual route, now curious about the strange events taking place up ahead.

He was still unable to see the hills as he got closer to them, finally turning onto the last dirt road and heading directly into the weird zone of darkness. He appeared to be approaching a dust storm of some sort. A charcoal curtain was draped across the road and the scene behind it was murky and indistinct.

As the truck flew through the boundary at a rather high rate of speed, Jonathan was thrown into another world. Fat snowflakes were swirling everywhere, bouncing off the windshield in swarms and fleeing wildly into the heavy mist. He looked around at the familiar expanse of sage, mesquite and creosote, but he could see only the bushes directly alongside the roadbed, the rest quickly fading away behind the veil of snowflakes and the thick layer of fog. The sun was shinning just a short distance back down the road, but he would never have known it from here.

He continued driving down the road heading deeper into the center of the storm, eventually turning onto a rough two-track road, the intensity of the snowfall increasing steadily as each minute passed. Snow was rapidly accumulating on the ground, piling up against the tall tufts of grass. He stopped the truck and opened the door. The wind was freezing cold but the wintry scene outside was incredibly beautiful and inviting. He emerged from the truck and his shoes alighted on a sparkling white surface of rippled banks and sculptured drifts, a strange sight in a land of perpetual sand and gravel. Emma did not wait for him and bounded from the truck while his back was turned on her, imagining that she had heard his voice telling her to exit, but the sound had only been the wind rattling the sideview mirror. No matter. They were going for a walk anyway.

Emma looked around and she couldn't believe it either. She took off running through the snow between the heavily encrusted bushes, kicking up a spray of snowflakes in her wake. Jonathan slipped his parka over his down vest and scrounged through the pile of clothing on the passenger floor mat for his insulated gloves and a wool hat. He followed Emma's fresh footprints across the barren landscape, imagining himself to be an arctic explorer striking off into some remote and uncharted tundra, a vision he had often entertained as a child when forging across the snowy fields and forests of his grandparent's farm in the upper midwest. He spotted Emma up ahead running madly, sweeping back and forth in long arcs, covering the ground as thoroughly as she could, looking for fresh scents or anything that moved.

The atmospheric lighting was fantastic, an eerie bluish-green glow enveloping everything, coming equally from all directions, backlighting the forlorn skeletal figures of shrubs standing on the frozen ground, accentuating the grooves and channels of the developing drifts that were being carefully molded by the invisible hands of the wind. The strange light reflected off the fields of snow crystals, but instead of glittering and sparkling in the rainbow colors of cut gems, the subdued lighting gave the surface the appearance of tiny metal flakes. The facets took on a range of shiny metallic hues from battleship gray and gun steel blue to coppery orange and ruby red rust, from specks of pure gold to the white oxides of heavily corroded tin, an intricate mosaic of smoothly polished surfaces.

After an hour or so of tromping through the snow, Jonathan wound his way back to the truck both to warm up and to explore new territory further down the road. He was clearly nuts. Nobody was stupid enough to be out on a day like this and even the main road was utterly deserted. Now he was even more alone and isolated on some barely visible side road where they'd never find him if he didn't come home that night. If the truck failed to start, he'd have to trudge quite a few miles just to get back to the interstate, and conditions were rapidly worsening.

The voices of prudence and sanity spoke to him at this point, their words as clear and distinct as any he had ever heard, but they held no power over him as he drove the truck further into the snowy wasteland. The deep ruts of the four-wheel-drive track were plastered over with a pristine coat of snow creating the illusion of a smooth roadbed, the deception betrayed only by the bumpy ride, the tires being guided not by the soft snow but by the underlying reality of large rocks and deep furrows. He stopped after another mile and a half and disembarked for another walk, having warmed himself thoroughly with the heater blasting the whole way. He saw no signs that anyone had been back there in a while. The surface of the road was as immaculate as the ground alongside it, indistinguishable from the surrounding area except for the lesser amount of vegetation within its boundaries.

He followed the road on foot into the bleak wilderness, continuing to watch for signs of previous activity, scanning the occasional windswept patches of bare roadway for the marks of earlier vehicle traffic, but he didn't see the slightest indication that a single soul had been out there all winter. The loneliness and isolation were exhilarating. The land was entirely his and he could walk around as if he owned the whole place. Confident that he wouldn't run into anybody that day, he marched along making tracks in the snow, watching Emma laying out a separate set of tracks of her own off in the distance. As he entered a broad basin encircled by spectacular stone edifices and precipitous mountains of rocks, he rounded a corner and was struck by the appearance of vehicles and people up ahead, previously hidden from view by an intervening hillside. The stiff wind and blowing snow had covered their tracks more thoroughly than he ever could have imagined. Staring in disbelief, he stood still for a moment. He didn't want to reveal his presence since he didn't know who these people were or what they were doing out in the middle of nowhere on such an inhospitable afternoon, so he refrained from whistling to attract Emma's attention. Emma was a good distance ahead of him and much closer to the unidentified intruders. She obviously knew that they were over there, but she was politely ignoring them. She didn't find people threatening—even strangers—and often payed no attention to them. No one had ever been mean to her in the past or tormented her in any way, so she loved people and just assumed that everyone was friendly.

Jonathan, on the other hand, had good reason to be suspicious of people, especially in a situation like this. Last year a friend of his had been hiking alone with her dogs when some guys started shooting at her with a high-powered rifle from across the valley, the bullets ricocheting off the rocks near her feet. She bravely marched over to them and verbally reprimanded them. They laughed out loud and told her that they were just target practicing. Confronting a bunch of strangers who were recklessly brandishing firearms was foolhardy in Jonathan's opinion, especially with no one around to witness the proceedings or testify on your behalf. She was lucky that nothing terrible had resulted from her actions. If the guys up ahead turned out to be similarly hostile, Jonathan was seriously outnumbered.

Emma was in particular danger. Shooting a dog would not put anyone in prison for a long time—and probably not at all. Given all the possibilities, Jonathan saw no harm in being cautious. He turned around and hastily backtracked down the road until he returned to a previous arroyo crossing where he took refuge in the shallow depression of the wash completely hidden from view. Luckily Emma noticed his about-face and quickly reversed direction to join him in his new quest. He took advantage of the channel cut by the arroyo to gain a position where he could spy on these people without being seen, his head still well below the general level of the surrounding desert. He stopped near a long line of bushes and retrieved the binoculars, while Emma continued on down the arroyo by herself. He positioned the instrument in the middle of a concealed opening and peered through the eyepieces. He saw several men in snowmobile suits mulling about a small collection of pickup trucks parked haphazardly in a tight group. The men appeared to be going from one truck to the other and talking to each other. He didn't see any tents or RV's, but some tents might have been pitched behind the trucks and hidden from view. The wind was too fierce and the visibility too low for them to be out hunting that day, and he saw no ATV's—or the flatbed trailers that were generally used to transport them.

The whole scene was very strange and Jonathan could not come up with an explanation. His imagination stepped in at this point and took over. The men seemed serious and animated, as if they were engaged in some important business. Jonathan wondered what kind of business that could possibly be. Then it dawned on him that he might be witnessing a transaction between drug dealers and smugglers. Perhaps he'd seen too many movies and tv shows with heavy drama and improbable plot twists, but he became gripped by the idea that he'd inadvertently stumbled upon the pre-established meeting place for the transfer of a large cargo of contraband hauled up from Mexico. From watching television he knew that drug dealers frequently chose the most remote locations to conduct their business. This strategy allowed them to see if they were being followed. The land was wide open and there was never anyone around, so the appearance of another person would be highly suspicious and immediately draw everyone's attention. In a more populous setting, they never knew who was watching them or possibly even spying on them. Jonathan understood this perspective because he also felt safe out in the middle of nowhere, for pretty much the same reasons. The sudden appearance of another human being always set off alarms in his brain and put him in a state of high alert, whereas in the city he paid little or no attention to those around him. In all respects, this would be the perfect place to execute such a deal, remote and out of the way, at the end of a road that stopped and went nowhere. The squall came up unexpectedly and with no cell service anywhere in the area, they would not have been able to contact each other and agree on some alternative plan for carrying out the transaction elsewhere. If Jonathan were a fugitive from the law and was forced to hide, this would one of the places where he would come to hang out, where he could live for long periods of time completely unnoticed by the authorities or other law-abiding citizens. He snuck back down the arroyo keeping his head down to rejoin the road in order to beat a path out of there. The idea that those guys might just be hardy souls enjoying a bit of winter camping was not entirely out of the question, even though he'd seen no signs of a campfire. Jonathan used to do that sort of thing himself when he was younger, but many years had passed since he last had the urge to suffer such hardships for no tangible advantage.

He remembered one time in particular, back in his university days when he was living in Wisconsin. After the fall semester exams were over and the Christmas holidays had passed, he decided to take a long weekend off during the winter interim and go camping with a couple of his college buddies. The three of them drove to the upper peninsula of Michigan and came to a place that had access to the Paint River. Jonathan knew about this place because he went there occasionally during the summer months to hike and fish.

They arrived in the late afternoon and found the place utterly deserted. To Jonathan's surprise the road he normally used to access the river was no longer there. The road was not maintained during the winter months and it was buried deep in snow. The county snowplows had merely dug out a small parking lot next to the highway, an icy flat-bottomed bowl ringed by huge chunks of ice and snow piled considerably higher than the roof of the car. The temperatures in that part of the country typically dipped to thirty or forty below zero in early January and few people had reason to go there. He worried that the car might freeze up while he and his friends were camping and they wouldn't be able to start it. They'd be stranded in the frigid weather until they located someone who could help them, not an easy task at that time of year.

They stuffed their packs with all the gear they could carry and shuffled single file down the trail on cross-country skis. They followed the route of the gravel road that led to the actual parking lot, the place where people normally left their cars during other times of the year, but all of that was now obliterated by several feet of snow and nothing more than a cutout in the trees. Jonathan knew the area well from his summer visits and he could picture the underlying reality in his mind, hidden beneath the monstrous drifts. He led the group down the footpath he had often walked on fine summer afternoons. No one would have ever guessed that it was there.

After another mile and a half the three of them reached the river. Jonathan stood on the bank sliding and shifting awkwardly on his skis. He could hear the rapids churning under a solid roof of ice and snow, seething and heaving only a few feet away, the muffled sounds rumbling and reverberating in the absolute stillness of the blanketed world. They chose a spot nearby sheltered by a sloping stone wall jutting out from the side of the canyon forming a small hollow underneath. The drive had been long and tedious and it left everyone little time before dark. They hurriedly set up camp, pitching tents and unrolling air mattresses and sleeping bags, stomping out small clearings in the drifts with their boots. They were going to need firewood for a roaring blaze.

One of his companions produced a rosewood pipe and filled it with black hash. It was the same stuff that had been all over town back then, supposedly imported by a biker gang in Milwaukee. Jonathan had seen a brick of it at a friend's house. An official-looking emblem was embossed in the middle and it came packaged in an authentic-looking muslin bag. They stood in a tight circle and passed the pipe around until the contents were exhausted, then struck out in different directions to scour the area for fallen limbs that were still stubbornly sticking out of the snow.

Jonathan ended up walking along the river. The sun had already set and the world was inundated with a blue light. Weakened by the cold and the strenuous effort required to get there, Jonathan felt shaky in the knees and overcome by the effects of the hash. He laid down in the snow and closed his eyes, listening to the sound of the water gurgling and trickling nearby. When he awoke, he faced a man and a woman standing in front of him, each with a similar shade of blonde hair. Both were tanned and muscular and scantily clad. A small waterfall gushed in the background directly behind them, the water gaily splashing through a verdant curtain of broad leaves. They said nothing, but the compassionate smiles on their faces greeted him as if he were a close friend. They turned around and started walking away, looking back over their shoulders and beckoning Jonathan to follow them.

The three of them walked down a narrow path through a dense forest, tracing the low bank of a rippling brook, emerging after a short time in a small clearing. A large house occupied much of the opening, carefully constructed so as to blend in with the natural environment. They entered the house and climbed some stairs to the roof. The view took Jonathan's breath away. He was in the middle of the most beautiful valley he'd ever seen. The flat roof was bounded by a low parapet and obviously meant to function as an observation deck. He went from side to side taking in the spectacular views in every direction. It was then that he noticed boxes of stuff scattered about. The boxes were filled with all the things from his childhood, old clothes he had cherished and worn to rags and finally discarded, but to his astonishment they were all here, all new again. He had loved to build electronic gadgets as a child and the wiring harnesses and circuit boards and metal cabinets were all here, thrown haphazardly into cardboard boxes. He'd had never been so happy in his life, to have all this stuff back again, everything that had ever meant anything to him. The items had all been gathered together in one place. This was his home, the place where he belonged, and it was such an unbelievably wonderful place. He never wanted to leave. He just wanted to stay there forever.

Jonathan awoke still lying in the snow, surrounded by the last glimmers of the fading blue light. The air was brittle, much as the icicles hanging from the eaves of a house. In fact, the whole world seemed to be made out of glass. His guardian angels had departed. The time had not come for him to leave this world.

He could tell that they were really in for it that night. The cold was tightening its grip, squeezing the last remnants of life out of the already dormant forest. Jonathan had just enough time to gather an armload of wood and stumble back to camp. His comrades already had the fire going, huddling around the faint warmth it provided.

The temperature kept dropping as the night progressed and there seemed to be no end to it. The roaring fire appeared to be just fakery, nothing more than the mere picture of a fire that was unable to warm him. Jonathan wanted to put his hand into the dancing yellow images to prove it, but his better judgement prevented him from doing so. He took off his gloves to drink from the canteen that he'd stuck into the snow at the very edge of the flames. He'd placed the canteen as close as possible to the fire without actually putting it inside the fire. He found that the water was frozen solid and he had nothing to drink. He tried to put his gloves back on, but they too were frozen solid. He crawled into his sleeping bag without removing his parka or snow pants, just taking off his boots. In Jonathan's mind, the sleeping bag was just another mirage, just more fakery, since it made no difference in his experience of the cold. He wasn't sure how much he slept that night, but there was no getting up until the blue light returned. In the morning he could not get his feet into his frozen boots. They were as solid as the rocks outside his tent. The three of them stayed another night then skied back up the trail to the highway. Amazingly the car started on the first try.

Jonathan had slept outdoors in the snow on numerous other occasions and a blizzard was always a special event and not something to be missed. He remembered a time in his life when he would have pitched his tent right there where those guys had parked their trucks and let the wind howl and the snow fall all through the night because he had more of an imagination back then. He pictured himself as an intrepid explorer, but now he saw himself differently, more realistically.

Jonathan and Emma left the strange camp of men behind in the falling snow, retracing their steps back to the general area where he had parked the truck. He hadn't bothered bringing his GPS since he knew the area well, so he was forced to navigate by sight, but the winter scene went on forever and he couldn't identify the white vehicle in the seemingly endless landscape of white shapes, particularly through the dense filter of falling flakes. He finally managed to locate the truck buried under a foot of snow. He opened the door and let Emma inside. Their presence did not appear to have been noticed by the strangers. Jonathan threw his pack into the back of the truck. As he latched the tailgate, he recalled a story that had been covered by the local news media a while back detailing the activities of drug cartels in New Mexico. The reporter had stated that gangs of criminals were entrenched throughout the state. They had set up store fronts and auto shops from Taos to Las Cruces to cover their operations. Perhaps it was a good thing that he and Emma had gotten out of there when they did. Jonathan quickly brushed the windows off with his gloved hands. The tracks his truck had made coming in were completely obliterated. He had no way of telling how long the other vehicles had been parked out there, perhaps for only an hour, perhaps for a week or more. What those guys were doing in such a remote location would forever remain a mystery to him, but as he slammed the door shutting out the wintry scene, he realized that the episode had taught him something important. No one could simply guess the truth. No one could possibly imagine an event that had not included them or reconstruct the proceedings of something that they had only observed from a distance. Jonathan decided that physicists should take notes whenever they ran into situations like this so they could review them when they got back to their laboratories and thereby put their work into perspective.

Physicists could not possibly know what was going on in the submicroscopic world. They could only study outlines and vague images and then try to ascertain the parties involved with little or no information, and this forced them to make up explanations in their heads. They resorted to ideas that originated from somewhere else and adopted ways of thinking that weren't necessarily valid. But if their reasoning was incorrect and they failed to understand what was actually happening, how then should they be thinking about the subatomic world? The question was far more complicated than they realized.

Jonathan started the truck and wheeled it around, desperately trying to identify the route he needed to follow, signs of the rough four-wheel drive track having mostly disappeared. He eventually reached the primary road, but instead of turning back toward the interstate, he headed in the opposite direction, toward the mountains. By now the storm had spread out in every direction and engulfed the entire area. After a couple of miles he found another place to pull over and once again struck out into the blizzard. A group of large boulders, each 10 to 15 feet tall, stood on a hilltop overlooking the surrounding area. He went over to visit them and hid behind each one for a brief respite from the freezing wind, touching each rock with his gloved hands as he watched Emma cavorting on the other side, appearing and disappearing from view as she explored the nearby terrain. Although Emma had a rather short coat, it was very dense and thick. She didn't seem to mind the icy temperatures at all. In fact, she loved the snow and cold and wouldn't stop running. The flakes, instead of falling to the ground, were being driven horizontally by the wind. Jonathan's exposed face was completely numb yet even in this insensate condition his skin was still somehow able to feel the painful pricks of the sharp crystals. He decided that he'd had enough of the cold and hastily abandoned the hilltop, dropping down into the valley and following the arroyo in order to stay out of the direct wind. Soon the sky darkened and the rocks on the hilltop all turned black, becoming evil shapes lording over the deepening gloom and standing guard, watching impassively as the night mercilessly devoured the land. He was grateful when the motor turned over and the truck lights came on, fighting back against the invasion of the blackness and driving it away, at least a little bit.

The truck crawled slowly along, plowing through the virgin snow as Jonathan struggled to maintain the tires on the concealed roadbed, spellbound by the crazy flights of snowflakes hysterically running away and escaping into the night. The lonely road kept winding back and forth, rising and falling, but the scene outside was monotonous, a flowing wall of brilliant spots of light just on the other side of the glass. Finally, he approached signs of civilization. Up ahead the interstate was full of red-flashing lights. He immediately assumed that a big pile up had been caused by the icy conditions and low visibility, conjuring up in his mind an ugly scene of wrecked cars and jack-knifed semitrailers, some of them flipped on their sides with their bumpers locked together. As he reached the nearest interchange and merged onto the roadway, he realized that the interstate had been closed down. The lights were due to a squadron of police cars and snow plows escorting the last drivers to safety. He fell in line with the rest of the group and crept along, mesmerized by all the beacons twirling on the rooftops, the flashing red and blue panels, the streams of snowflakes streaking across the beams of headlights, the wiper blades dancing on the encrusted windshields, the icy slush spraying from the tires and splattering onto the hoods and grills of the vehicles following behind them.

The deep snows that had fallen that day were surprisingly still here, many weeks later and well into the developing spring, resting quietly on the floor of this dark canyon, nicely preserved in a secluded museum that no one had yet visited. Jonathan wandered through the shadowy realm staring up at the towering cliffs, carefully winding a path between the numerous groves of Gambel's oaks, highlighted here and there by a magnificent cottonwood. Emma ran enthusiastically as Jonathan trudged slowly onward. Eventually the cliffs diminished and the world brightened. Jonathan emerged from the canyon and found himself completely on the other side of the hills. He noticed the afternoon sun had already begun to sink. Considering the late hour, he decided to return to the truck by crossing over a portion of the hills. He climbed upward and slowly gained elevation. The ground was covered with large blocks of stone tilted at every angle and wedged into every opening, countless tons of broken rocks strewn about everywhere. Emma managed a brisk pace by stepping nimbly onto the protruding corners of the rocks, but Jonathan struggled to make his way, floundering on the obstructive debris, until finally he reached a summit where he surveyed the land for a good distance in every direction. Everywhere the ground was a thick carpet of rocks. The force of the earth's gravity acted on each one of these rocks, but there was no corresponding motion. The desert was absolutely still and quiet in spite of the immense gravitational pull. For each rock, stone, and grain of sand, there was a second force pushing upward, exactly equal in magnitude to the force of the gravity pulling each one downward. This was convenient because it resulted in a free and unbounded sea being entirely frozen in place.

The two forces canceled each other out in the terms of a mathematical formula, but the equal sign did not cause the forces to mysteriously disappear, so Jonathan was looking at a tremendous amount of force, all right in front of him, surrounding him on all sides. He put the individual forces together in his mind and arrived at the total amount of force in the area. The calculations were straightforward and easy to execute, even for a student of physics. He needed only to estimate the total amount of mass involved. The combined forces clearly added up to an amount comparable to that of a large atomic bomb. An unexploded bomb was similarly quiet and serene, exactly as the desert was right now. Where were the forces at that point? Were they lurking inside the canister that housed the bomb? The ordinance was a metal cylinder full of forces, all tightly jammed together, packed like sardines, the handsome cargo crammed into the smallest of holds. By what clever trick had scientists managed to stow such a large shipment of goods into such a tight space? Jonathan looked around, imagining hordes of strange beings crouching in the shadows beneath the megatons of rocks, ready to leap out at any moment. He feared that he might step on them and injure them. He wondered if they squished. Being mechanical in nature, they probably crunched, bending and deforming as they strained under his weight, or perhaps they shattered like glass. Physicists always talked about forces as being things that existed in the world, but when Jonathan treated them that way and used the language that was appropriate for talking about things that existed in the world, he was plainly talking nonsense. He couldn't picture forces realistically because he couldn't think concretely about abstractions. Everything in this world had details because that was what existence was, but he couldn't see forces. He couldn't hear them, describe them or touch them. That was because they did not inhabit this world. They resided in another universe, a place without substance or weight, where everything was invisible. Physicists weren't talking about the concrete realities of the universe, but only about the relationships between a bunch of ideas. Surprisingly they never attempted to picture the truth. What they saw was something else entirely, a strange place, a far away never-never land that existed only in their minds. None of them ever wondered how they had gotten to this place, whether they flew over the rainbow or fell into a rabbit hole, sat on a magic carpet or climbed a giant beanpole, clicked their heels or snapped their fingers. After all, the mathematical world was not hard to find. They simply had to close their eyes, or stare blankly at nothing, or muse over the cryptic meanings of marks that had been hastily scrawled on a sheet of paper. This was an amazing situation, so Jonathan stopped to ponder its significance for a moment. Emma did not wait for him and disappeared over the hill.

He followed her down into the next arroyo and stuck to the channel for a while. The sporadic surges of water along the raceway had cut into the hills of graded aggregates and sculpted vertical cliffs out of the masses of small stones embedded in the matrix of sand and clay. The brief but violent flows undermined the cliffs in many places, particularly where the arroyo squeezed through narrow openings or where it turned sharply, carving out a hollow along the outside radius of each bend. These excavations often became quite deep, resulting in miniature caves that were used by various creatures for respites from the hot desert sun in summer and the cold winds in winter. Emma loved to explore these spaces and note the lingering scents of the previous occupants, but it was a practice that Jonathan did not tolerate because of the great danger involved. The huge mass of earth suspended precariously over the emptiness could come down at any moment without warning. He found ample evidence for this on many of his walks, large mounds of dirt and rocks from recent collapses, sometimes choking the entire passage. If the bank collapsed while Emma was under it, he would not have time to dig her out and save her. She would be buried alive in an instant and without a shovel he'd be forced to leave her there, lying within the grave that nature had provided for her. Years later the rushing waters would unearth her remains and scatter her bones along the course of the arroyo. A day would come when he would return to collect her bones and put them together once again, placing them in a shallow grave at a location of his own choosing, a beautiful place on a hill having a wonderful view of the surrounding desert, with a pile of rocks to serve as a monument to their time spent together. He rather not think about such sorrows. The desert was full of dangers and there was nothing he could do about it.

No upward force was exerted on the roofs of these caves to counteract the force of gravity, so Jonathan needed another concept here--the force of cohesion--to take its place. The stones were stuck in place by the weak cement of clay and sand. The force of cohesion pulled all the material together and held it in temporary defiance of the force of gravity. This was convenient because this force allowed Jonathan to pass by quietly, but what happened when the bank collapsed? The force vanished without a trace.

Physicists could put forces wherever they needed them according to the demands of their theories. If the results were not satisfactory, they could switch the forces around until they got what they wanted. A force was a geometrical construction. When Jonathan looked for it, it wasn't there. As a result, he could put it wherever it suited him.

Everyone believed that nature dictated the magnitudes and orientations of these forces and that physicists had no say in it. Physicists only observed what must be true. The forces existed through a series of sound deductions and the conclusions were inescapable. The forces were undoubtedly out there exactly as the physicists had imagined them, oriented exactly the way that they had drawn them on paper.

The electrons in an atom could not be stationary because in that case the electrical force between them and the oppositely charged protons of the nucleus would cause the atom to collapse. If the electrons orbited the nucleus, however, then the centrifugal force associated with this motion could exactly balance the electrical force. This was very convenient because just as with the rocks strewn about the desert floor, the precarious balance of forces created a stability, a permanence to the atom. The planetary model of atoms still didn't make sense, at least in Jonathan's mind, because the solar system allowed an infinite number of possible orbits and only a few of them were occupied by planets, comets, asteroids, and man-made satellites. The orbit of the earth could just as well have been something other than what it was, but there was no corresponding variability between the atoms of elements and thus the orbits of the electrons must somehow be strictly limited and precisely defined. This was puzzling. How did an electron know exactly how fast it needed to travel in order to achieve the required balance of forces? What mechanism controlled and regulated this velocity, since this also should be infinitely variable? Well, the electron assumed exactly the velocity that was required of it to make the equations work out. No one could actually measure the velocity of an electron as it orbited a nucleus. The value could only be inferred by making calculations. The equations could make the velocity into anything because physicists had no additional avenues of verification—other than more equations. The process of verification was simply the matching of numbers to one another.

The planetary model of atoms was obviously never meant to be taken literally. Now the electrons apparently jumped from one "orbit" to the next without passing through the intervening space. This could not possibly be what was actually happening. The model was nothing more than a schematic rendition of an unknowable reality, the coupling together of abstract ideas into long, artificial chains. No one understood what was really going on inside the atom. Even the subatomic particles seemed to have parts and these parts had further parts. The whole scene disappeared into the infinitesimal, a tiny world that was utterly beyond the reach of physicists.

So now in quantum mechanics, according to the equations, the electron could be found inside the proton at certain times. This had been impossible in classical mechanics because the force between them would become infinite as the separation went to zero. Physicists explained it away by saying that the proton didn't "feel" the force of the electron when the electron was inside of it. Jonathan stifled a muffled laugh as he labored up a steep grade out of the arroyo, shifting his pack on his shoulders and focusing his strength on each taxing step. The idea that protons could feel certain things but not other things was absurd. Emma was busy snuffling clumps of grass nearby. She jerked her head in Jonathan's direction at the unexpected sound of his voice. Her eyes twinkled and she had a quizzical expression on her face. She knew that Jonathan was happy to be sharing his day with her and this fact had been made obvious by his boisterous laughter. She never suspected that he was actually entertaining laughable thoughts in his head, thoughts that had nothing to do with her at all.

Many years of training in philosophy had taught Jonathan to question his beliefs, so as he reached the rim of the arroyo he asked himself: "Given all the stories he had been told by physicists, did he really believe in the existence of electrons?" No doubt the physicist would answer, "Yes, electrons definitely existed—absolutely and without a doubt." Jonathan wasn't so sure. Nikola Tesla hadn't believed in electrons and openly scoffed at them. Dismissing Tesla's considered opinion offhand was a bit rash, for after all, he had investigated electrical phenomena thoroughly and he was quite familiar with the subject matter. Furthermore, he'd been right about many things, including the practicality of transmitting alternating current over long distances. Jonathan certainly didn't believe in electrons as tiny elastic balls, just as he didn't believe in atoms as miniature solar systems, the way they were always depicted in books. To tell the truth, no one really knew what electrons were. It was kind of like asking the preacher if he believed in God. When the preacher answered, "Yes, God definitely existed—absolutely and without a doubt," then Jonathan would have to ask him, "What exactly do you mean by God?" The preacher would most likely define God through his actions. God was a human figure who sat on a throne and ruled the universe from a place called heaven, determining the outcomes of events and the fates of people. He was someone who heard your prayers and sometimes answered them. He sent souls to either one of two places after death. Yet the question remained: who was God? What kind of strange being would ever act that way? How was God created? Where did he originally come from? Jonathan would need to know such things in order to make an informed decision regarding the plausibility of God's existence. The preacher believed in a bunch of ideas in his head, merely describing a series of external behaviors, but what did those ideas have to do with determining the reality of a supreme being? Perhaps he was just stringing words together and merely believing in the logic of his own thinking.

If the preacher didn't know who or what God was, then the preacher was reduced to the statement that he believed in "something," just as the physicist believed in "something." Yes, Jonathan believed that "something" triggered the physicists' sensors. He believed that "something" caused electrical phenomena to occur, that "something" bound substances together, but if he couldn't really say what that was, then how could he honestly believe in it?

Both the preacher and the physicist believed in something they could only describe in words. Believing in something that could not be observed was a problem and ultimately their convictions rested on faith. The physicist insisted that his beliefs were different because they were based on observations, but of course he wasn't observing electrons, but only observing events that according to his way of thinking pointed to the existence of electrons. The preacher could likewise say that the events of the world according to his way of thinking pointed to the existence of God. The only thing that distinguished the physicist's beliefs from those of the preacher's was that the physicist relied on the authority of mathematics. He had faith that mathematics would reveal the truth to him, yet

41

when Jonathan compared mathematics to the reality of the world, Jonathan wondered exactly what authority that was. Jonathan looked at the desert and studied the nature of physical existence. Those direct observations forced him to question the legitimacy of mathematical concepts and the validity of mathematical thinking. In a non-mathematical world, mathematics had no such power and would only create illusions, forcing physicists to draw unsubstantial conclusions based on mere appearances. Physicists generated laws based solely on the comparisons of empty forms. The infinite particularity of the world reduced all such vacant concepts and irrelevant comparisons to utter nonsense.

A dust devil spun around in a circle much as the electrons spun around the nucleus of an atom, but in this case there was no force of attraction to counter the centrifugal force. While there was no danger of the whirlwind collapsing, what kept the air from flying away? What force held the whirlwind together? The devil should dissipate the moment it began, but instead these entities marched across the desert for miles. Jonathan had driven into a dust devil once on the interstate. It was crossing the highway just as he came up on it and he couldn't resist the temptation to find out what it was like. The devil slammed into the car and flung the vehicle sideways. He felt as if invisible hands had grabbed the steering wheel and yanked it to one side. He was quite impressed because the effect was much stronger than he had anticipated. He had read once that it was the 'Principle of the Conservation of Angular Momentum' that kept the swirling dust together, accelerating the rotation of the vortex as it stretched upward into the atmosphere. But mathematics could never be the cause of anything, since it didn't exist in the world.

Still Jonathan understood why people called these things "devils." Over the years he had observed them up close on numerous occasions. The pattern was always the same. The devil danced as it strode across the desert, arms waving over its head, hips swaying back and forth, waist gyrating in circles, knees bending crazily in one direction then flying off in another direction. The dust devil evoked the image of a human figure, perhaps suggesting a raggedy desert traveler or maybe some flamboyant individual wearing a floppy, wide-brimmed hat, a baggy, ill-fitting gray suit and flashy white shoes, sashaying down a busy metropolitan street all psyched up for a riotous night on the town. The body parts all changed position independently of one another, the chest at one moment stretching and bulging, the midsection twisting and turning, the head popping up and down then jutting out at ridiculous angles, the hips expanding and contracting, the arms and legs flailing wildly as the devil strutted its stuff. The devil displayed amazing dexterity and its movements were captivating, appendages darting in strange patterns that had hypnotic powers. First it jumped to one side then hopped back, then it twirled around and did another sidestep. What mathematics could anyone possibly use to describe such an uninhibited exhibition? The devil was anything but a simple, straightforward cylinder or a rotating funnel as the physicist might portray it, nothing at all like the column of warm air spouting from a fireplace chimney. The smoke that rose from the chimneys of houses never danced like that. The idea of a column of cold air being sucked down into the middle of a swirling vortex of warm air wasn't really any better. The ability of a diffuse flow such as the wind to assemble itself into a structured entity that persisted and had a clearly recognizable character was truly amazing. The devil's crazy performance went way beyond mathematics, yet the devil kept it all together and maintained its identity. The individual appeared to be high-stepping over the bushes and jerking around to some crazy beat. Jonathan wanted to hear the same music that the devil heard so he could understand it all and perhaps dance across the desert in the same fashion, mimicking the dust devil by employing the same patterns of moves, but by now the devil was at least a mile away heading for god-knows-where, absolutely refusing to quit the spasmodic boogie, carrying on with the same enthusiasm as before.

Jonathan wasn't seeing a pattern that no one had ever noticed before. Many others, people with direct and intimate contact with the desert world, had made the same observations. Jonathan might simply be portraying the mechanical actions of molecules in anthropomorphic terms, but it was also possible that nature was suggesting something to him here, perhaps pointing out the kinship of natural phenomena. Nature was telling him "I have made all sorts of things dance. I made the wind dance, much in the manner that I made you dance, dodging and high-stepping the obstacles that you encountered during your long journey across this earth. As you have deftly noted, the similarities were plainly visible for everyone to see."

Physics textbooks did not reserve a chapter for dust devils, not because they weren't interesting, but because for physicists, mathematics was the most interesting thing in the whole world and in this case the mathematics fit poorly—or not at all. The dust devil had to go away so that the mathematics could remain on the center stage. Physicists hastily concocted a lame explanation for these occurrences and left it at that. They had better things to do.

Unlike a dust devil, a house was a mathematical object, much more similar to the objects commonly studied in laboratories. Houses were not natural phenomena but rather they had to be built from scratch, assembled out of materials that were themselves mathematical objects, things such as bricks and boards and nails. The house was structured and ordered with regular spacings and uniform lengths and lots of perpendiculars, in stark contrast to the dust devil with its chaotic behavior and fluid motion, but the physicist could still observe the house interacting with its environment.

A house could be divided into a foundation, a set of walls, and a roof. When a hurricane or a tornado approached the house, the wind speeds increased and at some point the force of the wind tore the roof off the house. This value was a measure of the strength of attachment. If the physicist said that the roof and walls were therefore held together by a force, the physicist was creating an abstraction to account for the real situation. The roof was not held onto the walls by a force. The roof was held onto the walls by nails. The rafters were toe-nailed to the wall plates and also to the ridge beam. The sheathing was anchored to the rafters with special ring-shank nails, the shingles with galvanized roofing nails. The specific, detailed method of construction was necessary to understand the actual connection between the roof and the walls. Positing a force did not uncover this true relationship, but a force was something that could easily be assigned a number through a relatively simple process.

An explosive charge deposited inside the house would also blow the roof off. Or the physicist could bolt metal I-beams to each side of the roof and pull the roof off with a crane. Each of these techniques gauged the strength of attachment by measuring the force required to separate the roof from the rest of the house. The numbers produced by these methods might agree or they might not, but in no case did they establish that a force existed between the roof and the house. In each of these experiments, the physicist was only measuring the speed of the wind, the pressure generated by an explosive charge, or the tension on a steel cable. Just because a force was required to pull the roof off of the house didn't mean that a force was holding the roof onto the house. The force of attachment was merely a ghost hiding in the attic, lurking behind the boxes of unwanted junk. The physicist told Jonathan that there was something strange up in the attic, something Jonathan hadn't noticed before, but when Jonathan grabbed his flashlight and poked his head through the access door, he did not get the feeling of a strange presence. The attic was empty, so where was this force that the physicist spoke of? The force didn't reside in the attic. The force only haunted the minds of physicists.

The force that held the roof onto the house in the face of a strong wind depended at least in part on the shape of the roof and its orientation to the direction of the wind, factors that did not come into play in the other forms of measurement. Using the various techniques that were available to him, the physicist would get different numbers for the same roof or the same number for different roofs. A sturdily-built roof that had fallen into disrepair or weakened due to age might yield the same number as a brand-new roof that was poorly-built. The force did not indicate the type of roof that was in place or gauge how well the roof aesthetically matched the rest of the house. The force wasn't really about the roof at all. In fact, the force didn't tell Jonathan anything about the actual roof except for one thing, a general idea of what it might take to separate the roof from the house. Sure, he couldn't learn about roofs by studying forces and thus these forces did not grant him a genuine knowledge of roofs, but still the force told him "something." Jonathan questioned whether this "something" played the role that the physicist ascribed to it, namely, that it provided him with an adequate foundation for his understanding of roofs. The physicist could concoct other schemes for making measurements and piece together the results, weaving the quantities into a patchwork quilt of mathematical relationships, but where was the actual roof in any of this? The method not only fell short of an account of the roof in its infinite particularity, but it fell hopelessly short of any reasonable standard for knowledge. "No matter," the physicist replied, "I'll just forget about the actual roof and turn the roof into an abstraction, then I can deal with it perfectly well."

Physicists said that the atoms and molecules of the universe were all held together by forces. The solar system was also held together by forces. The problem was that physicists never figured out what these forces were or how they worked. They merely used mathematics to draw pictures of their influences and behaviors. They couldn't find explanations for them so they decided to skip that part, however, that was a mistake. Without knowing what forces were, they couldn't possibly proceed. Instead they settled for writing formulas that only described the outward appearances of things that remained anonymous.

Physicists later discovered a phenomenon they called "entanglement" where two particles were connected over certain distances for no apparent reason. The phenomenon was a source of great irritation to them because it didn't make sense in terms of their stock explanations—but a force didn't make sense either and they never explained that. Forces acted without intermediaries, yet that was perfectly understandable, so why was entanglement suddenly a puzzle? They felt that they had to explain how one particle could be connected to another particle over comparatively large distances with no apparent interconnection between them, yet that was something they already knew existed in the case of forces. Entanglement was not a force, but it might simply be another case of action at a distance. So how was the trick done? Good question. No one could say that they understood something until they understood what it was and how it worked. No, a force didn't warp space, because the structure of space was just a mathematical idea. The bending and molding of space was the ad hoc manipulation of an empty, non-existent, mathematical form that was nowhere to be found outside of the human imagination.

Jonathan once had a personal experience with entanglement, an incident that occurred several years after he first arrived in desert southwest. His parents had been steadily getting older and their health had begun to fail them. He hadn't seen them much since he had moved to New Mexico, so as the Thanksgiving holidays rolled around, he and Sarah decided to fly to Chicago to visit them for two weeks. When he first met Sarah, she had just bred her female blue heeler, a squirrelly, nervous and insecure dog who would stand right behind her and cause her to fall backward, with a magnificent male heeler who, although good natured, was so tough and strong that he was a bit scary to be around. They kept two of the pups, a female they named Rocky and a male they named Stewart. They decided not to kennel their two dogs as usual but instead leave them at the old farmhouse where they were staying at the time. They weren't happy with kenneling their animals anyway, locking them in small cages next to all the other barking dogs, dogs that were unfamiliar and threatening to them. They felt that the experience was too stressful, plus Jonathan couldn't stand the haunted looks on the dog's faces as he abandoned them and walked away. He had previously installed a dog door at the farmhouse so the dogs had the opportunity to go outside whenever they wanted and sleep on the bed together at night as usual. He put the lights on timers to make it appear as if he and Sarah were still around, and arranged for two women who lived nearby to come over and feed them twice a day and check on them.

Everything went fine for ten days. Sarah called the women regularly to make sure. Then on Sunday as Jonathan was watching football games on television with his father, he was overcome with a sense of dread. He left the room and went into the bathroom. He felt weak and fell to his knees directly in front of the washbasin, his hands clutching the sides of the porcelain sink to keep from dropping all the way to the floor. Terrible thoughts of Stewart inexplicably filled his head. He knew something was wrong. Stewart was in grave danger. He tried calling out to him, telling him in his mind to hold on, that he'd be back soon.

Jonathan recovered after a while and told Sarah what had happened. She tried calling the women but couldn't reach them. Jonathan sat in front of the television with his father but he wasn't interested in the games anymore. Later in the evening Sarah contacted the women. Everything was fine. They'd gone over to the house in the afternoon. She asked them if they'd seen Stewart. No, they'd hadn't seen him, but they were sure he was around somewhere. Jonathan knew right away from their report that his vision had been true and something bad had happened to Stewart. Somehow Jonathan knew that Stewart was dead. He and Sarah stayed in Chicago four more days, but Jonathan was miserable and he just wanted to go home.

When they finally arrived at the farmhouse, only Rocky came out to greet them. After quickly unloading the car, Jonathan immediately set out looking for Stewart, striking off in different directions and calling out his name, but he got no response. Winter was in full force in the mountains and patches of snow were scattered about under the trees. Sarah lit the fires in the wood stoves and cleaned up the house, while Jonathan stayed outside in the cold well into the night as a full moon rose over the hills. Several times he heard the sound of Stewart's collar as Stewart came running toward him and Jonathan spun around to greet him, thrilled and grateful that he'd been wrong after all, that Stewart had just been moping around in the hills, but every time he turned toward the sound, only silence and emptiness greeted him.

Physicists didn't believe in the entanglement of minds, another example of quantum effects in the macroscopic world, because they couldn't identify the causes or control the effects. Psychic entanglement was an extremely complex phenomenon and they hadn't figured out the conditions that must be met in order for it to happen. Jonathan often thought about people he'd been very close to in the past and sometimes he wondered if he was still connected to them. When they died, would he feel it, either in a vague way or in some intense vision, and know immediately of this person's passing? Had their ties faded completely or did remnants linger, and were these residuals enough to maintain a faint sense of entanglement in his consciousness? Perhaps entanglement depended on the flows of some form of dark energy that carried the bond along with it, much as the wind carried the sound of a person's voice. When Sarah was calling him, he might not hear her voice if the wind was blowing the wrong way, and yet, if the wind was just right, the movement of air carried the sound directly to him and he heard her voice with crystal clarity. His cell phone was often on the fringe of the reception area, particularly if he drove very far from the interstate, and sometimes he got a signal and sometimes he didn't. Sometimes the signal faded in and out at a given spot, even if he didn't move, and this didn't depend on which way the wind was blowing. So here Jonathan had two clear examples of long-distance communication depending on complex factors in the environment, and he might reasonably suppose a similar situation existed with psychic entanglement, only in this case the unreliability was generally taken as evidence that the phenomenon didn't exist.

Other factors might figure into his experience of entanglement causing further complications. What must the exact nature of his relationship to another person be in order for the possibility of emotional entanglement to arise between them? He might assume that he had an emotional bond with someone when in fact it wasn't all that strong. If he knew how emotional entanglement worked, perhaps he could control it, or at least influence it to a certain degree, but he couldn't dismiss the phenomenon simply because its patterns were beyond his ken. Jonathan knew that reproducibility was utterly meaningless in a completely non-mathematical universe, so his inability to cause the phenomenon to appear on demand was not the issue. The physicist didn't care about emotional entanglement. He didn't want to live in the real world because he couldn't master it and he couldn't understand it. The physicist could only think in certain terms and once he stepped outside of those boundaries, he was completely lost.

Jonathan had definitely heard Stewart's collar jingling behind him that night, but the physicist was convinced that Jonathan's mind had just been playing tricks on him. Jonathan felt that Stewart might have in fact seen him walking in the orchard and heard Jonathan calling out his name. He might have been happy that Jonathan had finally come back for him and he wanted to greet him, but Stewart had died several days earlier. Perhaps his spirit still lingered in the orchard and beckoned to Jonathan. Or perhaps Jonathan just imagined Stewart's presence in his mind. In any case, Stewart's ghost has long since departed the area and it no longer haunts the property adjacent to the old adobe house as it once did during the days and weeks following his disappearance. And now thirty years have passed away since Jonathan last heard Stewart's collar jingling behind him as he walked the fields and hiked the expansive piñon and juniper forests extending in all directions from the old farmhouse.

Jonathan wondered if the physicist was right and everyone lived in a bleak and barren world of mindless forces and mechanical transfers of energies. It was funny how people's minds could play tricks on them. Perhaps they started seeing things that weren't there and began to hear sounds in the night and imagine a reality that was very different from what was real. How could the physicist be sure that his mind wasn't playing tricks on him? Perhaps the physicist simply wanted a mathematical universe in the same way that Jonathan had wanted Stewart to come home to him, so that they could be happy together. The physicist was happy being with his cherished mathematical constructions, his constant companions in life, the ones he relied on for comfort and friendship. Or perhaps these things were not real and he was just imagining them as he walked the streets and hallways, hearing the calls of ghostly spirits that did not exist in the world. Perhaps his mathematical universe was just an illusion—not so much a deception of the senses, as a deception of the mind.

Leaving the hills behind, Jonathan struck out across the seemingly level plain, heading in the general direction of the truck. The overall flatness of the terrain created the illusion that he could see everything for miles, but strangely he couldn't find the truck anywhere. He always made a point of stashing the truck in a concealed location, a shallow depression or a deep arroyo, just to be safe. Visibility was high out in the open desert and he preferred that no one knew he was out there that day, but in fact a large portion of the land was completely hidden from view, including all sorts of troughs and bowls, valleys and pits, canyons and channels, both shallow and deep, lurking below the general elevation of the plain, far from trivial aspects that he would have to deal with on his way back.

Emma was glad to be freed from the hinderance of the vast bed of rocks and ran wildly across the soft ground. She was constantly in rapid motion so Jonathan got the impression that her activity level was already set on maximum, but she was cleverly holding herself back and pacing herself. Suddenly she smelled something and instantly turned up the tempo, streaking into a blur of motion, darting back and forth in a series of lightning quick moves, stirring up a haze of dust that rose above the bushes, the suspension of dirt rapidly thickening and spreading out concealing all further activities. After a few moments, Emma emerged from this earthy cloud with a rabbit in her mouth, its head and legs limp, the appendages dangling and flopping aimlessly as she strutted about proudly. She paused and stared at Jonathan with a triumphant look in her eyes. Her face was streaked with sap and dirt and she had the painted face of a fierce warrior, lending a terrifying aspect to her glare as she turned toward Jonathan with the hapless victim in her mouth.

Despite her obvious demeanor, the metaphor of a fierce warrior was incorrect. Emma didn't kill the rabbits she caught. They died completely of their own accord. Emma had a gentle mouth and she didn't clamp down on them. She was happy just to carry them around for a while and savor the sensation, then when she got tired of doing this she would put them down and go her own way, practicing her own version of the sport fisherman's "catch and release" policy. This was what she did, only the rabbits were already dead, having died for no apparent reason the moment she grabbed them, never understanding that this wasn't a necessary thing to do.

Essentially each rabbit was killed by the idea of its impending death. The fact that a rabbit could be killed by something as insubstantial as an idea seemed too farfetched to be true, but Jonathan always examined the bodies of the deceased and they never showed any marks or visible injuries. The most uncanny aspect of this whole episode was that the idea by which each rabbit died was a falsehood. Jonathan understood that false ideas could be more dangerous than true ideas and here was a good example. Each rabbit died from the false belief that it was about to be killed, something Emma never had any intention of doing.

Wars were waged by no more than ideas and as a result many people died. Ideas were responsible for all sorts of things, but ideas appeared in the world exclusively through the actions of the beholders of those ideas. The possibility of an idea acting by any other means was entirely out of the question. The idea that killed the rabbit didn't lurk in the shrubbery and then jump out and attack the rabbit externally. The rabbit could not die because of another rabbit's idea of death. Force was similarly an idea that could act only through the minds of physicists. Forces did not exist on their own and operate independently of the thoughts that had given rise to them. If only physicists didn't think so much, imagining straight lines that weren't there, picturing mathematical constructions that didn't exist in the world, they wouldn't have so many forces to deal with. If only rabbits didn't think so much, anticipating events before they happened, imagining outcomes that weren't real, many of them might still be alive today.

Emma happily paraded around with her trophy. She had just caught her twelfth rabbit. She carried the expired creature in her mouth as she explored the terrain in her usual fashion, savoring her hard-fought victory over the willy rabbits. Eventually the thrill wore off and her thoughts turned back to the hunt. She found a cool spot in the shade beneath the branches of a sagebrush and gently placed the dead rabbit on the ground, then carefully pushed the body with her nose toward the base of the sprawling plant, camouflaging this valuable and highly sought after object so that the hawks would not notice its presence as they circled overhead, thereby ensuring that the prize would be preserved for her brethren. Tonight the coyotes would find a free meal waiting for them, a generous gift from a kindred spirit. Emma didn't depend on rabbits for food. She knew that she would have an ample dinner tonight prepared by her trusted, lifelong companion. She'd rather donate the banquet to the less fortunate of her kind, the ones who didn't have human guardians to care for them.

—3—

As spring gradually unfolded, the budding yellow-green foliage of the mesquites infused the pale desert landscape with rich swaths of color and the ocotillos grew shaggy coats of small leaves to conceal their long, sharp thorns. Jonathan rose from bed in the dark of night so that he could reach his destination at the first light of dawn. This was now the most pleasant part of the day and not something to be missed.

Jonathan began breaking out his gear, a routine he had performed so many times before that he could accomplish it without being fully awake. He raised his eyes without thinking and was stunned by what he saw. The early morning sky had been set afire by an unseen torch. The peaceful black clouds that were only a few moments ago quietly floating above the eastern horizon had suddenly come alive with illumination. His first thought was that they looked like the blazing coals of a campfire, but he had stared into too many fire pits over the years not to know that the embers glowing in the dark of night were comparatively hazy and indistinct, the colors softly merging into one another and gradually receding into the unlit portions of the black coals. Some sunrises were in fact diffuse and indistinct, but this one was different. Here each briquet was outlined in exquisite detail, more like an LED display than a campfire, a crisp and clear mosaic of brilliant colors and exact shapes, the contours all precisely defined and exceedingly complex. Jonathan watched the light show develop, transfixed by each stunning new development. He finally turned away and finished assembling his pack in the orange glow of this grand exhibition, preparing for yet another excursion into unknown territory, loading stainless steel flasks of cold water into the main compartment and stuffing a rain jacket on top of them, along with a water bowl for Emma.

Jonathan ambled about his provisional camp looking for a place to relax, pouring one final cup of hot coffee from his king-sized thermos, eventually locating a comfortable spot nearby where he took the time to unfold a topographic map of the area that he had found while rummaging through a stack of books and papers in his library. Gazing down at the map spread out on his lap, he noted some of the main geographical features he would encounter today, surveying the swirling designs drawn on the large sheet of paper, planning the exact route he would follow through the hills and arroyos. Emma did not understand this and proceeded on without him, fully engaged in her usual routine of running around and sticking her nose into the confusion of twigs and branches surrounding each bush, sniffing out the burrows of geckoes, rabbits, and whatever other creatures that had found places to call home.

Jonathan checked his current elevation on the map. This number was the distance between where he was sitting right now and sea level. The only problem was that the seas weren't level. Not only were they plagued by huge waves and rolling swells, but the general outline rose and fell due to the gravitational pull of the moon, the strong and sustained winds of storms and hurricanes, the melting and freezing of the ice caps, the bulging of the oceans at the equator due to the earth's rotation, the variations in gravitational pull at different points on the surface of the earth. So elevation was the distance between this spot and what turned out to be something that didn't exist.

Everyone tended to think of the earth's crust as being absolutely rigid, but that too bended and flexed. Obviously, measuring elevation wasn't as easy as it had first seemed, but elevation could be determined by tying it to something else. Since atmospheric pressure decreased with altitude and atmospheric pressure was relatively easy to measure, elevation could be measured indirectly by measuring atmospheric pressure. The problem was that the atmosphere was itself an ocean of air that experienced swells and troughs, and these variations caused large swings in barometric readings. In fact, everything in the world rose and fell, expanded and contracted, jiggled and fidgeted, bent and flexed, increased and decreased, wobbled and vibrated, skewing every conceivable relationship. In an ideal world, none of this would happen.

When Jonathan tried to grasp the concept of elevation it slipped away from him. The fact was, he wasn't 5923 feet above anything right now, that is anything that could be identified, specified or defined. Something must be 5923 feet below his feet, perhaps a cave or a stratum of rock or a river of magma, but the distance was meaningless because this other object was not simply unknown, but unknowable. How was Jonathan supposed to conceptualize his distance between himself and a nameless, insignificant object buried deep in the earth where no one could ever find it? No one was talking about a real object below his feet. Everyone agreed that Denver was a mile high, but it was a mile higher than what? It was a mile higher than a concept. Mixing abstractions and realities together created deep and serious problems—not for making calculations, but for understanding what these people were talking about.

Much as the configuration of Orion in the sky, the measurements of elevation dissolved away into nothingness as Jonathan tried to trace them back to their origins, the ends of the imaginary strings dangling freely in a fantasy world of conjectures and visualizations that could not be easily connected to anything that actually existed. Everyone took the number 5923 feet to be an absolute fact, a calculation that was meticulously precise, a conclusion that could not be denied by anyone, but in truth, it was just a dream. The reference point was not an extrapolation of the surface of the ocean because the irregularities of the ocean could not be extrapolated. The expression "sea level" was just a figurative term that was not meant to be taken literally. Erecting conceptual bridges between the physical world and a universe of mathematical fictions was a rather risky business and not something to be taken lightly, yet no one seemed to give it a second thought.

However, sea level didn't need to exist. The mapmaker simply tied the numbers to each other. The fluctuations in water levels around the globe and over the years were haphazard and meaningless in themselves, but he could turn this chaos into a unique and stable surface, exactly the type of entity that his mathematics required, thereby establishing a suitable reference point for his calculations of elevation. He accomplished this task by lumping all the variations together and forming an arithmetic mean that replaced the infinite varieties of shapes and forms with a single representative, thus converting the endless stream of data into a single number that defined a permanent structure. All the imperfections conveniently turned into a unique geometrical shape and he had a surface as smooth as glass—a perfect sphere. This flawless object of mathematical precision was then used to establish the vertical distances that became the elevations of geological formations. Since the oceans rose and fell due to many factors and these factors were the result of numerous cycles of change superimposed over other cycles of change, the mean still varied depending on the sample area and the sample period, thus it was not a physical reality but only a calculation.

When Jonathan climbed a high embankment and looked out across the ocean, the details were lost and the water looked flat and smooth. He could easily imagine the ocean to be the perfect surface of the arithmetic mean. From this elevated perspective, the ocean appeared to be an ideal reference point for the measurement of elevation and he could see no problem with it, but could the titanic magnitudes of oceanic events simply be a matter of perspective? These sizable phenomena and indisputable facts could be dwarfed and essentially eliminated by climbing a tall embankment, but did they really go away? The tides, waves and swells were insignificant when placed next to the entire extent of the earth, but could Jonathan change the nature of the ocean simply by comparing it to something else?

Jonathan imagined himself standing on a shoreline facing an object that truly was a sphere. Instead of walking into the surf with his toes sinking into the soft sand, he walked out onto a hard surface as smooth as glass, unyielding and unblemished, perfect in every way—a sight incomparably more fantastic than the ocean itself. While this surface had all the properties that Jonathan would normally associate with a sphere, this object had nothing in common with the actual ocean. The ocean was not a surface and thus it had no mathematical properties. Instead the ocean was intricately sculptured by meteorological events and these determined the exact size and character of the swells. The topographies of the seabed then determined how these swells broke, and the variations in wind speed and direction determined the resulting configurations of the waves. The ocean was unbelievably complex, as every surfer knew all too well. The true spherical object was so radically different from the ocean that it had none of these qualities and therefore it did not resemble the ocean in any way. Jonathan could say that the ocean was approximately a sphere only by comparing the mental images of an unrealistic way of thinking called geometry with the questionable notion of a physical object having a pure form—a procedure that was most baffling and mysterious to the uninitiated. As someone who had gone swimming in the ocean many times, Jonathan understood that the statement of the ocean being approximately a sphere was absolutely ridiculous and an outrageous proposition that no one would ever take seriously. The imagination that was so crucial to the mathematical way of thinking was what enticed a person to see the ocean as a mathematical object and this had to be cultivated during long hours in dreary classrooms where the real ocean was conveniently very far away so as not to interrupt the physicist's beautiful dreams or contradict his rash conclusions.

The fact that the sphere disintegrated upon closer inspection proved that the ocean wasn't really a sphere, since something like that could never happen with a real sphere. A real sphere remained a sphere under all circumstances. An infinite particularity, on the other hand, existed simultaneously on many levels and therefore it could have multiple characters. Because the ocean was an infinite particularity, the ocean could be both a sphere and not a sphere at the same time. The ocean could appear differently from different perspectives and thus it could assume different natures that were incompatible with one another. A mathematical object was always the same, not matter how one looked at it.

From afar the ocean appeared to be a mathematical object, but when viewed up close, it definitely wasn't. An observation that had been so secure, a conclusion that had seemed so irrefutable, turned out to be utterly false. Instead of a simple, uniform sphere, the oceans were an indeterminate structure of infinite complexity. The oceans merely looked like a sphere from a great distance, but that was not what they really were. The geometry was just an illusion. The oceans weren't geometrical at all.

When Jonathan looked closely at the ground, what he saw was a vast collection of diverse objects: rocks and stones, grains of sand and clumps of soil, trees and plants, tangles of roots and fibers, an assortment of dense masses with all sorts of openings, gaps, tunnels, and empty spaces between them. The ground was definitely not a surface. A smooth wooden floor was more like a surface. A kitchen countertop might be considered a surface, but a surface was a geometrical concept that did not even remotely describe the exterior of the earth.

The problem revolved around the meaning of the discrepancies. To the physicist and geometer, these inconsistencies and variations had no real significance. This was like saying that the square peg was approximately the same size as the round hole, and if it weren't for those little edges, the square peg would fit. All one had to do was say that those corners—minor details really—were unimportant, then it was easy to disregard these slight variations. The square peg could be made to fit the round hole if one simply ignored a few rather trivial aspects of the square peg. The mathematical sphere didn't quite fit the earth, but it was close enough that the differences could be glossed over, and the reality of the earth could be made to fit the form of the geometrical construction, but as far as Jonathan was concerned the disparities were of the utmost importance because they revealed the utter failure of geometry to account for what actually existed and showed that the geometry of the earth was a completely false rendition of a reality that was utterly beyond geometry.

If the earth wasn't actually a sphere, but only sort of like a sphere, then geometry was only sort of like knowledge. If geometry had only the appearance of knowledge and no real substance, then it was an imposter and not what physicists claimed it to be, a genuine account of what existed. The geometrical form was not only inaccurate, but it fell woefully short of the truth. The earth was nothing at all like a geometrical construction and could not be understood that way, but viewing the earth as a mathematical object paved the way for all sorts of calculations and numerical analyses and that was the sole justification for the geometry—it led to more mathematics. The irregularities and complexities of the real earth could all be left behind as physicists entered a world of unrealistic configurations and misleading representations. The amazing part of all this was that everyone nonchalantly accepted a situation that was utterly unacceptable: the transformation of physical reality into mathematical forms. Everyone automatically adopted a mathematical way of thinking because they had been trained to do this from a very young age. They took it for granted that this was the right thing to do.

The essential features of the world could not be grasped through geometry because geometry was not a depiction of reality. If everyone actually lived in a mathematical universe, then geometry would provide them with a true understanding of that universe. Geometry would not only be a direct observation, but it would also be the reality behind that observation, that is, geometry would transcend observation rather than just be the appearance of something that when examined more closely wasn't real. The fact that some observations supported the image of a sphere while others contradicted it proved that the sphere was just an illusion.

Physicists employed a curious logical principle here. The principle stated that "if an object looked like something, then that's what it was." This kind of reasoning was encouraged because mathematical forms were mere appearances and this was the only way to make them more than that, to elevate them to a higher status. Instead of dropping mathematics as a valid form of reasoning, physicists decided to keep it and furthermore make it the centerpiece of their method. In truth, mathematics only painted a picture of reality, but more importantly, it produced a completely false picture of what actually existed. Strangely, these concerns were summarily dismissed as being of no importance. Appearances were taken to be facts and existence was understood to be based on mathematics. Concrete thinking on the other hand produced a very different view of the universe, completely at odds with the one generated by abstract thinking. Concreteness turned everything around, upending the physicist's most cherished truisms and contradicting his long-standing interpretations of events, yet thinking concretely was the only way to think correctly about a concrete world. Once the details were eliminated, objects could take on all sorts of weird forms and they could assume a wide range of alien identities. The physicist cast aside the details precisely because they interrupted his mystical trance and stood in the way of his ethereal visions.

Nevertheless the physicist persisted with his argument. He said that when he drew a circle around the earth, he was tracing the actual outline of the earth, and thus Jonathan was wrong about mathematics not copying the reality of the earth. Jonathan noted that the physicist was only drawing a circle around an image of the earth, an image that did not even begin to copy the reality of the earth. He was drawing a circle around something that was already an abstract representation of the earth, something that did not include the overwhelming complexity of what existed. If the physicist truly wanted to trace the reality of the earth, where would he even begin? He couldn't trace the piles of rubble that formed the hills or the swells and waves that rose from the oceans. He couldn't trace the swirling currents of wind or the seemingly limitless expanses of plants and animals. He couldn't outline the compositions of the soils or the crystalline structures of the minerals, or draw a picture of the arrays of molecules underlying all of these events. None of these facts were included in his image and that was why he liked images so much. Physics itself was just an image, a picture of something that the physicist couldn't even begin to fathom.

Materiality created all sorts of problems for the physicist. Not only was the earth not a sphere, but the electron wasn't a particle—that is, a sphere with mass. Light wasn't a wave, space wasn't curved, the shape of an atom wasn't the outline of its dipole moment. These mathematical forms were merely observations and deductions. They could not possibly be the ultimate realities of the universe. Jonathan's detailed analysis of macroscopic objects had shown him that the realities underlying mathematical forms were always non-mathematical in nature.

Since geometrical forms could truly be said to exist only in deliberately fabricated objects that had been constructed directly out of geometrical ideas, objects that had been artificially molded into mathematical forms, this proved that the world wasn't really geometrical. The physicist could not start with mathematics and then arrive at mathematics as a conclusion. The mathematics would have to come from the world in order for the world to be mathematical. The mathematics could not come from the human imagination because in that case the mathematics was purely imaginary, still everyone could see it in their minds and that was how they decided it was real.

Physicists could pretend that the oceans were spherical and that was all that mattered. It made no difference in the mathematics and mathematics was the only thing of any importance to them, but to Jonathan, an understanding of the oceans—or anything else for that matter—required him to determine what they really were, and if the oceans weren't really spherical and sea level didn't really exist, then he couldn't understand the oceans or anything that was based on them by pretending that they were spherical and that sea level was a reality. Mathematics was all make-believe and could not possibly be the road to reality.

In looking at the atomic world, physicists imagined that they were looking at mathematical objects, things that were fixed and determinate, but if in fact they were actually looking at infinite particularities, then the objects they saw might be radically different depending on the perspective they adopted through their selection of an observation point. Looking at something from a great distance washed out all the details and left them with nothing but a basic outline, much as the ocean from the top of an embankment. When physicists looked at the atomic world with their instruments, a great distance was still involved, even though the atomic world wasn't really far away. The great distance was a difference in size. The physicist could not possibly swim in the oceanic details of the atomic world. He could not experience the submicroscopic correlates of breakers crashing on the shores or swells rising and falling on the open water. All he could see was the general outline of this world, but that was alright with him because the details would have only gotten in the way. The fact that the details couldn't enter into the picture was actually a good thing. The physicist could never have dealt with them anyway, any more than he could deal with the staggering details of the macroscopic world. He could even go so far as to delude himself into thinking that the atomic world had no details, that it consisted solely of general mathematical forms and was absolutely devoid of content. The physicist ordinarily deleted this content anyway in order to mathematize everything, but with the atomic world this step was unnecessary because the content was unknown and could never be taken into account. He had nothing other than the outlines, but those were exactly what he wanted, those aspects of the atomic world that were purely mathematical in nature.

Emma had been anxious to get going and she had once again taken off. Jonathan finally stopped daydreaming and looked for her. She was nowhere to be found. He scanned the surrounding desert but didn't see her anywhere. He figured that she must be down in the nearby arroyo poking around in the bushes. He decided to take one last look at the map. He thought to himself that because the patterns on the map depicted the actual landscape, there was no reason why he couldn't just execute his entire walk today without ever looking up, the whole time attending only to the symbols on the map.

Knowing the direction he was traveling by using his compass and keeping track of the distance he had traveled by counting his measured steps, he could arrive at a predetermined location without ever referring to the real world, attending only to the isoclines that had been drawn on a piece of paper.

But he wouldn't get very far using this method. He would trip over the first rock he came across because it wasn't shown on the map. He would stumble into walking-stick chollas, twist his ankles stepping into coyote dens and rabbit warrens, impale himself on the sharp thorns of mesquite bushes, and get bitten by rattlesnakes coiled up in the cool shade of junipers, because none of these things were indicated on the map. He could make the map more complex and add more detail to it, but the map could never be a substitute for the real world. He could construct a map using the world as a guide, but he could never construct the world using a map as a guide.

So why was it that he could just follow a map when he was driving a vehicle? Because the highway system had already been mathematized and therefore the designs on the map and the actual network of roads could match each other. One could be substituted for the other because the overall patterns were the same. The variety of routes had all been placed into distinct categories—interstate highways, U.S. highways, state roads, county roads, dirt roads—and each one could be represented on the map by a different type of line. The road surfaces were smooth and flat—outside of generally minor blemishes and irregularities—the lanes were parallel and of constant width, the curves were gradual and uniform and they flowed evenly into one another, the shoulders met specified dimensional requirements, the junctions and intersections were standardized. The infinite particularity of the earth had been painstakingly removed, at least within the boundaries of the highway grade, making what Jonathan encountered on his camping trips somewhat predictable. The map represented all of these features quite nicely, and that was all he needed in order to navigate a course to his destination. The infinite particularity of the earth was still there, only it had been suppressed to a large enough degree that the mathematics now dominated the situation and he was only comparing the mathematics of the map to the mathematics of the highway system.

But the place he would visit today had not been similarly mathematized. If he added more lines to the topographic map and penciled in every minute variation in elevation, the map quickly became unreadable. Even if there had been some way to include the position and orientation of every rock and stone, he would still not have even begun to capture the true nature of the landscape. He would still need to take into account the fact that every rock and stone was itself a matrix of mineralogical sectors and these in turn were bewildering composites of tiny crystals. Each crystal had its own unique patterns of impurities and irregularities. The complexity of nature proceeded to unfold as he continued to zoom in on it. The crystals were arrays of molecules, the molecules were assemblies of atoms, and the atoms were structures of subatomic particles. It was naive to think that the composition of the landscape ended there. No one could see any further. The original map could not possibly show all of this, yet it was all there, all essential parts of the total picture.

Today the desert was silent and tranquil, but the outward serenity concealed a hidden world of activity. The vegetation was slowly growing in the warm sunshine. Jonathan thought about all the biological processes inside all the living cells, the molecules being created by enzymatic proteins and the molecules being broken down by digestive enzymes, the replication of DNA inside the innumerable nuclei. The immensity of this microscopic industry was staggering. The sheer number of plants currently within his view was mind-boggling enough--the sage and creosote, the grasses and Apache plumes, the wildflowers and cacti--yet each one was a cosmos in itself. The creatures of the desert were also miniature universes. The birds were fluttering and perching on the branches, beetles were marching across the pebbles, flies were swarming in the air. How could he put all of this on a map?

The dynamics didn't end there. The very features of the terrain were changing as he sat, the wind dislodging tiny grains of sand from the bluffs and depositing them in riverbeds and sand dunes many miles away. The ceaseless action of wind and water constantly sculptured the hoodoos, changed the courses of the arroyos, and molded the shapes of the hills. He caught glimpses of these striking transformations whenever he returned to a familiar spot following a series of downpours and he did not recognize the place where he had once stood. Beneath the large-scale remodeling of the landscape, the processes of mineralization and crystallization were slowly changing the very compositions of the rocks themselves. Any map he drew became instantly obsolete the moment he drew it.

The thing that most people failed to recognize was that the difference between the map and the earth was not a matter of degree. The discrepancy was not a matter of adding more resolution or more detail, because the gap between the map and the earth could not be bridged. The map could never be made to actually resemble the earth and the earth could never follow the unnaturally smooth and continuous lines on the map. So how then were the earth and the map related? They weren't. The map was connected only to a particular concept of the earth, the idea of the earth as a mathematical entity, an abstract topological object, and not to the actual earth itself, in the same way that a sphere was only connected to a peculiar image of the ocean and not to the actual ocean itself.

So what good was the map in terms of understanding the reality of the physical world? The map told Jonathan nothing about the real world—not in comparison to the totality of existence. The exact nature of the terrain he would negotiate today on his walk was not depicted in any way and could never be presented on any map, no matter how detailed. The contours of the rocks he would scramble over, the small lines of weakness in the ridges he would happen upon, the shallow cuts in the hillsides he would follow to reach the dry washes below—all these things he had to discover for himself. Regardless of how many miles he covered and how many days he spent here, he could never acquaint himself with every aspect of this small area or even begin to examine it in exhaustive detail. This tiny fragment of all that existed was itself absolutely infinite.

Jonathan realized that he was throwing the word "infinite" around recklessly, without saying exactly what he meant by it. Mathematicians used the word "infinite" in a very specific sense, in order to distinguish it from merely a very large number. The infinite was anything that could be put into a one-to-one correspondence with a set of numbers, such as the integers or the real numbers. A mathematician could therefore maintain that the world might be very large, but it was not infinite. Jonathan disagreed. The infinite applied to anything that could not be specified, to anything that could not be precisely defined. The infinite applied to anything whose boundaries were mutable, like the clouds in the sky, objects that were constantly multiplying and merging into one another, appearing out of nowhere and vanishing into thin air. The material objects of this world were not countable, first because they were collections of endless details and that made them indeterminate, and second because they were constantly morphing and mutating, transforming themselves, breaking apart and coming together, while numbers on the other hand, were fixed, definite, eternal.

A map could never uncover the truth because the truth was utterly incompatible with the map. This was a fundamental principle of the world, the one fact that everyone must know and hold foremost in their minds, that every aspect of this world was itself an infinity of details that could never be resolved into a final, definitive form. A map was merely a concept comprised of other concepts, and concepts were simple in comparison to what existed. Jonathan could never understand the world by looking at a map, yet this was what physicists seemed to be trying to do—and not only physicists, but every scientist in every field of inquiry. Neurophysiologists were busy mapping the brain and astronomers were busy mapping the galaxies. Scientists had turned the pursuit of knowledge into an expertise in mapmaking, the enterprise fueled by the blind faith that once the maps were in their possession, they would harbor a true understanding of the world they lived in.

When confronted by the hopelessness of their project, their plan was simply to draw more maps. They would then collect these maps and bind them into atlases. If Jonathan tried following such a course, he'd be stuck here all day, sitting on a rock and paging through tomes of sketchy pictures and abstract drawings, trying to keep track of the presentations on each page, struggling to put it all together in his mind. When he finally got up and started walking, he'd have to constantly keep referring back to these books because no one could possibly memorize all that data. The particulars of his journey today would be overwhelming and the endless connections between them mind-boggling. At some point he would have to put the armload of books down and look up, only to observe the vast multitudes of oversights and omissions, the infinity of infinities that were excluded from the diagrams—in other words, the real world.

Emma suddenly barged through the bushes at full pelt and at first Jonathan thought that he was being attacked by some wild beast. She had discovered a fresh scent and was wild-eyed, running with her nose just above the ground, darting first one way then another, finally reversing direction completely and backtracking, nostrils flaring, snorting loudly, sucking in the smells with a series of hollow, vacuuming noises. Jonathan watched her and thought about a new project: creating a map of this invisible network of scent trails crisscrossing the desert, laying out in detail the elaborate designs that Emma was now tracing with her nose. The map would be a snapshot of an ever-changing bounty of fragrances, similar to the world depicted by the topographical map only much more volatile. If he had such a scent-map in his possession, could he successfully navigate through this unseen world of odors without using his nose?

The facility with which people commuted between the world they lived in and the ideas they entertained in their heads was astounding. No one gave it a second thought, and here Jonathan had to include scientists as much as ordinary people—or perhaps even more so. The pathway was well-worn and easy walking, the transition so natural and straightforward that no one felt the need to pause at the gate and reflect upon where they were going.

Jonathan folded the map and rose to his feet. For the first time he noticed something partially hidden in the bushes along the arroyo a few hundred yards upstream. He walked over to his pack and put the map into the front pocket where it traded places with the binoculars. He pulled the binoculars out from their soft leather case and put the eyepieces up to his face. The black object appeared to be a pile of trash bags deliberately stashed beneath the branches of a mesquite. Hunters often deposited the carcasses of kills in such bags, including the hides and entrails, but upon further examination he noticed that the light-colored contents of one of the bags were strewn about on the desert floor. He concluded that the bags were probably somebody's household garbage. With the winds of spring about to be unleashed, the situation demanded his immediate attention, so he decided that his hike could wait until he tended to this emergency.

He gathered his gear and summoned Emma, who had once again returned to camp, confounded by the unusually long delay. He threw his belongings into the back of the truck and loaded Emma into the cab, then followed the road a short distance down into the arroyo. The flows in the arroyo had been divided among several channels that split apart and converged together in different pairings and combinations. He swung the truck onto the first of these and proceeded upstream, always choosing the channel that looked the most promising, at the same time dodging numerous large boulders and islands of tangled brush, mesquite thorns raking the paint as he squeezed between them adding their individual contributions to the existing patterns of mars and scratches. The bags came into view as he rounded a bend in the wash. He pulled the truck alongside the dump and got out to inspect it. The sun had already dissolved several of the bags in a number of places allowing the garbage to be released onto the ground. An assortment of kitchen waste was now scattered in a wide circle and bound for every corner of this sprawling tract of desert. Jonathan always carried an old, dirty pair of leather gloves behind the seat of the truck specifically for jobs such as this. He began stuffing the rubbish into fresh bags—another one of the many wares that he carried around with him—crawling into the thorny bushes on his hands and knees to retrieve paper plates and cups, aluminum cans and foil, plastic spoons and forks, and various kinds of food tins.

He was making good progress when an ATV sprang up over the top of the ridge and descended the reckless tilt of the canyon wall, swaying and rocking back and forth as the driver negotiated the rubble-strewn, makeshift ramp, the muffled exhaust note barely audible in the still air. Emma came up beside Jonathan and together they stared in disbelief as the contraption and rider approached. Jonathan took a few quick steps, opened the door of the truck, and beckoned Emma to jump into the cab. She did not challenge his request and promptly took refuge on the bench seat as the ATV pulled up alongside the truck.

The rider reached down, turned off the motor, and looked directly at Jonathan. Jonathan could not decipher his attitude through his baseball cap and mirrored sunglasses. When the rider spoke, there was a touch of hostility in his voice.

"What are you doing?" the rider asked. Jonathan thought that it was pretty obvious what he was doing. He was collecting all this garbage.

Jonathan replied in a similarly stern voice as if impatiently answering a child's silly question. "The wind is going to come up and scatter this rubbish across this entire area, so I thought I'd collect it before that happened. It was such a nice place, and I wouldn't want to see it littered with this mess."

The rider nodded. Jonathan wasn't sure if the rider agreed with him or simply acknowledged the fact that he had heard what Jonathan said. Jonathan suspected that the rider was the one who had put the garbage here and didn't like anyone poking around in his trash.

"You let your dog run around freely?" the rider inquired.

Jonathan replied, "Well, yes, I do."

The rider seemed miffed by Jonathan's response, obviously upset that Jonathan did not exercise more control over his dog. The rider stared down at the handlebars and hesitated for a moment, apparently reluctant to speak, then turned to face Jonathan.

"I've got some traps set just south of here. You might want to stay away from them."

Jonathan's mind started reeling. He'd never suspected that such danger lurked so near. He'd been coming to this canyon regularly over the past few years, oblivious to the likelihood of disaster. He and Emma could have easily gone in that direction. Emma was very fortunate that the trapper had come along when he did, before they set out on the trail. The crucial meeting was nothing more than blind luck.

"I appreciate the warning," Jonathan said. "But what about the other way, to the north?"

The rider was silent, as if lost in thought, or confused by the simple question. Finally he said: "No, nothing." Then he added: "I just ran into some javelina a little ways up the canyon here. One of them charged at me, came right up to the ATV. I had my pistol by my side, but it's illegal to shoot them."

"I didn't know that. That's crazy." Jonathan wouldn't hesitate to shoot them if they ever threatened him. No one would ever find out.

"They're everywhere now. I'm a hunting guide, and we've started seeing them up at the higher elevations. Soon they'll be throughout the whole state. They'll really mess up your dog."

A horrible image appeared in Jonathan's mind and pangs of fear surged through his chest. This guy was referring to all the places where he and Emma routinely hiked. They shouldn't even be out here. What was he thinking? He was walking around in his own little world and putting Emma in grave danger.

Jonathan and the rider suddenly became silent. Jonathan was stunned by the startling news and unable to come up with anything more to say. The rider appeared to have further business and needed to get going. They stared at each other blankly for a few moments, then the rider reached down, started the ATV, and signaled farewell with a slight motion of his head. The sound of the motor faded away as he shrank from view, his vehicle bobbing up and down and rocking from side to side. Jonathan was grateful to be once more alone with Emma in the peaceful desert.

Jonathan let Emma out of the truck. She jumped down and stared after the rider, disappointed that she hadn't gotten the chance to meet him. Jonathan stood next to her and peered down the arroyo along with her, even though the rider was long gone and there was nothing more to see. Finally Jonathan returned to his chore, gathering the last of the filthy items and mashing them into the tangled mess of the last remaining open bag. He toured the area looking for more stray garbage, insisting on a thorough cleanup of the area, when he found two freshly skinned carcasses, an adult coyote and a young coyote, side by side in the weeds. The bodies reminded him of museum displays, plastic models with clear skins showing the tendons and muscles and the arrangement of the internal organs for educational purposes. Jonathan realized that the traps might be a lot closer than the rider had indicated.

Emma also saw the corpses but showed no interest in them. She was busy sniffing what Jonathan now understood to be javelina digs. They had never run into javelina in this canyon before, yet the javelina were undoubtedly here somewhere. He didn't know much about javelina, but he knew that Emma could be confrontational at times and he doubted that she would fare well in a fight with these nasty beasts. Perhaps they'd better not hike here today. He called Emma and put her back into the truck. Since nowhere was safe, they might as well try a new place, a place where at least the danger was unknown and perhaps less immediate, a place they'd never hiked before. The day was still young and they had plenty of time left for a good walk. He climbed into the cab and joined Emma. They sat shoulder to shoulder, both of them wondering what they were going to do next.

Jonathan turned the truck around and backtracked down the arroyo, following the fresh tire imprints he had just made. He swung the vehicle onto the bumpy dirt road and continued onward, the tires crunching the gravel as he slowly wound his way through the hills looking for adventure. He came upon a side road that obviously hadn't been traveled in a long time and decided to try it. After a few hundred yards he understood why no one had used this road. Although the surface didn't look rough, at least by backcountry standards, the lack of vehicular traffic had allowed the grass to grow tall. A friendly facade now concealed a hefty layer of rubble underneath, including many large rocks. The truck pitched and lurched from side to side forcing Jonathan to slow down to a crawl. Even though the truck was barely moving forward, the hood jumped up and down, first on one side then on the other, mimicking a low rider's car cruising the streets of Española on a Saturday night, perhaps with an intermittent short in the control box for the hydraulics.

The rear end similarly jumped around, alternately bucking like a mule then hopping like a rabbit, the wheels skipping over the loose rocks causing the gears in the old differential to clang loudly. After a few more miles the routine got more than just tiresome. The constant jarring made Jonathan's body feel as if it were being shaken apart. He fixed his eyes on the landscape outside to gain some sense of stability, much as a landlubber on a ship stared at the horizon to ward off seasickness. He kept telling himself that he would find the perfect place just up ahead so as not to despair, although he began to question the merits of this grueling effort.

He finally rounded a bend and entered a long valley where the road smoothed out and cut through seemingly endless groves of magnificent ponderosas. He quickly found what he was looking for. The grassy area up ahead had no pullout or fire ring, but the ground was flat and an ideal place to camp, with a stunning stone wall for a backdrop, the tall rocks leaning over the opening to a narrow canyon. He parked underneath the trees and painfully emerged from the cab, standing unsteadily next to the truck, holding the door handle for support. "Ok, Emma! Let's go!" Emma saw immediately that this was a new place, somewhere she'd never been before, and couldn't wait to explore it. Jonathan's pack was ready to go, so he locked the doors and headed out.

He walked into the entrance of the canyon and followed the channel for a couple of miles, working his way through the foothills and climbing steadily toward the mountains. The canyon was bordered in numerous places by rows of towering, fractured columns of stone, all bundled tightly together. The reddish rock was tainted with areas of black discolorations. On the sun-drenched north side, the rock surfaces appeared to have been singed by a fire of biblical proportions, although that was clearly not what happened. On the shaded south side, the rocks had a shiny appearance and the black patches appeared to be oil oozing out from the interior and soaking through the porous material to the surface. A few of the columns were leaning heavily forward. Jonathan wasn't worried that they would fall over, but what did bother him were the sizable blocks of stone that sat on many of the tops. These seemed ready to slide off at any moment. He cast nervous glances at them as he walked past.

Jonathan eventually found himself in a remote and desolate region where the character of the land abruptly changed. The canyon suddenly narrowed and became populated with throngs of strange hoodoos. They all seemed to be watching him, tracking his movements with menacing and unsettling glares. He felt that he should at least acknowledge their presence, so he stopped and addressed the large audience of rock creatures, asking them for safe passage, but his words fell flat in the cool, dry air and the rocks stared back at him in silence.

The canyon opened up once more, vertical walls now set a good distance from the main wash, crowns of splintered rock towering above his head glowing yellow in the slanted rays of the afternoon sun. He scaled a short but steep pitch on his hands and knees, clawing at clumps of grass, grabbing handfuls of turf and pulling himself upward until he reached a level shelf. He stood up and began walking normally again. He discovered a well-established game trail hidden behind a thicket of trees. The trail led to an opening in the forest where the berm of dirt and gravel had been hollowed out by erosion. To his amazement, a collection of the strangest monoliths he had ever seen were standing in the opening, towering spikes of compacted rubble crowned with giant, oblong heads of stone, each spire maybe 15 or 20 feet tall. The massive stone heads were all oriented horizontally so that the tapered ends appeared to be the snouts of reptiles. The monoliths were all looking at each other, bunched together in a secluded alcove set well apart from the primary channel where it would be unlikely for anyone to casually stumble upon them. Jonathan scrambled along the loose gravel of the hollow to get a better look at them. The scene was utterly fantastic. He decided to take pictures of this collection of bizarre geological formations, unlike anything he had ever seen before, so he dug the camera out of his pack and captured their images from a number of different angles.

As he departed the magnificent shrine, glancing over his shoulder for one last look at the stunning rock creatures, he noticed that the sky had suddenly clouded over and the small patch of gray that was visible from the valley floor looked particularly menacing, indicating that a storm was brewing nearby. Given the unexpected turn in the weather, he elected to head straight back to the truck since he had a long journey ahead of him. As usual, he selected a different route for the walk home. Because he would be dropping down more than climbing up, he would tend to walk a little faster and arrive at camp a little quicker when compared to his outbound journey.

Approaching camp from a different direction forced Jonathan to connect with the valley where he had parked the truck well north of the vehicle. He released his GPS from his belt to check the mileage to his destination, just out of curiosity. The number was a bit puzzling. Why was a geometrical construction the first thing that came to mind when he thought about his distance to the truck? His true distance involved all the details of his route, the obstacles he would face, the encounters with wildlife, the mechanics of each step, the precariousness of each foothold. Why weren't these facts the first things that came to mind when he thought about his distance to the truck? The straight line told him nothing about the trials ahead, the various hills he would climb, the arroyos he would cross, the thickets he would circumvent, the rocks he would step over. The straight line conveniently disregarded the whole of existence, something unfortunately he could not do himself.

Jonathan spotted something up ahead, a slight movement high atop a ridge. After a brief moment of bewilderment he realized that he'd just caught a glimpse of Emma. Apparently she had gotten captivated by a fresh scent and had strayed way too far for her own safety. If she ran into trouble up there, he wouldn't be able to reach her in time to help her, but then for someone who could cover as much ground as quickly as she could, the distance was not so great. The simple fact was that her distance from him was not the same as his distance from her. Emma seemed very far away from Jonathan's perspective, but from Emma's perspective, Jonathan was only a short distance away. Physicists wanted distance to be the same for everyone because according to their way of thinking that was what made it objective. They accomplished this task by requiring everyone to execute the same type of calculations in every situation. Since distance could only be

60

measured through a standardized procedure, it became the same for everyone. Everyone got the same number, however, a hundred yards for Emma was still not the same as a hundred yards for him.

Instead of letting mathematics determine distance, Jonathan wanted to let the world determine distance, however, if his method took into account the practical aspects of crossing a certain parcel of land, the distance would be neither symmetrical nor universal, characteristics that the physicist deeply cherished. Each situation would have its own calculation tailored to fit the detailed circumstances of that particular location. The result would be much better than the physicists' distance since it would accurately reflect what was actually involved in moving from one place to another, but the mathematics would be a lot messier. For physicists, mathematics always came first, so distance had been conceived according to the ideas of geometry. This was the most satisfactory approach from a mathematical point of view, and much preferable to the unmanageable definition that Jonathan had just proposed, even if that definition was more appropriate to the actual world he lived in. Once again, Jonathan was putting the world first over and above the demands of the mathematics. He and Emma were separated by two completely different distances, but of course, he was referring to the realities of their individual existences and not talking about some ethereal notions involving nonexistent entities.

The physicist retorted that the real-life situation simply demanded a more complex mathematical formulation going well beyond the simple straight line, but complicating the mathematics wouldn't solve the problem. The problem was that the true distance could not be constructed out of mathematical elements. The distance was predominantly outside the range of geometrical considerations, incorporating factors that mathematics did not have the power to resolve, elements that could not be measured or quantified, aspects that were difficult or even impossible to define.

The desert was a very rough place and Jonathan could not easily run across it. Thus for his safety and well-being, he always walked carefully by putting one foot in front of the other. Emma, on the other hand, being low to the ground and having twice as many legs as he did, ran quite easily across the desert. Jonathan could calculate the speed at which Emma normally ran, compare it to the speed at which he normally walked, and come up with a ratio that reflected the difference in the distances between them. By doing this, he was making an extra effort to mold mathematics to fit the reality of his world, but this approach was only a small step in the right direction. His characterization of distance using the average time it took to traverse an intervening space ultimately fell into the same trap as the original straight-line definition. The true distance was also the effort required to reach the other location from the current location, and this depended on numerous factors, regardless of whether he was talking about Emma or himself or some other person. These factors included skill and stamina, the mental clarity to make the correct decisions, that is, identify the best route and execute the most graceful and efficient muscular movements, and the conscious objective not to put up psychological barriers and thus reap the benefits of positive thinking. Emma had an advantage over Jonathan in this department because she was more optimistic than he was and this fact tipped the scale even further in her favor.

Jonathan saw that distance was determined by many things, including vision, foresight, athleticism, coordination and attitude—not to mention the difficulty of the terrain itself, which involved a whole new set of parameters. If an arroyo intervened between him and Emma, Emma could negotiate this obstacle much more easily than he could, and this altered the ratio of the distances, favoring Emma even more. As Jonathan and Emma worked their way across the desert, all kinds of terrains intervened between them and all sorts of obstacles presented themselves, so the numbers would be different in each concrete example. Jonathan saw that distance was not as easy and straightforward to calculate as physicists had imagined and nothing at all like the simplistic relationships of Euclidean geometry.

The definition of distance in geometry never embraced the truth about Jonathan and Emma. The numbers that were generated didn't depend on the facts or circumstances of this particular spot or that particular spot, but only on the relationship between two mathematical points. With Jonathan's formulation, everything in the desert had to be taken into account. All of it participated in the calculation—every rock, every bush, every channel, every arroyo. Location made all the difference and all spots were not equal. Rather than being of the utmost simplicity, this version of distance was infinitely complex. That was alright with Jonathan because he much preferred a distance that embraced the whole of creation and granted every object its rightful existence.

Jonathan could see why physicists hated reality so much. He understood why they wanted to stamp it out, bulldoze it over, extinguish it, turn their backs on it and pretend that it wasn't there. Geometry was perfect for that purpose because geometry was about nothing. Geometrical constructions were empty and barren, however, that was not the world that Jonathan knew. Jonathan's world was filled to the brim with details and all of them were important. Jonathan demanded that the entire desert be acknowledged before he proceeded with any formulation, and this made his version of distance far superior to the physicist's version. His distance had real meaning because it was the real distance between him and Emma. Jonathan summarized the situation in his mind: if he let mathematics shape his ideas, he was studying mathematics. If he let the earth shape his ideas, he was studying the earth. If he let mathematics shape his ideas and then told himself that he was studying the earth, he was mistaken. The real distance between him and Emma was impossible to calculate, but the geometrical distance was the easiest thing in the world to calculate. Jonathan knew why physicists would never adopt his definition of distance: they had no interest in studying the earth or grappling with the reality of the natural world. So what reality were they studying? That was a good question.

Everything in the world shared Jonathan's fate of having to find its own course: the water flowing in the rivers, the wind surging through the canyons and squeezing between the hills, the lightning streaking down from the heavens, the mineral salts migrating and percolating through the rocks. So when physicists discovered that they could create straight line phenomena in a laboratory using particle beams and light rays, these highly unusual behaviors created a fortuitous situation for them, because it allowed for endless geometrical constructions, and all sorts of devices were built exploiting the straight line character of the phenomena, from cathode ray tubes to linear accelerators, from lasers to x-ray machines.

By building an optics bench, physicists were able to combine the geometries of wave phenomena with these linear configurations. Early investigators seized upon the opportunities, most notably Christiaan Huygens, and physicists had a field day constructing wavefronts and setting up optical arrangements with mirrors and lenses. Electromagnetic radiation became the centerpiece of scientific investigation largely because of this feature. The advantage lay only in the relatively simple arrangements of the laboratory. Given the fact that the straight line was the foundation for most geometrical constructions, the ubiquitous electromagnetic radiations opened the door to almost limitless applications, countless occasions for creating mathematical configurations within the phenomena themselves.

The physicist said that the reason why he couldn't construct wavefronts in nature or extract geometrical patterns from the chaotic confluence of light rays in the everyday world was because in nature light was composed of a broad range of wavelengths and trajectories, and each of these components was traveling haphazardly in every direction at once, but Jonathan immediately saw the error in his reasoning. When the sun shone in the desert, everything was illuminated equally. Only the geometrical analysis of light divided it into a confluence of waves, a configuration of points and lines, a multitude of separate rays each with its own identity. Breaking light apart in this way created a false image of light, an image derived from the forms traditionally envisioned in geometry. Physicists used light to create geometries for themselves, manipulating the phenomena of light by setting up artificial geometrical arrangements. They then tried to use these geometries to understand the nature of light, implicitly assuming that since light allowed them to construct geometries, light itself must be geometrical. Geometry became the essence of light, ignoring the fact that geometry could never be the essence of anything, that is, anything other than deliberately crafted mathematical objects.

In truth, light was a unity and not a multiplicity, meaning that the same light could be present in many places at once. There were no light rays but only light, a diffuse existence unbroken and undivided forming a uniform expanse of light. If light was not localized but spread out, and it contacted many different objects at once, then it was not like the motion of an object that could only be in one place at one time. In order to turn light into an object that moved, light first had to be interrupted in order to create a leading edge, broken apart so that the boundaries could then be used to establish positions and distances and ultimately the calculations of velocities. Light became understood in terms of its departures and arrivals and a mechanical view of light was developed. When light was broken apart, only then could light be said to travel. Where did the idea ever arise that light flowed like water, that it poured out from a source and then bathed everyone in its delicate, ethereal liquid? Perhaps that wasn't an accurate metaphor and light was actually motionless, just as it appeared to be. Waves, particles, points and lines were not the concrete realities of light, but reformulations of light in the terms of a mathematical analogue consisting entirely of nonexistent entities.

For a mathematical analysis of light to proceed, light needed to be understood in terms of the ideas of geometry. All analyses in physics began by setting up straight line distances, and in fact the entire world was seen in terms of distances. Every concept was automatically placed against this backdrop. Since laboratory setups were geometrical by design, distances were already built into the experiments and thus readily available, but thunderstorms and sunrises had no centers, no peripheries, no well-defined parts, so where should Jonathan place the tape measure? The physicist did not back away from his premise and insisted that Jonathan understand the world through lines and distances, even though a normal person might despair from the lack of reference points and the apparent meaninglessness of lines that were arbitrarily chosen. The physicist persevered because without these distances the cloud could not be compared to his laboratory experiments, so the physicist needed to find some sort of benchmark. Lacking a better idea, he latched onto the surface of the earth, since that was the only stable surface available to him, and then began casting meteorological events in terms of altitude. Perhaps he should forget about distance and think about something else.

When Jonathan turned his head up to the sky, he wanted to draw straight lines and measure distances within the clouds, but where did he begin? What struck him most about the clouds were their absolute irregularities, not only in terms of their outward shapes and appearances, but also in terms of their physiologies and internal designs, the complicated movements of water vapor and charged particles throughout their domain. The physicist wanted to introduce geometrical forms in order to try to understand clouds as mathematical objects, but a framework had to first be set into place before a mathematical structure of the clouds could be developed and unfortunately nature did not provide one and furthermore made it difficult if not impossible to establish one. Nature put its foot down and firmly commanded: "No mathematics here!" yet the physicist did not take the hint and proceeded anyway, imagining all sorts of ridiculous configurations that existed only in his imagination. He generated numbers and related the numbers together, but the world had no numbers and no mathematical rules to follow, so the physicist ended up playing games with empty markers on game boards that he had dreamed up in his head—or artificially constructed in his laboratory. The physicist wasn't studying the clouds at all, but only looking at his own definitions, engaged in a logical interplay of fanciful ideas that had little or nothing to do with the elusive reality of the sky.

Along with the majority of people in the United States, Jonathan had been brought up in the public school system where he was taught from a very young age to turn his back on nature by spending long hours sitting at a desk alongside the other students. It was here where he was trained to cast the world into abstract terms. His teachers told him that he really didn't need to look at the world outside because the thoughts of mathematicians were far more important. His education was not built around walking through the desert and learning to see things as they really were, because if he did that long enough he'd be able to put all the vain imaginations of his teachers into proper perspective. Instead he was locked in a room and forced to concentrate his attention on blackboards where the schematic relationships of a mathematical style of thinking were drilled into his head. The teachers regularly caught him dreamily staring out the window and they always attributed this to a lack of self control. They never suspected that he actually knew what he was doing. He wanted to be out in the world, the real world, and not sitting in a chair trying to focus on something that had no relevance to anything whatsoever. Each day Jonathan mused over the emptiness of pure abstractions, teetering perilously on the edge of this dark star chasm, desperately clinging to his sanity while he stared into the abyss of nothingness. He looked around at the other students and watched helplessly as the black hole of mathematics sucked their brains into its gaping maw. He always gave the teachers the answers they wanted, but inwardly he rebelled. He was never going to allow himself to become another brick in the wall. His teachers would find that out soon enough.

Jonathan survived the long ordeal but this type of upbringing caused disturbing disabilities in his classmates. They ended up feeling more at home with their thoughts than they did with being out in the natural world, because they had unwittingly been coerced into trading in their natural lives for artificial existences. They became more adept at solving puzzles and playing games, more comfortable visualizing abstract ideas and negotiating chains of pure reason than crossing formidable terrains and finding their way across vast wildernesses. They grew up believing that mental constructions were more important than physical realities, or even worse, that mental constructions were physical realities.

Jonathan could still remember one day sitting in class. The teacher thrust a piece of paper toward him, placing it on the desk in front of him, in essence saying: "Look at this,

Jonathan, this is what the real world looks like." He turned his head away and stared out the window. He thought to himself: "What in the hell was this person talking about? The pencil lines were nothing at all like the real world. The connection between the two was non-existent—or at best thin and convoluted. A long series of mental acrobatic leaps and flips were required to reach the world from this measly piece of paper. No paltry sketch, no simple drawing, could ever capture the richness of all that existed, or encompass the diversity of the vast universe. Listening to his teachers talk about mathematics, he understood that these people had no interest in the real world. They were hooked on the narcotic of abstractionism and they couldn't get enough of it. They scribbled on paper day and night. He didn't understand the attraction of the symbols or feel the euphoria of a proved theorem. He didn't need to solve the problems they handed him because he knew that the answers would never lead him to an understanding of the world around him. He just wanted to get the homework over and done with so that he could get back to the more important task of trying to comprehend what actually existed—or at least start looking at it. The world was fascinating and Jonathan wanted to know what it was all about. Mathematics was neither here nor there, an exercise in futility. He didn't care if he got the answers right. The teachers were asking the wrong questions.

Mathematics wasn't knowledge because it wasn't about anything, yet mathematicians often imagined that they were investigating an independent realm that existed in some sense, apart from everyone's thoughts, because mathematics was a shared vision. Everyone memorized the same definitions and by using the same rules of logic they all followed the same lines of reasoning and ended up with the same conclusions, giving everyone the impression that they were studying some sort of objective reality, and of course they were, a reality of definitions comparable to the reality generated by a Monopoly board. No one needed to think this way. The world did not demand mathematics, but casting phenomena in mathematical terms put the physicist on familiar ground. Everything in mathematics had clear and precise explanations and that was the way he wanted the world to be, so even though actions in the physical universe couldn't possibly parallel the interconnections between mathematical ideas, merely associating these events with mathematical ideas made them seem more understandable. Long ago the idea had dawned on physicists: why couldn't these mathematical explanations also be the explanations for the world as well? Physicists felt safe and comfortable in a rational universe of mathematical relationships, so they decided they would make that world everyone's world.

Mathematics was dangerous because it surreptitiously enticed people to start thinking in totally unrealistic terms, linking vacant abstractions with merely conceptual ties and then presenting these meaningless connections as somehow representative of what actually existed. In order to think about the world, people had to think in concrete terms. They had to embrace the world in the fullness of its being and thoroughly explore the innumerable details, variations, nuances—the qualifications that were necessary to every statement. They had to specify the infinite possibilities and endless complications involved in every action, the factors that constituted the real world and made it what it was. This required them to acknowledge the fundamental limitations and ultimate failure of concepts in capturing something that was utterly dissimilar in nature and structured along entirely different lines. Mathematics was the opposite of the truth—that is, the truth of the world. If a person was serious about gaining a knowledge of the world, then that person couldn't waste a lot of time studying mathematics. But people wanted to put themselves on the center stage and not let the world hog the limelight. Their understanding would be based on their own creations, and of all the things they had created, they loved mathematics the most.

Although mathematics was fine when considered on its own terms, it certainly wasn't the reality of the world. Jonathan continually found himself sidetracked in class, led in different directions, lost in strange clutters of symbols that didn't seem to matter. Where was his world in all of this, the world he knew and loved? He looked out the window. The world was still outside, waiting for him. He couldn't wait to put the book down, straighten his desk and shoulder his pack in order to immerse himself in a landscape of infinite possibilities, a refreshing break from the tedious lines of reasoning that led him nowhere. Instead he wanted to let nature be his guide, his companion, his teacher. He didn't want to play games. He didn't want to turn away from the world.

Jonathan was much older now but nothing had really changed for him. Every day he confronted nature full in the face and his wonder of it never ceased. The world never got old and tiresome the way that equations often did. Emma shared Jonathan's enthusiasm and this was why they got along so well together. She never argued with him about where they should go today. As far as Emma was concerned, the destinations were all good and the days were equally filled with excitement. They both sat shoulder to shoulder on the seat of the truck staring at the panorama outside the windshield, curious at to where this new road Jonathan had just discovered would take them and what they would encounter around the next bend. They shared the same sense of adventure and in this appreciation of the unknown they were soul mates, kindred spirits, bosom buddies. Jonathan did not seek predictability and certitude, the mainstays of mathematics and physics, and he would not trade what he had for the dreary attributes of empty abstractions.

Everyday life was the reference point for Jonathan's understanding of the world because his existence in the world was self-evident. Physics had originally started out embracing this position. The world did not need explaining because it was right there in front of us, the most certain truth we could get our hands on. As mathematicians invented more and more outrageous constructions and convoluted lines of reasoning, and physicists applied these ideas to progressively more enigmatic gadgetry, the new formulations and setups were first cast in easy to understand, everyday terms to make them more accessible. The continued escalation of mathematics gradually reversed all of this. People started living in a world of pure forms, a world of the imagination. Mathematics took over everything and became the new criterion for understanding. Instead of explaining mathematics in the language of common experience, common experience had to be explained in the language of mathematics, and ideally in the most esoteric mathematical terms available. Knowledge now resided in the farthest reaches of theoretical speculation where no one but a chosen few were able to go. Truth existed in the most obscure and impenetrable fabrications of abstract thought. The mathematical ideas suddenly came alive with new powers and human experience was stripped of its privileged position. Physicists treated mathematics as if it were a reality and began to envision a hidden realm lurking behind daily events and familiar objects, another world embedded in the things they saw and felt around them. Eventually they concluded that they were living in the holographic mirage of another dimension and this fantasy was far better than any science fiction movie. They trusted equations. If the equations said that something was true, no matter how ridiculous it was, they believed it, because the authority of mathematics was irrefutable.

But when Jonathan looked at the process closely, he saw that mathematical calculations were actually the most precarious form of reasoning there was, with unlimited opportunities for mistakes and oversights, because physicists were converting the world into fictions, stripping away realities to set up the measurements of fanciful concepts, then trying to correlate the outlines of all these bare forms. There were so many ways to go wrong with this. The procedures were highly complicated because the world wasn't presented to them in the form of numbers and abstractions. Many transformations and substitutions were required, each one with the potential for leading them astray. They constantly had to keep tying it all together. So why then was the person who made the calculations the one everyone turned to when they wanted answers? What made people think the calculator knew what he was talking about? Yet he was the one who was universally held in the highest esteem, and everyone listened attentively to his words.

Calculations were portrayed as the highest form of knowledge when in fact all of these imaginations, fictions and fabrications took everyone away from the truth. Those who leaned heavily on this form of reasoning were assuredly going to fall into errors, as they had done throughout history. People had tried to calculate all sorts of things. The calculations were almost always wrong to start, but subsequently the figures were brought into line with other calculations, because this correspondence was what made the calculations correct. Physicists looked at their mistakes afterwards, discrepancies that hadn't been apparent when they first set up the problems, only to discover that the numbers they had originally cranked out were way off. So now they had to go back and make adjustments, which was easy to do because none of it was real. They could change the formulas any way they wanted. They were bound only to the rules of logic and the theorems of mathematics, and the ways they applied these rules and theorems were entirely up to them. So they went back and fudged everything. They rearranged the terms and eventually they got the numbers to agree, then they vigorously applauded the technique. They said that this was the only way to go, the way that everyone should approach every problem—just make a whole bunch of calculations, even though the procedure had failed before and failed again until the brainiacs finally got it right. Nothing had changed today. Physicists blindly turned on their instruments and started calculating, even though the numbers were as questionable now as they had been in the past.

Instead of patting the human calculator on his head and saying "There, there, you go have fun now. You go play with your numbers in the next room because we'd like to have a serious discussion about reality and there's no place for you at this table," everyone put the calculator on a pedestal and took his proclamations as gospel. They'd find out later that he had no idea what he was talking about, yet that wouldn't deter them in their admiration of his arcane skills. They offered excuses for him: "It's ok, everybody made mistakes."

Calculations were tedious and when taken on in large numbers they became mind-numbing. Jonathan had read somewhere that prior to the advent of computers, quantum physicists had taken large sheets of paper and started writing in the upper left-hand corner in the smallest script they could manage. The tiny letters and symbols slowly marched out across the giant sheets of paper. The efforts gave Jonathan a headache just thinking about them. The rules for proper syntax were all prescribed and they only had to be memorized in order to write the text. The physicists just had to follow the rules and keep making substitutions. After this Herculean effort, the physicists often found that the numbers didn't pan out and they had to start the whole process over with a different set of assumptions. He'd heard a rumor that the intense concentration required to carry out such difficult operations, if carelessly maintained over a long period of time, led to an increased incidence of brain cancer in mathematicians. Whether or not this was true didn't matter because Jonathan had ample reason not to pursue such a course. The calculations were meaningless in themselves and the reasonings led nowhere. He wasn't willing to sacrifice himself for a cause that had, from what he could tell, no redeeming value.

The writing of mathematical equations was a technical skill, similar to playing chess. A master chess player might not do so well in other areas of mental engagement. He might get caught up in a bad relationship, bungle his personal finances, ruin his career, or otherwise make a mess of his life. He might also manage his private life very well, but that outcome was not connected to his ability to outwit opponents by moving miniature figures on a checkerboard. Although this was a remarkable ability, it didn't mean that he understood his place in the universe or comprehended the meaning of life. The special talent he possessed was not to be confused with wisdom or any general powers of perspicacity—in fact, it wasn't necessarily connected to anything. Back in college, Jonathan had found out that thinking about things philosophically was a lot harder because the philosopher wasn't playing a game. He wasn't acting out the part of a robot by strictly adhering to a set of fixed protocols.

Leaning heavily on many years of specialized training in the nuances of mathematical thinking, physicists invoked their adeptness at manipulating formulas and applying theorems to qualify themselves as interpreters, assuming the job of translating the indecipherable symbols into commonly understandable words and phrases—a task for which they received no training. Well, what sort of training could they have possibly received? This dual function monopolized the dissemination of scientific truths to the remainder of humanity in much the same way that holy men had traditionally brought the word of God to their people. Priests typically used their self-proclaimed and highly privileged relationship with the sacred world to maintain control over what the laity believed to be true.

But the meaning of the exceedingly abstract and intricate mathematics was problematical and there was little agreement among physicists as to how to make sense of it. Physicists handed everybody prepackaged bundles of concepts that few people could appraise or criticize adequately. They said that everyone had no choice but to accept these non-sensical ideas because the equations had spoken. People could only shake their heads in amazement at the crazy world of quantum mechanics because there was no doubt about its veracity. This was what mathematics did to a person. It caused the person to think in strange ways and with a straight face proclaim ideas that made no sense, but that was certainly not a reason to question mathematics—or was it? Had physicists, Jonathan wondered, duped themselves into creating a preposterous and convoluted world of formal relationships linking empty and vacuous terms, laboriously devising a world of imaginary entities that winded up being conceptually bankrupt? Had the bubble already burst, and the fallacy of this approach already become apparent, at least to a few bystanders?

Jonathan reached the ponderosa grove where his camp was situated, but twilight had not yet begun. The time was too early to retire, so he decided to scale the ridge on the other side of the valley to check out the view from there. Emma was thrilled at the chance for one last venture into unexplored territory.

They both enthusiastically climbed the side of the canyon making their way up the increasingly steep pitch. Jonathan looked at the unbroken layer of clouds above his head and found the sky to be a soothing, pacific shade of gray. The presence of the clouds had made the afternoon cool and comfortable, but as he neared the crest, the clouds began to emit occasional rumbles, followed by long silences. The clouds seemed to be speaking to him, muttering soft words in low, confidential tones of voice, each time pausing to consider the next phrase with great care. He listened attentively to the strange words but failed to decipher the cryptic messages or the subtle warnings they contained, until a bolt of lightning rifled the top of the ridge. The unexpected event stopped him in his tracks and obliged him to ponder his further advance. Despite the powerful shock of the blast and its close proximity, he still thought that the lone shot was just an aberration. The clouds really didn't look that bad. The abrupt electrical discharge must have been a fluke. He dismissed the authenticity of the threat and decided to simply ignore it. He looked up once more at the serene clouds. Satisfied with his evaluation, he continued on his way.

Jonathan picked his way through the rocks, laboring heavily as the difficulty of the terrain increased, until finally he caught sight of the top of the ridge. He paused briefly to catch his breath, but flinched violently when an intense flash of light blinded his eyes. The earth shook, causing him to lose his balance and nearly pitch backward down the slope. The dazzling surge of electricity was accompanied by a devastating detonation. The bolt had apparently gone right over his head. He didn't see where it hit the ground, but the impact must've been close. This second blast resolved the issue for him. Conditions were simply too dangerous for him to summit the ridge right now. The best plan was to retreat to a lower elevation as quickly as possible.

He whistled for Emma but soon realized that she was nowhere to be found. He yelled out her name. A draft of cold air swept over the ridge bearing the distinctive smell of hailstones and mountain snows. Jonathan became impatient as the sudden urgency of his predicament weighed upon his mind. The same clouds that had been so innocuous just a few moments ago had suddenly exploded into a dangerous storm centered directly over his head. He and Emma needed to go now. He whistled repeatedly for her, since every second they lingered out in the open increased their risk of getting hit. He waited nervously, wondering what had happened to Emma, watching the clouds thicken and darken in an amazingly rapid transformation. He puzzled over her strange disappearance, then it occurred to him. Maybe she'd been up on the ridge when the lightning struck. Panic gripped him as he realized that she might have been killed by one of the blasts. He had no choice now but to endure the hazard posed by the impending storm and go up there to look for her.

He started climbing into the heart of the mounting squall, deliberately choosing not to think about the risk he was taking. As he turned his head upward to gauge his objective, Emma popped into view, looking down at him from the top of the ridge. She was oblivious to the danger and couldn't understand why he had stopped. She pranced about encouraging him to follow her and to continue their walk, but he vetoed the suggestion and firmly commanded in a loud voice: "Emma! Let's go home!" She immediately obeyed and disappeared from view as she circled around, then reappeared directly above his head. She meticulously negotiated the rocky surface, steadily coming toward him as he about-faced and set off down the side of the canyon.

Since the storm had originated on the other side of the ridge, Jonathan imagined that he would easily escape its wrath by simply dropping back down into the valley, but the storm quickly overtook him. The lightning was now striking in every direction and he was walking into the storm as much as he was walking away from it. His progress down the rocky hillside was annoyingly slow until the gradient tapered off as he approached the valley floor and he started running, something he almost never did due to the danger of tripping and seriously injuring himself. His wild behavior excited Emma because for once she wasn't alone in her enthusiastic racing about. This was one of those rare occasions where they found themselves engaging in this pursuit with equal dedication.

A fierce lightning bolt exploded to Jonathan's left. The blinding flash and thunderous roar were nearly simultaneous, the sound lagging behind the flash by only the smallest fraction of a second, a delay that was barely perceptible but still noticeable, putting the strike at about 150 yards. Jonathan happened to be looking in the right direction when it hit and observed the bolt slam into the wall of the valley, inclined at about 45 degrees, issuing from a brand new focus in the widening circle of blackness. A fresh storm cell had been born in the ravaging sky and this latest nucleus of electrical energy was heading directly for them.

As the cell swooped down the scene became pure madness and they were both running for their lives. The rain intensified as Jonathan finally approached the truck. He nervously fumbled with the keys. They leapt out of his hands and landed underneath the truck. He released his pack and got down on his hands and knees in the wet clay and bristly pine needles to reach for them, and then using both hands this time, he managed to unlock the door, chiding himself out loud for always insisting on locking it. Emma needed no encouragement. She sprang onto the seat as soon as the opening door allowed it, followed by Jonathan's pack and then Jonathan himself. He slammed the door with force, ensuring that the storm would not find a place to enter. The interior of the cab was rank with the smell of a wet dog.

Without warning the dimly lit world outside burst into an ethereal blue radiance so dazzling that it washed out every facet of the scenery and they were plunged into an immaculate and all-encompassing aura of the strangest light that Jonathan had ever seen. Neon, ultraviolet, florescent, incandescent—there were simply no words to describe it. The crack of the explosion strangely passed over them in silence, the compression of the air outside apparently off the scale of audible sound, but then Jonathan could hear the thunderous roar materialize a short distance away as it radiated outward. The concussion bounced off the canyon walls and came back to them splintered into a cluster of deafening thunderclaps. Lightning had just hit the ponderosa next to the truck, the same spot where he had been standing only a moment ago. He pictured himself still out there, looking up with childlike wonder at the mysterious clouds overhead. Unbeknown to him, a cosmic ray originating from a distant point in the galaxy had loosened some atmospheric electrons, the beam caught by some as yet unidentified receptacle within the clouds. The electrons were then thrown towards earth, knocking more electrons loose along the way, picking up momentum as they went, much as an avalanche of snow surging down a mountainside. For some unknown reason the electrons accelerated to higher and higher velocities, sending a beacon of antimatter out into space. The electrons began generating unusually intense flashes of x-rays, again for reasons that were not clear. The cosmic freight train then slammed headlong into the poor pine tree where it was deflected slightly to one side by the force of the impact. Had he still been out there gazing at the clouds, the runaway train of electrical energy would have plowed squarely into his upturned face. A short time earlier Jonathan had read the latest theory of lightning presented in a scientific journal. Lightning was more complex than anyone had ever imagined.

He did notice the tree when he first parked the truck. The top had been burned off, the bark split open in a number of places, the underlying wood blackened on all sides. The hapless victim appeared to have been struck by lightning on multiple occasions. None of the other trees in the area showed any signs of such previous impacts, which was strange, because many of them were taller than this particular tree. What caused

70

independent storms separated by long intervals of calm to each single out this one individual, when there were so many other targets to choose from?

Perhaps the tree sat atop a vein of some mineral that conducted electricity more effectively and thus it was rooted in a more efficient ground. The mineral pocket would have to be very small not to include trees only a short distance away. Or perhaps the tree marked the location of an extraterrestrial mother ship buried beneath the surface of the earth and the tree served as an antenna, collecting atmospheric electricity to power a subterranean colony of alien creatures. That would certainly explain the bizarre nature of the storm. It was the type of storm often featured in science fiction movies to signal the approach of bewildering entities wielding inscrutable powers. The tree was a mystery to him and he'd never know the answer. He simply had the misfortune to park next to it.

The electrical tension in the sky rapidly dissipated now that Jonathan and Emma were safely ensconced in their protective shell and the clouds unleashed a fierce torrent of rain. Jonathan couldn't see anything outside the windows as sheets of water cascaded down the windshield and a dense fog collected on the inside of the glass, so he sat patiently listening to the rain hammering on the hood, trapped inside the tiny cubicle with nothing more to do. He glanced over at Emma and she reciprocated with a deadpan stare. They both knew they'd be stuck here for awhile. Emma immediately curled up into a compact ball and decided that she would use this time to take a nap. Jonathan decided to think about what just happened, and search for an explanation.

The sky had been cloudy all afternoon so there had been no differential heating of the ground to create thermals, and by late afternoon the heat from the sun had been waning anyway. Therefore the storm must have been created by the latent heat already stored in the clouds, heat that had been gathered during the morning hours. This heat did nothing all day, lounging within the peaceful confines of the clouds until it was released by the condensation of water vapor into rain droplets as the general atmosphere cooled down. The rapid darkening of the clouds signaled an increase in cloud density, as water droplets precipitated out of the vapor, as well as a concomitant increase in cloud height caused by the sudden release of heat. Or so the story went. The account still left a lot of questions unanswered.

The atmosphere had been quiet and serene all afternoon, the vaporous water molecules happy and content with their peaceful existence. Something then triggered a revolution among the molecules and a widespread panic ensued. The statement that the air was 'unstable,' meaning that atmospheric conditions were ripe to precipitate a radical change in cloud behavior, was merely an empty reference, just a bunch of words strung together. The clouds were a massive horde of tiny individuals, a mixture of conformists and mavericks, a mosaic of obscure agents with complicated agendas. Here and there molecules of water joined hands to form droplets, but at other points they turned their backs on each other and stubbornly remained aloof. At some points the diverse crowd moved together as one, but a certain percentage of the molecules didn't participate in the general mood and struck off in other directions seeking different goals. Specks of dust picked up electric charges and carried them around like backpacks, yet many of the particles chose to run free and unburdened. Jonathan could sketch arrows over all of this and draw circles around the different areas of the clouds, but these abstract forms were just convenient fictions that didn't exist. The storm had been created by the complex social behaviors of its inhabitants. It was not the sum of these activities, because the manners of conduct were so diverse that they could not simply be added together, yet they were coordinated somehow to form a whole that had a character of its own, apart from the activities taking place at each point. These activities could have simply amounted to nothing, just a uniform yet energetic and chaotic blandness. Why did all this enigmatic hustle and bustle turn into a structured entity?

The storm had a huge cast of actors, an unbelievable wealth of atoms and molecules all playing uncredited bit parts. No one could really grasp the size of the assembled masses because the number was well beyond human imagination. The particles existed independently of one another and each one followed its own course, oblivious to the actions of its distant relatives located in other parts of the atmosphere. The particles interacted primarily by pushing and shoving one another, contacting no one but their immediate neighbors.

Physicists often portrayed molecules as the helpless victims of forces that were beyond their control, passive elements manipulated by the abstractions of theories, their behaviors determined by mathematical concepts with unique characters, because this was the essence of an experiment. Physicists had found ways to influence the particles and then they observed their responses. In the laboratory, particles were the recipients of external causes, but once set free in the world without the experimenter to guide them, they had to act on their own. The molecules needed to create the forces and energies themselves and then use these tools to make everything else happen. No other agent could possibly be manipulating the forces and energies other than the particles themselves. With only particles and motion, the molecules were burdened with the responsibility of doing everything for themselves, and what they created was unbelievable.

The physicist pointed out that in the laboratory the responses he got from molecules were reliable and therefore predictable, supporting his claim that these elements were passive. He could control the behaviors of these molecules and get them to do all sorts of things, but Jonathan countered that people were also easy to control. In fact, you could get them to do almost anything—and without having to hold a gun to their head. If you offered them money, they would perform every sort of job for you. If you played upon their fear of death, they would enlist in your church and proclaim their faith in front of the entire congregation. If you granted them the opportunity for fame, they would humiliate themselves in bad movies. If you catered to their sense of self-image, they would dress in the latest fashions and drive around in the most stylish cars. If you employed the proper advertising and marketing techniques, you could entice them to buy things they didn't really need. People weren't nearly as unpredictable as they'd like to believe.

But out in the world, the molecules were incorrigible builders, constantly assembling new forms, putting together structures and then tearing them down, single-mindedly forging ahead much in the way that squirrels burrowed tunnels, unable to stop even when they had more tunnels than they could ever hope to use. The instinct to dig was so well ingrained in the psyches of squirrels that they couldn't help themselves. They dug simply for the sake of digging, just as Emma ran for the sake of running and mathematicians wrote equations for the sake of scribbling. There was no point to any of it. Jonathan was as much a part of this circus show as anyone else, walking simply for the sake of walking, laboring each day to march long distances simply for the sensation of movement. When he parked the truck, he was already in the middle of nowhere, already alone in the wilderness. There was nowhere further to go and nothing more to find. He was already where he wanted to be, so why did he keep walking? He could just as well have pulled out the folding chair and sat down, remaining right where he was, satisfied with admiring the scenery at that particular location. One place was as good as the next. Why did the molecules build a thunderstorm when they could have just relaxed and been satisfied with what they had?

The urge to build things was found at the most fundamental levels of material existence. The irresistible impulse then worked its way up through each new layer of organization. The atoms built molecules and the molecules learned to put together ever more complex arrangements of atoms. They bundled enzymes together into tiny assembly plants, machine-like complexes that cranked out proteins at phenomenal rates. The proteins built single-celled organisms and these individuals collaborated on more projects. They all got together and joined forces to create something higher. The proteins and lipids became the organelles of cells and the cells formed structures and societies that became the tissues of organisms. As the process continued, the molecules learned from their successes and failures, creating ever more elaborate designs, tiny ion pumps and protein motors, elastic strands pulling nuclei toward opposite poles, tubular conduits of circulatory systems and fibrillar networks of muscles and nerves. The creatures they created then built roads and aqueducts in much the same fashion, and onward and upward the construction proceeded.

Scientists construed all of these works as the result of blind forces haphazardly pushing objects against one another, but it seemed to Jonathan that some kind of awareness had to be present all along. At each new level of organization, this consciousness grew and carried on the traditions of its forbears, ultimately constructing an entire mathematical universe out of ideas. But the sentient beings did not stop there. The mathematics was further employed in the design and construction of devices and machines, first lights blinking and wheels turning, then screens glowing and speakers vibrating out the sounds of human speech. Yet the molecules had been behind it all along. They had caused everything to happen, in the same way that people had built cities and empires and then filled them with the richness of human activities.

One thing was clear from all of this: the existence of the world was not based on simplicities. Rather, the world rested on an abyss of complications, an infinity of particulars that was as incomprehensible as the phenomena themselves. Molecules were boundless agents of action and motion, bottomless pits of details and possibilities. No concepts could corral them and no abstract quantities could describe them. Jonathan didn't feel that the metaphor of molecules as tiny independent beings endowed with impulses and objectives was unjustified or inappropriate.

If molecules were controlled by forces and energies, then these entities would have to exist apart from the molecules, otherwise the particles were just controlling themselves through fictitious intermediaries. Forces and energies were nothing more than descriptions of the collective behaviors of the vast multitudes, gross generalizations of the group dynamics, vague summations of things that didn't add up. But how did one particle know what the others were doing? How did they all get together and build a structure as intricate and complex as a thunderstorm? The effort required harmonies and conflicts to mesh, necessitated a unified course to appear out of nothing, called for a collective consciousness to materialize and build overarching structures out of chaos. One particle could not possibly understand the overall plan, or foresee the end result, or work toward the common goal, yet that was what happened. The storm had been a collaboration, a magnificent work of art, a stupendous engineering project, a monster that lived and terrorized the townspeople.

Particles surged down invisible corridors much as commuters raced down the concourses of stations in the heart of the city to catch their appointed trains. The particles then fanned out onto crowded streets, merging with the existing flows of pedestrians, blending into the unified throngs of diverse individuals, yet maintaining their unique destinations. Other particles clashed head on with opposing factions, created scuffles and turmoil, crashed barricades and broke through invisible boundaries. Violence erupted. Riots ensued. How could anyone predict the course that these events would take? All Jonathan could do was sit back and watch the events unfold. That was all anyone could do. The unruly mobs took control, ran roughshod over the communities of ordinarily placid citizens and all hell broke loose, the sky went black, and no one was safe from harm.

Consider the billions of people around the world right now engaged in tasks, driving cars down congested city streets or walking lonely paths through forests, cooking meals in towering apartment complexes or huddled around wood fires and barbecue pits in small villages, gathered in stadiums watching baseball games or sitting in large concert halls listening to a small group of individuals play music. A tiny whiff of cloud also had untold billions of inhabitants, and they were a motley crew of dust particles, chemical pollutants, fine sprays of water, atmospheric gasses, pollen grains, rare elements, charged ions, ice crystals, engaged in every sort of task and group activity, playing roles and performing functions, acting together or standing alone, all participating in something greater than themselves, the larger society of molecules.

The molecules had all reached their destinations and done their jobs within the thunderstorm, putting on a spectacular show up on the ridge. Now that it was over, Jonathan wanted to stand up and vigorously applaud the elaborate production, just as if he had witnessed a rousing presentation of one of his favorite operas in the theater, afterward falling helplessly back into his plush seat to contemplate what he had just seen. The molecules in the opera house had also put together a stunning show, first assembling themselves into incredible living organisms, stretching chords of fibers across openings and fabricating bellows to force air through the orifices, then singing delicate arias and delightful four-part harmonies, the music a final product of processes that were unfathomable. Jonathan imagined the flux of ions across membranes, the contractions and relaxations of cellular walls, the seemingly endless numbers of tiny muscles fibers all flexing in unison, the vibrating air molecules bouncing off one another, the collisions transferring complex patterns of motion to the next in line, passing long-winded messages down established networks of communication with the speed and efficacy of gossip racing through a rural community.

Jonathan had an array of concepts in his mind to explain it all, but what conceptual system were the molecules a part of? They had gathered together without thinking about it, assembled themselves according to no preconceived plan, followed no established architectural blueprint. So how then had the feat been accomplished? Where was the mathematics behind it, the equations that guided the course of events? How could mental constructions be at the root of everything he had just seen and heard?

When Jonathan asked the physicist for an account of the performance on the ridge, the physicist handed him a written synopsis of the plot. Jonathan had just seen the storm play out in full, but he could find no way to sum it up in a few words or capture it with a string of symbols or portray it using a few simple diagrams. There were so many events taking place simultaneously, so many characters all speaking at once, such an enormous stage spanning more than he could possibly encompass. He could not keep track of all the essential facts, follow the unfolding of each subplot, or see how the action sequences were integrated into a single storyline portraying the ephemeral life of not just any storm, but this storm in particular.

The physicist's explanation of the storm was comparable to a movie review. A movie had a fixed structure, much as a laboratory experiment. Jonathan could watch a movie as many times as he wanted and the scenes were always the same and they were presented in the same sequences, just as the physicist could perform the same experiment over and over again and get the same results, but each thunderstorm was absolutely unique and it could not be duplicated or replayed. Furthermore, the physicist could not give a synopsis of a thunderstorm in general any more than Jonathan could give a synopsis of a movie in general.

If Jonathan always summed up each movie with the same generic outline, the same stereotypes engaged in the same empty dialogues, the same events taking place in the same drab town, he was forced to strip the synopsis of content until it fit all situations. This was the nature of mathematics and therefore the nature of a scientific theory. A thunderhead had updrafts and downdrafts, heating and cooling effects, the condensation and evaporation of water, changes in pressure, but Jonathan had just seen the storm play out in exquisite detail and it was much more than that. The storm was a fantastic show, a grand display of power, a wild beast on the rampage. No mathematical description could copy every detail or capture the erratic temperament of this one individual.

The extreme variations between storms were not just minor differences that could be disregarded for the sake of simplicity. Over the years Jonathan had been trapped in scores of thunderstorms, yet he had never seen anything like this, the way it came out of nowhere in such a short period of time. Each storm had its own personality, and if the physicist wanted to see all of these storms as being essentially one and the same, then he'd have to squint and make the images fuzzy and indistinct so as not to look at the specifics or focus on the details. He'd have to block out more aspects of the scenery than he was taking into view. The price was too high to pay. Jonathan wanted to see everything and account for every last bit of it, because nothing about the storm was trivial and unimportant.

Jonathan had read a few books on cloud physics, but applying the laws of thermodynamics to thunderheads was disappointing and yielded no useful information. The scientific method tried to shatter the intact phenomenon into pieces, then make the broken fragments look like replicas of laboratory experiments. The problem was that the phenomenon didn't naturally come apart that way. The mathematical patterns that had been set up in the laboratory weren't the real configurations of a thunderhead. The thunderhead didn't even have fixed and well-defined parts and that distinguished it from the mechanical behavior of ideal gases in a vessel. The storm was a shifting yet integrated whole, more like a living being with invisible circulatory systems and shrouded organs, specialized regions of activities with specific functions. The flows of bodily fluids— the mixtures of air, moisture vapor, ice crystals, dust particles, and raindrops—all fanned out and coalesced, driven along transparent arteries by unseen pumps. Hidden streams of electrical charges gushed through neural networks of invisible axons, dispensing these corpuscles into shadowy reservoirs with tenuous boundaries. Localized structures materialized out of thin air and completed their life cycles in a matter of minutes, only to vanish as suddenly as they had appeared. The physiology of the thunderhead could not be seen with the eye or dissected with the cold scalpel of reason. As with all living organisms, every piece of cloud tissue served multiple functions and contributed to the overall design of the cloud in diverse ways.

The same criticism applied equally well to all occurrences in the desert. The numbers that were cranked out by formulas were useless to Jonathan because he was not out in the desert to play games with numbers and he could find no other value for them. He shrugged his shoulders and stared blankly at the ground when the physicist confronted him with the data. Who cared about any of that? He wanted answers, answers to real questions about real things. Why was all of this happening? This grand spectacle stretching before his eyes in every direction, incessantly transforming itself at every moment into endless sequences of unfathomably complex arrangements. The puppet shows that so captivated the attention of physicists were nowhere to be found, and the laws that governed these artificial performances were not the laws of the desert. The simple rules of physics could not account for what Jonathan witnessed everyday on his treks through the desolate wilderness. There was so much to see and he could never take it all in. That was why he kept coming back. He could never exhaust the endless supply of experiences, the perpetual variations on themes, the infinite creations of this fantastic world of his, the countless works of art to study and admire. What did physics have to do with any of this? Physicists did strange things. They collided streams of subatomic particles together and analyzed the spray, then mentally tried to reconstruct the whole universe from this totally bizarre event. What made them think that they could ever do this? Physicists could not even tell Jonathan what would happen next, so he had to find that out for himself. How could they claim that the method worked?

Understanding the ways that variables were connected together in equations was not what Jonathan needed to know. He wanted to know what just happened to him up on the ridge. How could he possibly understand it? Just because physicists used the physical world as a foundation for their mathematics didn't mean that the mathematics was therefore the foundation of the physical world. Mathematics and the physical world could not be more opposed to one another: a spiderweb of ethereal thoughts in the mind versus an infinite particularity of corporeal bodies in the world.

Physicists built a laboratory inside an airplane and then directed the pilot to fly the airplane into a thunderhead. Once the instruments generated the numbers, the investigators were locked out of the real world, stranded on their island of mathematical relationships. They could only make further calculations and try to correlate the numbers in new ways. Physicists then inserted wild speculations about what must have taken place outside the airplane, but the associated ideas came more from the mathematics than from the clouds.

Suddenly the truth dawned on Jonathan and he understood exactly what had happened to him up on the ridge. Previously he had angered the lizard creatures by taking their pictures. He should have known better. Rock formations like that were often inhabited by powerful spirits. These spirits flew into the sky and took on the forms of menacing storm clouds. He recalled that the clouds first appeared at that moment and then they followed him all the way back to camp. The spirits camouflaged themselves and waited for an opportunity to strike, concealing their true purpose until the time was ripe. Jonathan could still see the clouds coming over the ridge and hear the noises they made. The muffled words weren't subtle warnings at all, but rancorous threats whispered only to emphasize their seriousness: "Listen carefully. We are coming to kill you. We graciously allowed you to find the monoliths, but then you showed no respect for your elders, and now you must pay for your insolence. Your little friend must also die with you."

As soon as the spirits announced their presence the attack immediately ensued, the clouds ambushing Jonathan and Emma at their most vulnerable moment, fully exposed high atop the ridge. The spirits gave it everything they had and nearly succeeded in killing them both. The spirits then quickly fled as soon as they realized that they couldn't get

Jonathan and Emma once that they were safely inside the truck. They narrowly escaped this time, but Jonathan vowed never to take pictures of hoodoos again. The practice was way too dangerous and he might not be so lucky the next time. He was just grateful for a second chance.

Jonathan peered out the window into the darkness. The rains had tapered off and only a light sprinkle remained. He opened the door to let some fresh air into the musty cab. Emma, who had been pretending to be asleep, twirled her head around ready to spring into action, her eyes alertly following the motions of Jonathan's hands waiting for the signal to disembark, knowing that the order was forthcoming. She was well rested, and anxious to resume her tracking and hunting out in the field.

He stiffly turned and slid off the seat, his muscles having tightened after sitting in one position for too long. Water had pooled around the truck and he nearly slipped as he put weight on his foot, not anticipating the full extent of the muck. Water still sheeted across the light slopes, adding a final amount to that already collected. He gave Emma the signal to exit the truck, a sweeping arc of his hand away from the open door toward the emptiness of the night, reinforced with the words "Ok, let's go!" She leapt off the seat and splashed through the puddles, mud spraying in all directions, her feet slapping the soggy surface as she departed his small illuminated world and disappeared into the impenetrable blackness. She was determined to find out what had happened while they were hiding in their metal compartment, but she was not going to analyze concepts and think about things as Jonathan had done. She was going to discover the truth for herself.

—4—

In certain parts of the arid southwest, long stretches of interstate highway crossed unpopulated and desolate expanses of desert, providing the only means of access to these areas. Beyond the sparse traffic along these isolated ribbons of asphalt, there was little or nothing in either direction—sometimes for a hundred miles or more. The widely spaced exits all conformed to the standard configuration that was familiar to everyone, an overpass connected to the interstate by a pair of exit and entrance ramps, only in this case there was no crossroad at the top. Drivers who availed themselves of these traditional fixtures of interstate travel found themselves in the middle of nowhere. The pavement did not extend beyond the boundary of the interchange where a sign was usually posted advising drivers of the situation. After a thorough inspection of the surrounding area, everyone could see for themselves that the sign stated the obvious: "No Services Ahead."

From this point a rutted dirt road typically struck out across open country and disappeared into the distant hills, beckoning the adventurer in his mud splattered four wheel drive vehicle, but no one else. This questionable route was often the only means of further progress outside the interstate. There were no more signs, no placard of mileages to proper destinations, because there was nothing up ahead other than unmarked and often vague geological formations, casually known to locals and others who frequented the area as coyote butte, rattlesnake hill, wild horse mesa, burnt cabin flats and skeleton ridge.

Jonathan left the interstate behind at one of these exits, rattled the truck up a washboard grade for a few hundred yards, then guided it onto the first side road. This secondary road was narrow and bumpy but passable, until it dove into a steep canyon. He pounced on the brakes and skidded a few feet on the gravel, creating a cloud of white dust. He saw that last summer's rains had carved a deep depression on the righthand side of the roadbed and the overall condition of the road did not look good. He feared worse up ahead. The afternoon was well underway and he did not have time for major road repairs. The arroyo crossing was too close to the interstate to allow Emma to run freely while he laboriously carried rocks and threw them into the holes and ruts, so he bailed out of his plan and backed up, swerving from side to side until he regained the main road. Pressed for time, he stopped a short distance further down the road at an old garbage dump and parked alongside the edge of the same steep canyon that had blocked his original attempt to access the area.

Jonathan shouldered his pack and paralleled the chasm to check out a side canyon, looking for a place to cross, but he found no weaknesses in the canyon walls. He hiked back up to the landfill and tried the other direction. After a time, he located a game trail that cut a diagonal through the rocky terraces and descended the whole way down to the wash, but once on the bottom he found himself trapped by impenetrable walls and was forced to trace the course of the arroyo back toward the highway. Considering Emma's usual range of activity, this path took her dangerously close to the trickle of barreling trucks and speeding cars on the interstate. As the highway bridge slowly came into view, the ridge finally petered out and he scaled the end point of the spur up to level ground. After a quarter mile or so he joined the side road that he'd previously abandoned with his truck and followed the dirt track on foot as it meandered toward the uplands, finally escaping the annoyance of the light traffic in the background.

The sky was filled with gray and black clouds marching across the desert surface, merging with the hills and overrunning the peaks. The hordes of dark beings had him surrounded and appeared ready to unleash volleys of cold raindrops at any moment. Rain had fallen prior to his arrival and the smell of creosote was overpowering, the fragrant oils having been released during the serious drumming the leaves had received by the heavy droplets.

No one had dared negotiate this side road for quite some time and Jonathan found no signs of any cows, so he and Emma had the whole place to themselves. The freedom was delightful. Jonathan's spirit soared as he strutted across the open expanse, reveling in nature, watching Emma scamper about exploring the low hills. He resolved to return soon without her to spend an afternoon working on the road, because after that first crossing the road was fine. He had worked on all of these crossings two years ago and the efforts had mostly paid off.

The principle was simple. The runoff from summer storms scooped up surface debris as the water surged down gradients, digging out small channels in the loose sand and gravel. These ruts, once established, collected more and more water and concentrated the flow of the runoff even further, accelerating the erosion and expanding the once minor conduits into sizable trenches. Putting obstacles in the way of the water caused the water to abruptly slow down and deposit its burden instead of picking up additional material. The ditches automatically filled in with sediment and the road repaired itself. Thanks to Jonathan's previous labors, two of the formerly impassible arroyo crossings were now only slight impediments to backcountry traffic. If he could only get past that first arroyo, he would be able to access the entire area with relative ease.

Jonathan found himself attracted to the desert for several reasons. The desert possessed a special symbolic meaning for him because the quest for knowledge had begun in a desert, an arid region of the Mediterranean, a landscape perhaps not unlike this. As physics emerged from the diverse mixture of philosophical speculations, the discipline gradually took on a life of its own. The enterprise gained momentum and acquired new perspectives, ideas were put together and elaborate experiments were devised, wherein physicists pulled up stakes and moved indoors. Although physicists sometimes claimed that nature was their laboratory, for some reason nature didn't build laboratories. Perhaps this glaring omission from nature's vast repertoire was just an oversight on nature's part. Given the grand diversity of life, the marvelous geological formations, the spectacular spinning galaxies and gaseous nebula, the grandeur of the night sky, the making of a laboratory seemed well within nature's reach. Either nature lacked the motivation for such an undertaking, or the idea simply had never occurred to nature. Apparently nature had no use for laboratories and didn't see any reason to populate the landscape with them. What physicists meant was that they were measuring something that hadn't been entirely fabricated, in contrast to the bulk of experiments where the phenomena under study were completely artificial. But they still needed a laboratory, and this they had to build for themselves.

The reason why nature didn't build laboratories was because a laboratory was a microcosm of mathematics. The whole structure was chock-full of mathematical forms, from the geometrical shape of the building to the geometrical arrangements and precise alignments of the apparatuses. Nature didn't build laboratories because nature didn't understand what mathematics was and couldn't make things according to mathematical principles. In fact, nature had nothing to do with mathematics. The laboratory was the special place where physicists introduced mathematics to the world. The laboratory was the vehicle by which they brought their strange ideas to a land of irregularity and chaos.

Physicists had both the freedom and the ability to put laboratories wherever they wanted them: telescopes high atop mountain peaks, particle detectors in underground vaults, barometers in planes flying through clouds, radiometers in satellites orbiting the planet, but there was always a laboratory involved, because the physicists' world existed within those boundaries. This was never considered to be a problem because the truth was everywhere the same: inside, outside—it didn't matter. Gradually their focus shifted away from the mysteries of the desert and the laboratory became everything, the totality of existence. For them, the laboratory became the New World, a previously undiscovered domain waiting to be conquered and exploited.

Thus physicists no longer had any reason to come out to the desert, and in fact, not many people did. Perhaps they felt that it was too harsh and inhospitable. Occasionally Jonathan ran into a rancher or a hunter, very rarely an equestrian or a hiker, but he never ran into a nuclear physicist engaged in the work of physics. He found that strange because the desert was made up entirely of atoms and molecules. Physics was supposed to be about all this, the world he lived in, so why weren't they out here working on the mysteries of the universe, or at least examining all of these strange things and wondering about their true natures? None of this mattered to physicists and they were never going to come to the desert and attempt to explain it because they knew they couldn't. The world was not the proper subject matter for physics. Physicists performed experiments in closed rooms and then with a lot of gesturing and hand waving intimated that the results also applied to all of this.

When Jonathan reflected on the matter further, he saw perfectly well why physicists had abandoned this vast territory. This was no country for physicists. The land was unpredictable and lawless. No one could forecast what would happen next. The sun might rise in all its glory or remain hidden behind a layer of impenetrable clouds all morning, but no one could be certain until the day actually dawned. To everyone's dismay, lightning might explode from the most innocuous cloud floating in the most placid sky, while at the same time the approaching black thunderheads remained strangely silent. A violent hailstorm might sweep down from the mountains without warning, or the ominous clouds certain of rain would mysteriously disappear causing the sun to beat down on them once again. The trail might end unexpectedly, or continue farther than anyone could ever have imagined. Danger might lurk around the next bend or behind the next bush or under the next rock, or perhaps he and Emma would happily see no sign of trouble the entire day. Jonathan could not tell what would happen next because there were no rules, no calculations he could make, no way to know for sure what the outcome of the experiment would be. This was about as far away from physics as one could get. The desert was wild and rebellious, capricious and erratic, entirely free from the constraints of civilization and the rules and regulations of a structured society. For the rationalist, logician and mathematician, the desert was the utopian antipode. It was the hideout of eccentrics and outcasts, the repository of baffling occurrences, the proven sanctuary from the burden of logical necessity. The physicists had fled to the safety of their precise and predictable world, the one that they had made for themselves, the place they now called home.

Emma stopped and looked at Jonathan from a distant ridge, verifying that he hadn't changed directions. She independently conducted her own tour of the landscape, yet checked on Jonathan's position at regular intervals. Jonathan stopped and looked back at her, seeing that she was alright. If he had a dog that always walked alongside him looking up at him and attending to his every move, faithfully waiting for his next command, he would take it to the pound. Obedient dogs were in great demand and he was sure that it would find a good home. He wasn't looking for a military cadet trained to mindlessly snap at the barks of a drill sergeant. As his backcountry companion, he'd rather have a biker chick, a dog that was independent, rebellious and self-assured, full of surprises and fun to be around. Sometimes Emma stonewalled Jonathan, ignoring his commands—in a sense giving him the finger—and Jonathan always thought that it was cute. He related to it. When Jonathan was her age, he had given the finger to lots of people, including his parents. He had always thought that they deserved it. Deep down inside he still loved them. It just meant that he wasn't going to let them mold him or dictate to him or push him around. Jonathan would never try to push Emma around. He had the utmost respect for Emma. He knew that she was determined to be herself and he was never going to change that.

Back when he was younger, Jonathan ended up working part time and attending college on and off for nine years. At the end of that long stint he could have taken a degree in mathematics or physics, but instead he took a degree in the humanities—not philosophy or the fine arts—just humanities. Jonathan knew that a degree in the humanities was worthless and that he'd never get a job with it or ever find any way to use it to his advantage. He knew exactly what he was doing. He was giving the establishment the finger. Jonathan never regretted what he did. He would have made a lousy academician. Sooner or later he would have been forced to confront his mistake and rebel against the system. He simply had the foresight to choose sooner rather than later. So he walked away from it all and never looked back.

The reason why Jonathan and Emma got along so well together was that they were both fiercely independent. They had the kind of relationship that wilderness travelers had always sought. Jonathan knew full well the difficulty of finding a companion who didn't interfere with his experience of nature, who didn't come between him and the wilderness he loved. Emma was never a burden to him. She didn't talk to him. She didn't waste his time with moronic prattle. She didn't distract his attention with commentaries and observations. Jonathan could focus on the path ahead, soak in the scenery, meditate on the beauty of nature, knowing that she was with him, doing more or less the same thing in her own fashion. He would have been upset with her if she had paid little or no attention to the world around her. He would have thought that something was lacking in her character, a blindness, a self-absorption. Emma saw Jonathan relentlessly exploring the terrain, constantly checking out every hill and arroyo, always poking around the rocks and bushes, and she understood that this was why they were here. His clear example corroborated her personal viewpoint. This was the way life was supposed to be and all was right with the world. Knowing this made Emma very happy, and perhaps her happiness was what motivated Jonathan to a large extent, and vastly increased his estimation of the great outdoors.

Jonathan understood that Emma was not the perfect dog, but he found that she had many good qualities. He strove to embrace her virtues and forgive her faults. Years ago when Jonathan was looking online for tips on training a puppy, a man with a black lab commented that his dog would start running around with its nose to the ground, more interested in scents than him, and that this behavior was completely unacceptable, something that should never be tolerated by anyone. Jonathan thought to himself at the time that this person should consider getting himself a different breed. This was what labradors did, and any attempt to break them of this habit was cruel and inhumane. People had made them this way and they couldn't help being obsessed with smells. Jonathan didn't want to change Emma because he loved her the way she was. He knew that she had to discover for herself what was important to her so he cut her a lot of slack, but that didn't mean he let her walk all over him. Everyone had to do things they didn't want to do and Emma was no exception. Emma was smart. She understood this and developed a good relationship with Jonathan. She saw that Jonathan did things for her, so in return, she did things for him. For the most part, they had worked it out.

The road made a sharp turn and veered away from the mountains, heading back toward the interstate. This unwelcome deviation forced Jonathan to abandon his convenient walkway, but he'd gotten weary of it anyway. He'd gladly swap the ease of strolling down an untraveled road for the challenge of bushwhacking over rough terrain.

The brush thickened as he entered a shallow basin, the branches camouflaging the rocky surface. He was already surrounded by a host of creatures when he first looked up at them. The dense creosote had masked the approach of the javelinas and he had never heard a sound or had any warning of their presence. The squat animals came at him from several directions, much as the rush of water flowing around obstacles. He could see only the leading edge of the onslaught and had no idea of the full extent of their numbers. He found the faces of the javelina ugly and disgusting and they caused him to recoil involuntarily. They reminded him of the orcs and trolls of fantasy literature, a race of beings that were despised and fought off whenever they attacked. He swiveled about and started running in the opposite direction. He hadn't marked the whereabouts of Emma until this moment, but he found her only a short distance away, leisurely following in his footsteps, preoccupied with scents and burrows as usual. She hadn't noticed the javelina yet, but snapped her head at the sound of Jonathan's feet grinding the gravel in a frantic dash. He called out to her: "Emma! This way!" She assumed that he'd spotted some game and was leading the charge, so she eagerly joined forces with him. Emma couldn't possibly comprehend the fact that Jonathan was actually leading her away from a confrontation with wildlife, because that kind of behavior made no sense to her. Why would he ever do such a thing? They were obviously out here to hunt wildlife, to pursue it and chase it, and not run away from it.

Jonathan pulled up as he approached a gully that was too steep to sprint across and stood still. Emma figured that the game had gotten away. She looked at him as if to say: "That's alright. It happens to me all the time. At least we gave it a good try." However Jonathan was not searching the area ahead of him. Instead he was scanning the area behind him. He did not see any signs that the javelina were following them so apparently he had been successful in eluding the creatures. Emma was still oblivious to their presence, which was good. Jonathan crossed the gully and maintained a brisk pace, just to be safe, nervously glancing over his shoulder to reaffirm the situation. Confident that they had lost the javelina, he circled around to regain his original route, along with his casual attitude toward hiking in the desert.

Jonathan climbed a small hill and noticed some interesting stones on the ground. As with most desert explorers, he was a bit of a rockhound. He stopped to examine the stones more closely. Wherever he went, Jonathan was always picking up stones and taking them home with him. Some of the stones were polished smooth and others were crystalline in structure, but he also collected larger rocks with strange shapes, cavities or inclusions that appeared as if they were eyes looking back at him, juxtapositions of contrasting minerals welded together into an interesting mosaic, or splashes of molten lava frozen on the surface of an underlying matrix, a freeze frame of the moment of impact. He often put them into the pouches of his pack and lugged them up and down the hills along with his gear. Sometimes he forgot about these weights and took them back out into the desert again on his next walk, needlessly adding to his load. The collection varied as Jonathan visited different localities, at times comprising many little stones, at other times just a few larger rocks. He didn't keep track of them and never found a reason to count them, because he was only interested in the stones themselves.

In order to count them, he had to attach the same label to each stone, regardless of its size and shape, regardless of its appearance or structure. They were all "ones." Calling such a diverse array of objects by the same name was counterintuitive. Jonathan stooped and picked up a tiny specimen of clear quartz and put it in the palm of his hand, then asked himself "What is this?" He said "This is one." He then walked over to the arroyo and rolled an anvil sized chunk of rhyolite onto its side for a better look at it and again asked himself "What is this?" Again he said "This is one." Really? The objects weren't at all alike. He wasn't making any sense. A person had to understand the idea behind it first. Repeating the same word in different situations was a clever plan devised to stamp out all traces of individuality. Mathematicians and physicists wanted to erase the eccentricities of the world and cause all objects to conform to one universal standard. To achieve this end they would simply ignore the creativity of nature, cover up the truth of existence, and live happily ever after in their alternate reality of empty words and hollow concepts. Objects were stripped of their identities so that they could be assigned new ones, identities more suitable to the roles they would play in their new lives as the residents of a mathematical universe.

Jonathan could only replace an object with a place holder by disregarding the object for what it was. He had to learn to ignore reality. This was not an easy thing to do. Reality naturally took center stage and a person might rightfully ask: "What else was there?" Somehow he must distract his gaze momentarily, whisk the object away using a clever slight-of-hand trick, and then put the plain number there in its place. He now saw the rock not as a rock, but as a mathematical object.

The switch only seemed natural because he had been indoctrinated into this way of thinking at an early age, taught to count before he had learned to do much of anything else. The object was gone, but now he had the number instead. Now he could start playing games with numbers. For this he did not need to bring the rocks and stones back because they no longer played a part in any of this. He could just put them down on the ground where he had found them and forget about them, because he had something better—he had mathematics. This exchange was never suggested by the complex nature of the stones or by the contexts in which the stones had occurred. Rather the switch was based on an idea Jonathan had entertained in his head, and in fact a very bizarre idea: naming different things by the same name. Understanding the motivation behind this weird scheme made everything clear. Since numbers were concepts and not physical objects, he could do all sorts of things with them that he could never do with rocks and stones. He could invent calculus, for example.

A primitive caveman did not collect stones and then come up with the sums of numbers by observing the behavior of the stones. Addition involved manipulating the stones, getting them to mimic a set of preconceived ideas. Handling stones and studying them was not how someone discovered what sums were because counting and adding were not about stones. When a person added stones together, the stones were just pawns in a game, and it was a game that the stones had not invented.

Once Jonathan tied each stone to a "1" using an imaginary string, he could make the stones dance to his equations because the numbers were pulling on the strings. He could act out "2 + 2 = 4" with the stones by pairing them off and then piling them together, because he had created a puppet show. That didn't make the equation a law of nature or give it an independent reality. He had set up the game in a certain way and he had sworn to play by the rules. He then had to accept the outcome of the plays.

When Jonathan was a young child, he spent much of his time horsing around with two of the neighbor girls. One was several years older than him and the other was a year younger. Sometimes the girls would pick on him. The older girl would wrestle him to the ground and pin his wrists with all her weight while straddling his chest. The younger girl would produce a freshly dug earthworm from her pocket and attempt to force Jonathan to eat it. Jonathan would press his lips tightly together and swing his head from side to side trying to avoid the disgusting creature as she pressed it to his face. With the encouragement of the older girl, the younger girl would tickle and prod him, poking his ribs, trying to get him to open his mouth. Finally they would give up, laughing heartily, and he would take that opportunity to run for his life. They knew how to get rid of him when they were tired of playing with him.

The three of them would sometimes play Monopoly on the front lawn of his house. On one occasion Jonathan made several unfortunate roles of the dice and the older girl took all of his money, ending the game for him. He protested that this wasn't fair because this was not the way the game was played. Taking up the challenge, the older girl lunged upon the directions in the upturned cover of the box and quickly located the relevant passages, taunting him as she read them aloud with exaggerated inflections in her voice. Her arm then shot forward with the piece of paper waving in his face so he could verify for himself that what she had just told him was correct. Infuriated, he snatched the directions from her hand, marched over to the neatly trimmed hedge in front of his neighbors' house, and flung them into the bushes. They floated over the top of the hedge and sank into the narrow gap between the impenetrable tangle of branches and the imposing brick wall of the house where no one would ever get their hands on them again. He had never agreed to play by those rules. Someone had simply shown him how to play and he never actually read the official rules. He stomped into the house to see if dinner was ready.

Jonathan could not deny that he had landed on Park Place and therefore had to fork over a large sum of Monopoly money to his heated rival because all of the action on the board was strictly controlled by definitions. The definitions could have been otherwise. They did not exist independently or express a truth about physical reality. Similarly, Jonathan could not deny that "2 + 2 = 4". The one-to-one correspondence he had previously set up between stones and ones made the formula "2 + 2 = 4" seem like an absolute truth, one that was independently verified by nature, but that was only because two and four were defined in such a way that the conclusion was inevitable. The ideas fit together and these relationships would never change. The illusion of reality was caused by the idea of "correspondence." This clever notion allowed Jonathan to leap back and forth between numbers and objects, but he hadn't gone out into the world and found this correspondence already existing in nature. He had put it there in a move so sly and disarming that it went unnoticed by everyone.

Jonathan had observed stones in many locations and under every set of circumstances and he had never witnessed an act of counting among them or caught them engaged in any way in this kind of behavior. Counting externally imposed a set of ideas on them without their consent or active participation. Counting was not a natural occurrence and it did not happen spontaneously. Counting was not a part of the world and it was not something that needed to be understood in order to comprehend what existed. From the point of view of what existed, counting was meaningless and should simply be ignored. When the physicist turned all of this around and made counting the most important thing in the whole world, he was leaving the physical world out of the picture and concentrating his attention on a world of his own design, a highly imaginative dream world where nothing was real. Jonathan wanted to know what stones were, and in order to answer that question, he had to refrain from counting them because their essence lay not in numbers. Counting was nothing more than a means of bringing mathematics into the world and the method merely used stones to achieve that end. When Jonathan embraced the stones, he saw that no counting was involved in their existence, and this was the only issue of any importance to him. Mathematics played no role in the nature of things and could only be added to existence in a careless and wanton act that disregarded everything that made things what they were.

Of course, Jonathan could count not only stones and rocks, but anything whatsoever: one set of tires, one leafy bush, one handful of popcorn, one imaginary triangle, one rock aggregate, one cloud in the sky, one thought of dinner. The ones added up to seven, but the objects that corresponded to these ones didn't add up to anything, and in many cases they weren't even objects at all. The motley collection didn't unify in a meaningful way even though Jonathan could still mentally carry out the algebraic operation.

What was the sum of a rock and a leaf? A rock and a leaf certainly did not add up to a number. These objects could not be put together in such a way that they resulted in a crooked line scrawled on a sheet of paper. The sum in this case combined the idea of a rock with the idea of a leaf, ideas that did not fit together in any obvious way, but what about actual leaves? Leaves came in all sorts of shapes and sizes. Was a leaf from a plains cottonwood equal to a leaf from a desert willow? Given their different shapes and colors, the basis for such an equality was difficult to explain. And didn't the size and weight of the rock mean something? Shouldn't a large rock produce a sum that was different from a small rock? A gemstone was certainly worth more than an ordinary chunk of sandstone. Shouldn't it command a higher value? The more that details were added to the formulation, the more absurd the concept of addition became. If nothing equalled anything else, then the arithmetic was impossible to carry out. First the meanings of all the details would have to be specified by rules, but laying down the endless definitions in unequivocal manners would be a daunting task. Instead everyone had to start with mathematical concepts because physical objects did not lead up to mathematical concepts. Concrete objects could not possibly be transformed into ones. They could only be replaced by ones.

The idea of putting two disparate objects together to form a sum reminded Jonathan of the late-night talk show comedy routine where the host tried to imagine what the offspring of two well-known celebrities would look like. Portraits of each celebrity were first displayed side by side, then certain facial features from one were mixed with certain facial features from the other, resulting in a hilarious rendition of their supposed love child. Jonathan imagined a flat rock with a ragged margin containing pockets of light-absorbing pigments distributed along a network of fluid-filled tubular inclusions, or a hefty leaf with greenish mineral sectors embedded in the epidermal layers instead of the organic molecules known as pigments. A mishmash of individual details destroyed the original integrities of each object and produced only hogwash, nothing at all like the logical, intelligent outcome of an arithmetic sum in mathematics.

The only reason Jonathan could add objects together that had nothing in common with each other was that he had started out with the mathematics. He formed the concepts "one," "addition," and "every entity can be transformed into a 'one' without regard for its reality" in his mind, then imposed this structure on the natural world, but if he started out with nature, he clearly saw that counting was wrong. He couldn't possibly count concrete objects because that meant integrating them into a whole. He could only count "ones," but in the real world, there were no "ones." Physical reality was a denial of arithmetic and a resounding refutation of mathematical reasoning, but this problem could easily be overcome by establishing the mathematics first and then holding onto it tightly, even in the face of extreme adversity. This forced him to let go of the physical world instead. The objects didn't have to add up to anything other than a number because the mathematics was not about the objects. Mathematics was purely self-referential so it could stand on its own without any outside assistance from the real world. He could count absolutely anything, with the caveat that counting didn't matter and the result had no significance. The physicist insisted that the sums weren't imaginary because he was counting objects that actually existed, but the process of adding these things together was completely imaginary and moreover a violation of the sovereignty of an existence where every object and event was absolutely unique. In contrast, "one" was a perfect unity that was absolutely identical to every other "one." Ones were indistinguishable from one another, while objects were incommensurate with one another and therefore incomparable. Thus they could not be added together in any meaningful way.

Jonathan saw that the act of counting imparted a false harmony to the objects being counted. Ones were integrated through identity and this quality was transferred to the objects in the process of counting them although it had no such authority to do so. The Three Musketeers were primarily united by their allegiance to one another, captured in the remarkably illustrative slogan "All for one and one for all!" but they were also united by being "three." Each musketeer had to deal with the unique circumstances of his own life and meet his own personal destiny and unlike 'ones' these facts were not interchangeable.

The Four Horsemen of the Apocalypse seemed to be riding together as a group and acting in the name of a common cause, although their missions were vasty different and they were united only under the general banner of unspecified death—in addition to being united by the mathematical banner of "four". The Seven Sisters of the Pleiades appeared to have similar magnitudes and they were clustered together in the night sky, but counting did not rely on such associations and conferred unity with no regard for the actual relationships between objects.

The problems associated with counting concrete objects applied equally to all forms of abstract thinking. A while back Jonathan had been talking to his neighbor about diets and their relationship to health. He pointed out to her that every food possessed a toxicity which accumulated in the body over time, so it was important not to eat the same food everyday. The neighbor argued that if that were true, then eating a broad diversity of foods did not solve the problem, because the sum of the toxicity was not thereby reduced. Each food contributed its own share to the overall toxicity and thus nothing was gained by altering one's food intake. Since every food participated in toxicity, a person might as well just eat one type of food and live on nothing but say watermelon or butter or whatever that person liked most in the world, to the exclusion of everything else. Of course, the problem was that different foods did not possess the same toxicity, thus each food posed a unique danger to the consumer. Each food had a specific brand of toxicity not shared by other foods, so by varying one's diet, the different toxicities did not add up. No single toxic chemical was ever allowed to stockpile in the body to the point where it began to interfere with normal metabolic processes or block essential biochemical reactions.

Jonathan found it interesting how turning a concrete situation into a set of abstractions changed everything. The abstractions created a new reality of their own. The ideas had highly original identities and these novel attributes allowed a person to arrive at conclusions that were contrary and in many cases the opposite of what could be deduced using more concrete thinking.

Abstract thinking replaced a multiplicity of individual examples with a single entity. Instead of facing a mixed bag of discordant elements, everything was dressed in the same uniform with the same name tag on the lapel. Once toxicity was stripped of its concrete existence, it could be viewed as a simple, independent entity—or in other words, a quantity. The new state of affairs forced everyone to think about things differently, but more importantly, it created a totally false view of the world. Turning toxicity into an abstraction made all foods essentially the same. This turned toxicity into a matter of addition and subtraction. The details were still out there and they were what constituted the true situation. The toxic accumulation of substances encompassed a wide range of specific compounds, each one corrupting a different set of molecular interactions in the body and causing a different set of symptoms to appear. These facts had to be dealt with and understood, but abstractions failed to accomplish this task, instead sweeping all the details aside. Everyone wound up tossing around abstractions in a theoretical funhouse, which was unquestionably an entertaining pastime, but unfortunately it had nothing to do with the real world.

Toxicity was just a word. It could be the subject of a discussion but not the object of an investigation. The word could be studied and analyzed grammatically and semantically, but not objectively and experimentally. Toxicity could not be found in the world because it didn't exist in the world. The emptiness of the term allowed all sorts of conflicting definitions and the lumping together of unrelated examples.

What disturbed Jonathan the most was not that the neighbor had instinctively turned to mathematical reasoning to resolve a matter, but that after a considerable period of reflection and deliberation, she still found the argument convincing. She was confident that she had successfully refuted Jonathan's position and thus she could continue her eating habits in the full knowledge that he was wrong and she was right, even though she had turned the whole situation into a travesty of the factual relationships between foods and the human body.

Jonathan felt that a small amount of abstraction was harmless enough and was in fact necessary, since every concept was an abstraction and no one could lead an entirely non-conceptual life. As long as a person did not stray too far from reality, everything was fine—well, more or less. The danger lay in the fact that abstractions invariably led to more abstractions. Getting caught up in this ongoing process was pretty much unavoidable and everyone quickly found themselves lost in ugly wastelands of meaningless expressions, deluding themselves that their sure-fire conclusions were the absolute truths of the universe.

At even the most rudimentary level of mathematics—the simple act of counting—mathematics transferred its own character to the objects and events of the world. Replacing the pliability and variety of concrete items with inflexible ones instilled an unyielding uniformity to these items, a commonality that came from the mathematics and not from the items themselves. Expecting such a wanton act to have no effect on the nature of things was baffling. The mathematics was so utterly different from the real world, a way of thinking where all substance and content had been deleted, in contrast to a reality that was composed exclusively of details. This meant spanning an infinite abyss between polar opposites, trying to understand what existed through something that had nothing to do with existence. Enlarging the domain of mathematics by adding more concepts did not transcend this character, but only served to compound its effects and amplify its falsity. Employing mathematical methods radically transformed physical reality, making it impossible to understand that reality through the ensuing mathematics. How could anyone believe that this was not the case?

Despite Jonathan's carefully reasoned analysis, the physicist claimed that he had absolutely no problem counting atoms and molecules and then tying these numbers to his calculations of mass and energy. Since the method worked and the numbers matched, the physicist maintained that Jonathan was wrong in saying that there was a problem with counting. Jonathan pointed out that first of all the physicist didn't actually count atoms and molecules. Instead he arrived at these numbers through an entirely different process, that is, through the measurements of quantities that were then divided up into equal portions, each allotment attributed to an entity that was presumed to be identical to all of the other entities. This was clearly a mathematical construction and as with all such constructions, it worked perfectly. All of the determinations took place in a conceptual universe that neglected the unique personalities of the participants. Bulk quantities were distributed among fictitious entities in a neat interplay of ideas that supposedly represented the true state of affairs—but what exactly was the true state of affairs? Even in the physicist's highly schematic account of events, atoms could exist in a multiplicity of states. He separated these types into gross categories that most likely included a wide range of deviations and abnormalities. He similarly reduced the specific actions of individual atoms and molecules into basic outlines and cartoonish renditions. Who knew the wide variations in behavior that atoms and molecules were capable of exhibiting? Some interactions might be quite rare, but these occurrences were lost in the din, averaged in with the more commonplace interactions, everything added together to produce a uniformity that was not characteristic of the actual events. The physicist's picture of reality was derived entirely from mathematical considerations. Even if the physicist could count atoms and molecules, they would not add up to anything.

Jonathan could measure the sides of his kitchen table and calculate the surface area. He could use this number to determine the volume of paint he would need to deposit a uniform layer of liquid on the surface. He could say that the fact these calculations worked proved that mathematics was the language of nature. The only problem was, the table was not an object of nature but rather a mathematical object that he or someone else had fabricated. By building a flat rectangular surface and adding a layer of uniform thickness on top of it, namely the volume of space between two parallel planes, Jonathan was introducing mathematics to the world. The mathematics worked because he had brought the mathematics to the table. He constructed items that had mathematical properties and then said that this mathematics had consequences, that it implied further mathematical properties, however, he had made all these ideas incarnate by molding material substances to conform to them.

As soon as the physicist set up an experiment and started constructing axes and distances and scales of measurement, he was imparting mathematics to the situation, and it was this mathematics that he was going to study when he performed the experiment. The physicist retorted that the experiment was more than just the mathematics, because at some point a material object was going to respond in some way to some external influence and this response was not dictated solely by the mathematics. What the physicist failed to mention was that this object had already been mathematized by the setup, by the way the object was handled and presented within the apparatus. The object was turned into a fabricated mathematical form, much as the table in Jonathan's dining room. No one was talking about objects running wild as they did in nature. The object was being controlled in mathematical ways by strictly defining trajectories and velocities, by setting precise values for energies and momentums, by introducing unnatural purities and uniformities, so that the object acquired a mathematical disposition—at least for the time being. The physicist than captured the object in its moment of weakness, before the mathematics evaporated and the object returned to its natural state of confusion and disharmony by mingling with unknowable foreign elements and being barraged by haphazard and unfathomable influences. The order and the conformation to laws that were seen in experiments were fake when viewed from the perspective of the world. Physical laws, by being mathematical in nature, could only exist by virtue of the relationships within mathematics itself, and these had to be introduced externally. Otherwise the physicist had to somehow maintain that the mathematics was already there before he fabricated it, an essential part of the universe that had existed prior to the experiment, but this was a dubious proposition given that mathematics was made up of definitions and chains of reasoning and that mathematics was nothing more than the way that certain ideas could be fitted together.

People had lived on the earth for untold millennia, moving around and interacting with objects other than themselves without ever thinking of space as a three-dimensional manifold, because that was not what it was. A coordinate system was not an observation. No one could observe something they first had to construct in their minds. Jonathan could stare out into space as long as he wished, but he would never see three perpendicular lines intersecting at a single point. Mathematicians argued that it was a necessary assumption, but they did not finish the sentence. They failed to add the crucial final qualifier, which completed the thought: the assumption was necessary only in order to turn basic geometrical concepts into a coordinate system, but in doing so they were not visualizing space, they were just thinking about mathematical forms and juggling mental constructs. No one ever needed to mathematize space because space itself was not mathematical. Everyone understood what space was because they were surrounded by it all the time. People had interacted with space long before mathematics had been invented and they had continued to do so without having the slightest knowledge of mathematical reasoning.

If mathematics was truly a necessary assumption, then people wouldn't be able to function until they had learned mathematics. If the world was in fact a mathematical system, then people would have to know mathematics before they could successfully deal with it, but obviously that wasn't the case. People today were living more and more in a mathematical world, surrounding themselves with the mathematical objects that science and technology had provided for them, but primitive people had lived in a thoroughly non-mathematical universe. Nothing they did required mathematical knowledge and they were never called upon to engage in mathematical thinking. An understanding of geometry wasn't required for cooking a meal over an open fire, skinning a carcass out on the open range, or setting up a village along the river, but it was required to bisect a line segment, find the area of a circle, or draw a perpendicular.

Geometry wasn't necessary for any of the usual activities of human life, but it was absolutely necessary for all of the activities of physicists and mathematicians, however these people were not engaged in normal human affairs. Mathematics was only necessary in order to do mathematics and that was as far as it went. If a person wanted to construct mathematical objects, then that person was going to have to understand mathematical ideas in order to achieve that end, but for people who were not in the business of constructing mathematical objects, mathematics was utterly meaningless and irrelevant. Mathematics was only necessary in a given context, but physicists assumed that this context was valid in every situation because they were only thinking about themselves. The real world never entered into the picture. Physicists lived in their own little bubble, an abstract world of their own making. Nothing else mattered and no other factors needed to be taken into account. If mathematics was necessary for them, then mathematics was necessary for everyone.

These days nearly everyone lived in a world of man-made geometrical shapes—cylindrical cooking pots, rectangular picture frames and tv screens, planar walls and floors, not to mention rulers, compasses and protractors, all constructed from straight lines, perfect circles, sharp corners and flat surfaces. The mathematical world also included basketball courts, football fields and baseball diamonds, planes, trains, boats and automobiles, books and word processors, calendars and clocks. Mathematics was literally everywhere, and in these objects various geometrical forms, including the axes of the cartesian coordinate system, were made visible, so everyone naturally thought that geometry was real and coordinate axes were the fundamental parts of all objects, but when Jonathan picked up rocks and sticks on his walks, their profound irregularity hadn't the slightest hint of geometry and the application of three perfectly straight and perpendicular lines was awkward and inappropriate—or as Jonathan often said, irrelevant. He oriented these objects in different ways, searching for natural axes, but at no time did the objects ever suggest these artifices. The desert revealed to Jonathan the utter absurdity of geometry. In the laboratory, geometry was bestowed with great significance by constructing apparatuses with geometrical shapes and configurations, which themselves engendered more geometrical patterns in the artificially created phenomena. Geometry entered the spotlight and all eyes focused on it, creating the illusion of a geometrical universe. Under these circumstances, three dimensions was a natural fact and believers in the supremacy of mathematics were triumphant. They could hold onto the illusion as long as they didn't wander too far into the desert and suddenly realize the folly of their actions. It was easy to forget that all of these things were mere fabrications based on mathematical ideas.

Mutually perpendicular axes were of no use to Jonathan on his travels. He wandered freely in every direction, paying attention to real landmarks and geographical features rather than imaginary lines. He could not locate these lines because they originated neither from the world at large nor from the objects around him. Jonathan could make the grid appear by pulling out his handheld GPS unit. The device was in constant communication with a fleet of space vehicles orbiting the earth in precise formation, the satellites continually broadcasting microwave ranging and navigation signals to receivers on the ground. The whole setup was basically a sophisticated geometrical construction. The trees and rocks of the desert and all the geological features of the earth were not a part of this geometry, so anything could be located at a given point and what it was didn't matter. The device was navigating the grid and not the earth. Similar to the cart in Jonathan's grade school experiment, the concrete details of the desert were totally irrelevant and the only thing that mattered was the general outline of the grid. The little calculator was constantly cranking out maps and charts and numbers, but Jonathan had to look up, away from the device, in order to see where he was at.

The fact that a hill or an arroyo was found at a given location made Jonathan feel that he was navigating the desert with his GPS unit, but he was doing so only incidentally and none of these encounters were implied by the structure of the mathematics. This mathematics actually told him nothing about the desert. Geometry could not incorporate the details of the desert because these details had not been placed in geometrical patterns or laid out according to any preconceived mathematical plan. The desert could never produce the interplay of ideas on which the Global Positioning System was based and it stood outside of every mathematical formulation.

Jonathan could locate north-south and east-west lines radiating out from his current position but only because the GPS device brought these ideas along with it and introduced these ideas to the world. The device could not demonstrate the existence of a coordinate system because it created an independent reality that was set apart from the reality of the desert. Without the GPS, all these lines were unavailable and unknowable. Without the GPS, the coordinate system disappeared and he had no way of finding it. He was forced to navigate using the hills and arroyos as reference points, just as he had done before the Global Positioning System had been invented.

The thing that Jonathan noticed about the real world was that he didn't make it up. He didn't walk around and say to himself that he was going to arrange the trees in certain patterns, put a tree over there and another one over here. The trees existed of their own accord and he had no say in it. He did not make up reality in his mind. The physicist, on the other hand, by letting definitions create realities for him, lived in a dreamworld of his own making. Allowing the magnetic pole to set the directions of his axes did not mean that the resulting directional grid was an independent reality of the world. The existence of definitions in geometry was completely unlike the existence of trees in the forest, or clouds in the sky, or rocks on the ground, or bears in the mountains. Jonathan did not trip over his imagination or run away from it when it turned on him and attacked him.

90

The magnetic pointer of a compass played a role similar to the dice in a Monopoly game. The players let the dice determine where they landed on the board, just as the navigators let the pointer determine which way they positioned their axes. The real world made a token appearance in a make-believe game world and suddenly the game was real and nobody was making up the plays anymore. The game now shared a crucial aspect with physical reality: both were self-determined and people were no longer just imagining what they saw taking place before their eyes. Everyone had to sit back, watch events unfold, and accept the outcome of the plays. The game board had brought a set of definitions into existence, but these definitions didn't exist outside of arbitrary acts of creation. The game was real, but it was a purely fabricated reality.

The hardest thing for the physicist to accept was the fact that perpendicularity didn't exist. Plato also had a problem with this, but the issue had since been resolved—at least outside of the sciences. Jonathan could use the idea of perpendicularity to make an object with rectangular sides, just as he could use the idea of a weight attached to a handle to make a hammer, but these ideas existed only within the fabricated objects. Perpendicularity was an abstraction, constructed out of elements that similarly didn't exist. The physicist was so convinced that everything he imagined was real, that mathematics was the way of the world, he couldn't see that all of his mathematical ideas were just fictions, nothing but a particular way of thinking about things and not an accurate representation of any kind of reality.

Jonathan heard the physicist exclaim: "But perpendicularity is such a useful idea and I can do all sorts of things with it. If perpendicularity is just a fantasy, then how can I possibly do anything with it? Didn't the indisputable fact of its utility imply the existence of that which was being utilized?"

Well, a hammer was also a useful idea. Jonathan could do all sorts of things with the idea of a hammer, much as a geometer could do all sorts of things with the idea of perpendicularity. He could tie the idea of hammering to other ideas, those of bending, flattening and straightening for example, or knocking free, breaking apart, and pulverizing, or driving, nailing, and pounding, in the same way that a mathematician tied perpendicularity to parallelism, squares and rectangles—and right triangles. But Jonathan never envisioned the hammer as being an ultimate reality of the universe. C'mon, it was just a hammer. What was so special about the idea of perpendicularity that made it fundamentally different from the idea of a hammer?

Many of the ideas surrounding perpendicularity and hammering could be acted out in the physical world. Jonathan could bring the idea of a hammer into existence by making an actual hammer. He could also bring perpendicularity into existence by making a wooden box with sharp corners and straight sides placed at right angles to each other. But perpendicularity didn't exist in the world until he made it appear. Jonathan had walked this earth for thousands of miles and he could tell you for a fact that there was no perpendicularity anywhere out here—unless it had been deliberately put here by humans beings. So Jonathan mused: "Let's replace the physicists' misguided aphorism with a new one: 'Existence cannot possibly be based on something that doesn't exist.'" And then, he said, "Let's add another one: 'A person cannot use fabricated objects to account for the world because the world had not been fabricated by anyone.'" Jonathan felt that such maxims might at least lead the physicist in the right direction.

The problem was, of course, that everything that existed was concrete, and as such it could never be the basis for anything else. The Greek philosophers originally searched for something that could account for everything and naturally they chose things that existed, or rather simple abstractions from things that existed—fire, earth, water, air—but water could only account for things that were wet and fire for things that were hot. Anaximander realized that nothing existed that could possibly be the foundation for everything that existed, so he invented something he called the **apeiron**—the "boundless." Obviously there was no such thing as "the boundless," but in a strange breach of reasoning, he decided to just say that it existed. In a most amazing move, he said: "Let's pretend. Let's make believe that it's real." Jonathan understood the motivation. The existence of "the boundless" would have been a great thing because then the world would have made sense and we could all have slept well at night. Existence could be based on existence as it should be—as it must be—and we could all have what we wanted: an explanation for the world. We just had to overlook one little detail. Our explanation wasn't based on anything but our imagination. Anaximander didn't do what he did because it was the right thing to do. He did it because he had no other choice.

Jonathan took Anaximander's example and challenged the physicist with it: "Can't you see that what you've created through mathematics and geometry isn't part of the world?" Continually solving problems and thinking about mathematics had put the physicist in a state of dazed confusion. Jonathan wanted to interrupt his trance and bring him back to an awareness of his surroundings. He needed to stage an intervention in order to stop the madness. The physicist had been hypnotized by all the sparkling ideas and their promise of theoretical salvation, and he had lost his senses after staring too long at the glittering treasure chest of mathematical jewels, the innumerable facets capturing the unnatural glare of the artificial light of reason, particles of matter spinning in the void and the curtains of space rippling in the invisible winds of time. Constructing the world out of mathematical ideas had become his Frankenstein monster. The physicist could not create the world out of formal relationships and colorless silhouettes any more than the biologist could create life out of dead and inanimate matter. The very idea itself was sheer madness.

Jonathan held one truth in clear consciousness: mathematical ideas did not come from the world. They did not belong here. They were not what nature had used to create the world. So how could they be the ultimate realities of the world, the foundations of all that existed? The physicist had locked himself in a concrete cave, a place where the intense light of the real world had been filtered and diminished. When he crawled out of his windowless prison, his declaration of a mathematical universe was based not on what his blinded, squinting eyes now told him, but on the shadowy forms of his laboratory, recounting the ideas he had entertained while he was lurking in the dark recesses of his scholarly dungeon. Jonathan listened to the physicist's fantastic tales of a magical land and was filled with incredulity. He suspected that the physicist was delirious with a fever and beset by strange visions and hallucinations, seeing apparitions that couldn't possibly exist. Jonathan wanted to make the physicist whole again and bring him back to his senses so he could join Jonathan in his quest for the truth and walk with Jonathan on his long journey across this remarkable landscape, witnessing inexplicable events and beholding creations of sublime beauty. "Tell me once more how you plan to account for all of this? With mathematical equations?" Jonathan stifled a muffled laugh and turned his back on the physicist, kicking a few stones along the way.

Nature hadn't been built out of ideas. Jonathan hadn't been walking along one day, alone with Emma out in the middle of nowhere, enjoying all the splendid works of nature, when he found that nature had made a hammer. It was just laying on the ground among all the rocks. He picked it up and said to himself: "Wow, look at this! Now here's a great idea! Why didn't I think of this? It's too bad that nature thought of it first." Actually, it would have been great if nature had thought of it first. Nature could have helped everyone out by providing them with all the tools they needed to live a good life, but nature chose not to do this. Jonathan felt that this criticism wasn't fair because nature wasn't worried about anyone's welfare or concerned about their problems. Nature never considered what Jonathan required and what might make his life better. Nature had given him a brain so that he could figure all of that out for himself, and build a world for himself based on his own ideas. But how did Jonathan know that a bush was not an idea? He could say that it was not a mathematical idea, but it might be some other kind of idea. Maybe nature just thought about things differently. Maybe nature reasoned in strange ways and was obsessed about its own needs and ambitions.

Had nature just been going along enjoying the fruits of its labors, basking in another fine day, observing all the interrelationships it had brought into existence thus far, when suddenly it thought to itself: "You know, I have a great idea. Why don't I make something that has tiny manufacturing plants that convert sunlight into sugars. I could put these workshops in the epidermal layers of leaves, and these leaves could be supported by a network of branches in order to spread them out and capture as much sunlight as possible, and as an added bonus, I could make the whole assembly grow and multiply and populate the landscape. Now there's an idea for you!"

Jonathan knew that the more accepted explanation—both in the scientific community and in the public domain—was that the plant had evolved. It didn't have a particular plan or overall design in mind when it started out. Things just went along and the plant that Jonathan encountered standing on the open range today just happened. The same was true for the beautiful bluffs that bordered the arroyos. The rains came down, the water flowed, strata were exposed, the aggregate was sculptured, the colorful bluffs appeared. They weren't formed according to an idea. Nature didn't envision them and then find ways to make the vision come true, but that was the way that both the hammer and mathematics had come into the world.

The bluffs might serve some purpose, providing a place for owls to carve out their hollows and hatch their young safe from predators who wanted to steal the eggs, but that purpose was not built into the bluffs from the beginning and it was not the reason why the bluffs were formed. The bluffs just happened to acquire a purpose inadvertently. The same held true for the bush. The bush didn't grow berries so that it could feed the birds, but it turned out that the bush ended up fulfilling this function. The berries did serve a function that appeared to be deliberate, but the bush had created berries for another purpose altogether, to provide nourishment for its own seedlings when they sprouted. The birds took advantage of this unforeseen opportunity. They said to themselves: "You know, we're gonna eat those berries. We're going to utilize those berries for our own purposes because they were just there for the taking." The birds sounded like a bunch of physicists. The physicists said to themselves: "You know, we're going to take all of this stuff that was never meant to be mathematized and we're going to mathematize it, because it was just waiting for us to come along." The world wasn't created to be put into mathematical relationships and made to appear mathematical, but physicists realized that the opportunity was at hand so they took it. The mathematics they got, just as the owl dens dug into the bluffs and the bird food hanging on the bushes, was inadvertent. The physicists were just studying the conditions and situations that they had brought on themselves through their own actions, by engaging in a very strange way of thinking entirely unrelated to nature.

Perpendicularity was just an idea and the world had not come into existence through ideas, but as an idea, perpendicularity could be combined with other ideas in many different ways. Mathematicians could mentally fabricate coordinate systems with any number of axes, each axis perpendicular to all the others. From the purely mathematical standpoint of simply combining ideas together, they could freely choose between these systems, but in terms of making calculations, they needed to use the one with three-dimensions. Therefore they concluded that our world had three-dimensions. By the same reasoning, two plus two could conceivably equal any number, but in our universe they equaled four. In another universe, any number of perpendicular lines might intersect at a single point, but in our universe that number could only be two or three.

But what did these other higher-dimensional coordinate systems have to do with anything? Did physicists honestly think that one of these constructions could possibly be the reality of some other world? Comparing the calculations of this world to the hypothetical calculations in some imaginary world did not define this world or set it apart. Nothing existed which could be compared to this world so the calculations could not be otherwise. The definitions of the concepts were what forced the calculations to be made in certain ways. Using fictitious worlds to distinguish this world created the impression that none of this was necessarily so. Physicists could envision things being different, but they were looking at physical reality solely from the point of view of the infinite possibilities of mathematics. None of these other possible worlds were realistic. As far as physicists were concerned, the world we lived in was just a bunch of definitions and of course these definitions could be otherwise. Definitions were products of the imagination and thus they could be changed freely, but the realities of this world were firmly established and they could not be other than what they were.

Mathematicians created higher-order manifolds by taking a straight line and transforming it into something completely different: an ordered set of numbers. The numbers represented the coordinates of each point along the line. Whenever Jonathan was confronted with the idea of putting another line perpendicular to the first three lines that he already had, his mind stopped dead in its tracks. He couldn't do this because the idea made no sense to him. With the addition of this extra line, he could no longer picture the result in his mind. The ideas that he had started out with no longer fit together. Physicists attributed this to the structure of space, however, asking why four perpendicular lines didn't pass through a single point was like asking why two plus two didn't equal five or why a circle wasn't a square, but adding another number at the end of an ordered set of numbers was the most natural thing in the world and made perfect sense. Of course the mathematician could just keep adding more numbers to the existing sequence of numbers, turning the original triplet into a quadruplet, then into a quintuplet, then a sextuplet, and so on.

But the mathematician was not talking about straight lines anymore. He was talking about ordered sets of numbers. This was an entirely new way of thinking and the new ideas fit together differently than the old ideas, the shapes of geometry. They had their own logic and their own peculiar relationships. The two formulations were not at all equivalent. Mathematicians had simply devised methods whereby they could go from one system to the other through formal correspondences. If they transformed the lines of Euclidean geometry into ordered sets of numbers and then started writing equations for these new mathematical entities, they found that they could write equations for the sets with the additional numbers in them that were similar to the equations they had originally written for the triplet. They told themselves that all of these new entities must also be lines, even though they had ceased to be lines the moment they were converted into numbers, because numbers could not also be lines except in the most abstract way, a relationship that did not maintain the full meanings of the concepts. Instead of saying that everybody lived in a four-dimensional universe of space-time, physicists should be more careful with their declarations and say that we all lived in a stiff batter of a finely-ground flour, or a rigid framework of molecular-like complexes, which was what their equations were actually telling them.

Jonathan thought about being surrounded by the points of a coordinate system and wondered if this array ever reflected light, because once in a while he caught the flash of something in his peripheral vision and thought that it might be the glint of this crystalline latticework shinning in the sunlight. Sometimes the air had a peculiar haze and maybe the coordinate system, an invisible spiderweb of tenuous interconnections on most days, was vaguely showing itself, making its faint presence known. Perhaps the coordinate system could be glimpsed by the trained eye of someone who knew that it was there and was consciously looking for it. Jonathan decided to keep a sharp watch for it.

The points of the coordinate system didn't cause anything to happen in the real world because the coordinate system didn't exist in the real world. On the other hand, it did cause everything to happen in the physicist's conceptual universe, a faraway place populated by strange inhabitants, situated somewhere between Middle Earth, Narnia, Avalon, and the quixotic realms of Heaven and Hell. If physics was based on observation, then the one object that was key to everything else should be observable. After all, physicists had built all sorts of spectrometers, electron microscopes, dish antennas, particle detectors and photovoltaic sensors to prove that their concepts were more than just fantasies, that their ideas referred to things that were real, yet given the one object that underpinned all the other entities in physics, the one reality that made everything else a reality, they had never made a detector to observe it.

Jonathan did not walk through a tapestry of straight lines or wade into a structured ocean of tiny pinpoints on any of his walks. Mathematicians had just made all of this stuff up. He was not really moving within the framework of a coordinate system and he had no idea what that would be like. He was only walking through his own ideas and not walking through the ideas of other people, especially those of mathematicians.

The confusion resulting from mathematical thinking was commonplace in physics. Mathematicians could not seem to hold onto the realities of their conceptual world and they jumped back and forth willy nilly, directed by voices in their heads telling them to substitute one thing for another simply because the two things could be tied together by vacant abstractions. People who were unable to maintain their grasp on reality had traditionally been put into institutions for their own protection, as well as for the safety of those around them. Where were the authorities when we needed them? Why were these mathematicians being allowed to run around freely, endangering the minds of others with such outrageous replacements and substitutions? If everybody started thinking in these terms, the society that people had created and the world they knew were both in grave danger of crumbling into disjointed rubble. Jonathan fumed at the logic of mathematicians and their callous disregard for content and substance. A physicist similarly hopped around wildly from one bare form to another without a care, much as a madman dancing along the streets and shouting incoherent nonsense at each hapless pedestrian who came along, combining dissimilar thoughts with free associations, saying one thing when he meant another, seeing shapes and designs that weren't there, connecting objects and events together by threads that were so thin and tenuous as to be incomprehensible to all but a deranged mind.

Einstein's genius had lain in the fact that he saw clearly how the concepts of physics were just toys and he could play with them freely, and thus he began a new era in physics that mirrored a corresponding movement in the arts taking place at about the same time, championed by Picasso along with many other cubists and abstractionists. Art was not bound by nature and it could become detached from reality without suffering any serious consequences. Art could be centered instead around the unfettered imagination and this created many more opportunities for artistic expression.

Einstein essentially did for the sciences what Picasso had done for the arts, namely, free it from the burden of what existed. Time and space could now be anything Einstein wanted them to be and he could connect them together in any way he saw fit, and by this clever maneuver he was able to win the game played in laboratories with lines and quantities to the satisfaction of physicists everywhere. The fact that these lines and quantities didn't copy the realities of the world didn't matter. Physics was art and it could create its own realities, painting pleasing images for an audience of mathematically-minded individuals to praise and admire. Abstractionism took over and replaced the former realism, releasing everyone from the constraints of the natural world. Nonrepresentational shapes and symbols now occupied the center stage and basked in the limelight while the actual details of material existence slipped out through the back door and disappeared into the blackness of the alley. The inexhaustible trivia of this humdrum world had gotten very boring anyway. Life was so much better in a purely conceptual world of abstract shapes.

Jonathan had once read that a journey of a thousand miles began with just one step. Long ago mathematicians had decided to pack up and leave this world and Euclidean geometry was their first stop on the way out of town. The mathematicians had many more destinations ahead of them. The journey continued and after a time the mathematicians looked back on the path they had taken. From this vantage point, Euclidean geometry looked like home to them, even though it had been itself a great leap into the imagination. It still reminded them of the world they had left behind. Each step they had taken was another abstraction leading them farther away from the point where they had started, and soon a great distance intervened.

This walking away from reality granted physicists an opportunity to employ a clever trick. The degree of abstraction involved in simple geometrical constructions such as lines and circles was small enough to allow these virtual realities to be made into fabricated realities, but as physicists kept assembling abstractions by adding them together and placing them on top of one another, the disparity between the resulting conglomerates and the world gradually increased and the possibility for turning these ideas into fabricated realities evaporated. The dissimilarity became so great that the bridge between ideas and realities could no longer be spanned and the mental constructions had to remain forever nothing more than imaginations. The problem with higher-dimensional manifolds was that the process of abstraction had gone too far and the result no longer related to the material world, except in a highly complicated and convoluted manner where geometrical concepts were connected to other mathematical concepts. Adding one more perpendicular line did not seem like a higher level of abstraction, but the result could only be obtained by viewing the lines themselves as abstractions, that is, as sets of algebraic coordinates in a fanciful system of mathematical ideas. Because abstractions were products of the imagination, the continued application of this mental operation quickly led the physicist into pure fantasy and after a while there was no longer any possibility of a fabricated reality based on these ideas. Freely generating more abstractions based on the previously generated abstractions created the illusion that the earlier formulations had contained an element of reality within them, but this was only from a further conceptual point of view, and thus it was based entirely on the comparison of ideas. The appearance of reality was relative to these additional higher-level abstractions and certainly not true from the standpoint of the world.

The problem could be traced back to the straight line itself. The misconception that the line was real had led to all sorts of errors in thinking, the most egregious of which was a belief in the existence of mathematical forms. The line was a complete fiction and its purported ties to the world were nothing more than further ideas that were added to the line itself. The idea that creosote leaves were the endpoints of a line segment was certainly one way of thinking about creosote leaves, and it was an idea that cognitively tied the leaves to this improbable phantasm, but as the mathematician played with the ideas of lines, and not only assembled lines together but transformed them by incorporating these ideas into higher-order frameworks of other types of ideas, he began to see lines merely as constellations of abstract concepts in a rich universe of abstract thoughts. He could do things like draw a straight line into another dimension, a dimension that had no depth and immediately curled back upon itself. Of course, he couldn't construct a material object based on such an idea, but he could still say that the idea represented a facet of reality as long as he could tie the numbers generated by experiments together by invoking this idea. In scenarios such as this, the interplay of ideas and objects had been raised to a higher level of abstraction and the connections became entirely cerebral. After a while the formal correspondences required to establish a bridge to the physical world became impossible to maintain because the thinking no longer allowed it. The thought processes unfolded only with regard to the constraints and potentials of the ideas themselves, yet additional ideas were necessary to reestablish the ties to the world, and finding these ideas was no longer possible. The ability of someone to fabricate a simple straight line did not mean that such a line revealed a truth about the world because the line was not about the world. Since the line wasn't real, it could not play a role in the mechanics of natural phenomena or serve as an explanation for what was actually happening. When compared to the world, all mathematics was equally unreal and by its very nature divorced from what existed. All of it was abstract and the level of abstraction didn't really matter because in judging the relationships of these empty forms to the infinite particularities of material existence an unbridgeable abyss necessarily intervened between them and this was not a matter of degree.

There could be no world in which two plus two did not equal four. Since the world did not support the equation, the world could not refute it either. The only way to make two plus two not equal four was to change the definitions of two and four. There could be no world where an object was both a triangle and a square at the same time. There could be no world where four perpendicular lines passed through a single point. The concepts simply wouldn't permit it. The only way to get around these limitations was to change the definitions, something mathematicians had done with complete success.

There also could be no world where the angles of a triangle did not add up to 180 degrees. The physicist could not postulate a world where the angles added up to something other than 180 degrees, say in a curved universe that corresponded to the curved surface of a globe, because putting ideas together and playing with mathematical relationships had nothing to do with material existence. The physicist was only thinking about a schematic of formal relationships, but such empty correspondences were not, and could in no way be, equivalent to an actual world of physical objects. The world wasn't based on ideas. The physicist could not dream up a universe out of his imagination. This notion arose from the false belief that the universe everybody lived in was based on the ideas of Euclidean geometry, when Euclidean geometry was just a system of mental constructions that had to be fabricated in order to make them real. The measurement of the angles of a triangle employed the same concepts that had created the triangle in the first place, so there was no possibility of conflict between them. The sum of the angles was not a matter of observation and not subject to experimental tests. There was no way the angles could not add up to 180 degrees just as there was no way that two plus two could not equal four, because the result followed logically from the definitions of point, straight line, intersection, and the manner in which they had been put together to form a triangle. A triangle drawn on the curved surface of a globe was different from a triangle drawn on the flat surface of a plane because the idea of a curved line was different from the idea of a straight line. Ideas could be assembled in many different ways, but ideas could only lead to more ideas. Creating actual worlds out of these ideas was ridiculous. The notion that the inhabitants of a curved universe saw a curved line as being straight made no sense because people could not live in an idea and there were no inhabitants of a curved universe. Although Jonathan could imagine a curved three-dimensional space in his mind, he could not actually go there and become a resident of this fantasyland—so dream on.

A rock with the shape of a triangle was still not a triangle. A triangle was a concept in geometry that was made up of other concepts, and a rock on the ground was a physical object that was made up of atoms and molecules, so the comparison was superficial and meaningless. The relationship between a triangle and a rock could never be more than mere appearance. The triangular-looking rock was not special and it was no different from all the other irregularly shaped rocks that were strewn about all over the place. A person could not understand a triangular-looking rock by using the concept of triangularity because triangularity did not exist within the rock and therefore the triangular-looking rock did not exist because of triangularity. The rock appeared to be triangular for no particular reason. The rock ended up that way by chance as all the rocks were fractured and pulverized by geological events, but the triangle in geometry existed by reason of the essential features of its component parts.

Yet connecting the triangle in geometry to the shape of a rock on the ground was the most natural thing in the world. If they looked the same, then why couldn't the physicist say that they were the same? He could bridge the chasm between deliberate mathematics and inadvertent mathematics easily enough if he simply forgot about the true natures of things and disregarded all questions surrounding existence. Well, Jonathan declared: "I'm afraid to say the difference is monumental, all-encompassing, and undeniable. The two forms of mathematics are not at all alike, and they cannot be switched around or substituted for one another."

The physicist dwelled in the void left by the removal of all content and substance. A physicist was so accustomed to the emptiness of mathematical forms and the blank outlines of formal relationships that if a rock looked like a triangle, then the rock was a triangle, because he had stripped the triangle and the rock of their essential meanings and they had become nothing in themselves. The world lost its solidity in order to make room for the fluidity of mathematical thinking and now everything flowed into everything else and all things were related. Previously impossible connections were not only plausible, but the most natural state of affairs. Brute existence no longer stood in the way of the creative impulses of the symbolic artist known as the physicist and he painted his canvas with abstract shapes and incongruous lines fashioning a portrait that did not at all resemble the world he lived in.

Jonathan came to his senses. He was tired of looking for stones. He'd thoroughly covered the area and picked up every specimen of any interest to him. Emma had no patience for Jonathan's obsession with stones, or for his proclivity for engaging in theoretical speculations, and had continued on without him. She had long since disappeared in the direction he had previously been traveling. He'd wasted enough time and needed to get going anyway. After a short distance he spotted Emma up ahead merrily running like the devil, thoroughly occupying herself with her own passions. Emma saw that Jonathan had finally abandoned his ridiculous quest and sensibly resumed the hunt, so once again they were working together as a team, tracking game and chasing rabbits.

After climbing up and down several more hills, Jonathan came upon a small, boxlike arroyo with nearly vertical sides and a relatively flat bottom blocking his path. The channel entered a chute up ahead which emptied into a basin where the rocks offered him a place to cross. He carelessly maintained his gait as he began stepping onto the descending footholds. The rocks were spaced a little farther apart than he had anticipated and he gained momentum with each stride, having to reach out with his foot as far out as possible to gain the next rock while dropping down to a lower level at the same time, forcing him to accelerate his pace. About two thirds of the way down he missed his target slightly and his ankle buckled. Losing this expected and much needed support, he careened out of control the rest of the way down, a bit like a twin engine airplane that had suddenly lost one engine. He almost made it to the bottom when his foot slipped at the most inopportune moment causing him to fall backward and he went down hard on the last tilted rock face. He slid into the sandy arroyo much as a baserunner sliding into home plate, similarly acquiring a few strawberries on his thigh. Jonathan lay motionless on the ground for a few minutes. The fall had knocked the wind out of him and he was in a slight state of shock. The scraped skin on his leg burnt fiercely and his ankle throbbed, but he hadn't hit his head on anything so apparently he was alright. He should have been wearing a helmet for a stunt like that.

He sat up and unbuckled his pack, letting it fall to the ground behind him. He looked around, assessing the situation. The sheltered spot was quite enchanting, a picture postcard of the desert southwest complete with yuccas, flowering sotols and fruit-laden cacti, all intermingling with large populations of white, smoothly rounded limestone boulders—a nice place to stop and rest for a while. He noticed that Emma had continued on without him. She was running around somewhere on the other side of the arroyo having fun on her own, blazing a fresh trail through the sage and creosote. She finally realized that he wasn't following her so she backtracked to check on him, thinking that he had changed directions, but instead she found him sitting quietly on a rock at the edge of the basin. She hadn't returned out of concern for his welfare, but was seeking an explanation for his dereliction of duty, since he wasn't leading the expedition as he should be. She noted his position and took off again, disappearing over the rim. Emma saw Jonathan largely as a littermate and she had little compassion for him. Well, she could plainly see that he was alright. If he wanted to rest that was fine with her, but she would keep going in the meantime. She'd check back on him later.

Jonathan pulled out a tube of arnica from his first aid kit, then removed his shoe and sock and rubbed the ointment onto his sore ankle, thoroughly massaging the cream into the tender joint. The ankle wasn't swollen, so it was probably just a mild sprain. He felt stupid, because he always assured Sarah that he was never reckless on his treks. He never told her that he'd taken a number of nasty falls over the years because then she'd worry about him. Things like this happened occasionally. They were a part of wilderness travel.

Jonathan replaced his shoe and sock and tentatively stood up, gradually putting weight on his injured foot. The ankle hurt, but it was definitely not broken. He took a few baby steps, then rested, then took a few more baby steps. He might limp all the way back to the truck, but other than that he was fine. Still, he should give his ankle a little more time to recover before striking off across the desert and putting more stress on the weakened tendons and joints and possibly doing further damage to them.

Casually moving about on the level floor of the arroyo kept his ankle from tightening up without putting undue strain on it, so he decided to perform an experiment while he allowed his ankle a little more time to recuperate. He slowly walked off the distance between the rock he'd been sitting on and the base of a small juniper tree standing downstream, placing his shoes heel to toe and counting the number of times he did this. He wore size 11½ shoes and they were very close to being one foot in length. The distance was 42 feet. He then measured the width of the basin by selecting a boulder on one side and a large mesquite tree on the other side. The distance was 27 feet. The physicist at this point made a bizarre statement. He said that the two distances were connected to one another. They were related even though the rocks and trees were all sitting quietly, each minding its own business, paying absolutely no attention to the others. They weren't interacting, they weren't touching, they weren't engaged in any group activity, yet the physicist insisted that they were all joined together.

According to the physicist, the trees and rocks were all parts of an overarching entity that spanned the entire universe. Physicists called this entity "space." Space was something that tied everything together. Jonathan could reveal the hidden bonds between the two distances by enacting further measurements. He walked off the distance between the juniper and the second boulder, then walked off the distance between his seat and the mesquite. The four objects were now connected together by distances, and the original two distances were now linked to one another by the additional two distances. The fact that he could measure the distance between any two objects whatsoever weaved everything into a community of like individuals. Two distances, which originally had nothing to do with one another, were now part of a single entity. The two unrelated events of walking off different sets of footsteps, each having to be conducted independently of the other, were now related not only in the general form of the measurements, but in being two attributes of a single structure. The universality of the technique created the universality of space.

The fact that astronomers could see the distant galaxies with their telescopes proved that the universe was connected, but it wasn't space that connected the galaxies with the astronomers, it was light. If all the light in the universe were suddenly taken away, everyone would be lost in absolute darkness and the universe would be utterly unknowable. In the infinite blackness of space, everyone would be completely alone because space itself did not connect things together. Without this light, the physicist could know only what he could touch and feel, and knowledge could be obtained only through direct contact. Even if only the smallest space intervened, the effect was the same as if the largest astronomical distance intervened, meaning that distance had vanished.

Physicists couldn't use light to furnish them with a concept of space because in talking about light they were talking about something other than space, switching to another topic —the classic bait and switch. Nothingness was not a good concept for space since it did not have the qualities necessary to enable calculations. What could anyone possibly say about nothingness? Nothingness produced only silence. There was nothing anyone could do with nothingness except contemplate its emptiness. The calculations required a connected universe, so the mathematicians needed to have everything tied together, a universe where everything stood in definite relationships to one another. If space was absolute nothingness, then it could not possibly be mathematical in nature.

Most people had trouble dealing with the concept of absolute nothingness, but physicists saw it as a grand opportunity. Nothingness could be anything they wanted it to be simply because it was nothing in itself. They could fill the void with exotic designs and weird shapes and nothing prevented them from carrying out these intentions. The universe still wasn't connected through space because the lines they used to connect it weren't real. Something would have to actually tie the points together, something more than just imaginary lines. The space occupied by the universe was only connected in the minds of physicists. They could measure all sorts of distances, stretching their rigid measuring rods in every direction, putting their lines everywhere, introducing their geometry to every corner of the universe, but space wasn't doing any of this—physicists were doing this. Pure and unadulterated space insulated everything, enveloped everything in nonexistence, erected a barrier that then needed to be bridged, formed an obstacle that then needed to be overcome. Whether with beams of light or the straight lines of geometrical constructions, something other than space had to provide a structure for space.

While someone could cut a drafting triangle out of a sheet of clear acrylic, cast a wheel out of metal, fabricate a wooden cabinet with straight edges and perpendicular corners, no one could make a material object that corresponded to M. C. Escher's "Waterfall" or the impossible stairs in his "Ascending and Descending." Was it the nature of materiality or the structure of space that caused this impossibility? Or was it just that the ideas did't fit together? Escher had simply been drawing lines on a piece of paper, positioning them so that he could connect the endpoints together, and not making an actual staircase. The connection between the bottom of the stairs and the top of the stairs in Escher's drawing wasn't real. He had created an optical illusion that was mere appearance. The connection was only a symbolic correspondence. All relationships in mathematics were purely symbolic and thus they led to similar nonsense.

Why was it possible for someone to draw a perfectly straight line? Did this act depend on the structure of space? Jonathan could easily imagine a world without straight lines. In fact, that world already existed, because it was the world he now saw all around him—the seemingly endless desert. But he could not imagine a world where he couldn't draw a straight line even though he had the idea of a straight line in his head. Euclidean geometry was a visualization and this system of images could be acted out in the physical realm, but why was it that some ideas worked and others didn't? An idea worked if it was suitable for the purpose at hand. Jonathan couldn't make a hammer out of a balloon and a string, not because of the structure of the universe, but because the idea of a balloon fastened to a string was not compatible with the idea of a heavy weight attached to a rigid handle. The ideas didn't match. They didn't fit together. There could be no universe where a balloon

101

and a string made a good hammer. None of this had anything to do with the nature of material reality, but only with the ideas themselves. The confusion arose when ideas were seen as existing independently of the person who had thought them, when they were projected out into the world and they became no longer ideas, but realities in their own right.

But what if Jonathan couldn't conceive of a straight line at all? Not that the thought simply hadn't occurred to him yet, but that there wasn't the idea of a straight line to be had anywhere, that it was impossible for this idea to ever arise in anyone's consciousness under any circumstances. That raised the question: were there concepts that Jonathan couldn't think? What if a straight line didn't make sense to him and he couldn't figure out what the words meant? What if he couldn't visualize a straight line in his mind? Would that be a statement about space—or a statement about consciousness?

Saying that the structure of space made the thought of a straight line possible implied that physical reality was at the root of consciousness, that consciousness depended on the outside world for its ability to function, that consciousness was structured along the same lines as the universe, that consciousness could only copy what was presented to it. All of these conditions were clearly false, but what determined the bounds and limits of conceptualization if not the physical universe? If consciousness was self-determined, then the universe had nothing to do with the ability of consciousness to form thoughts. Physicists were not imagining straight lines because the world allowed it, but because consciousness allowed it. No one had gotten the idea of a perfectly straight line from the world anyway, so where did the idea originally come from? Jonathan could think of things that didn't exist, things such as perfectly straight lines, and he could think of things that couldn't possibly exist, things such as Esher's staircase. His mind wasn't limited by what he saw around him, but that didn't mean his mind had no limits at all. Jonathan wondered how many thoughts had not yet been thought. He asked himself: how far could thinking go? Could a person think a universe that had no thoughts of its own?

The other source of ideas was commonly known as the human imagination, a faculty that arose directly from consciousness, but one that certainly also had its own limitations. Without knowing what could not be imagined, the limitations of the human imagination could only be defined by what could be imagined. Consciousness could not go outside of itself, so who knew what lay outside of consciousness? A whole mental universe perhaps, a limitless realm of unthinkable thoughts, a vast world of inexpressible ideas that were forever unknown and unknowable. Perhaps herein lay the true conceptual basis for the physical universe.

Jonathan saw that his foot was fine. It had fully recovered thanks in large part to the arnica. He'd used arnica many times before, whenever he tweaked a knee or a hip, and the salve always worked wonders for him. He saw no reason to abandon his hike today. As he was gathering himself together, Emma came looking for water. He broke out her bowl and watched as she guzzled down the full amount. He reached for the water bottle to replenish her empty dish but she immediately turned her back on him and ran away before he could unscrew the cap. She had caught a whiff of a new scent and decided to follow it down the arroyo. Jonathan packed up her water bowl and the rest of his gear and headed out after her.

The arroyo was a narrow channel with vertical sides completely submerged below the general level of the desert. The eight to ten foot walls were composed of compacted aggregates, vaguely reminiscent of the slot canyons in Utah but not nearly as spectacular and beautiful, yet still providing a somewhat similar experience. A rabbit startled Jonathan as it came bounding around the corner, streaking up the canyon with the intensity of a bullet shot from a gun, literally flying through the air and touching the ground only every ten feet or so. He jumped sideways because he thought that the rabbit was going to hit him, but at the last second the frantic animal darted and ran straight up the canyon wall, somehow knowing that there was a small channel carved out by runoff at that exact spot, the only path that was not absolutely vertical. Emma came dashing around the corner with a similar trajectory, presenting a correspondingly streamlined profile, racing at high speed but not actually flying through the air as the rabbit had been. She was too late to see which way the rabbit had gone, but she obviously knew the way because she turned at exactly the same spot where the rabbit had turned. Jonathan laughed because there was no way Emma could possibly pursue the rabbit given that it had shot straight up the canyon wall. Certainly she was going to stop once she realized that the wall was too steep for her to scale, but to his surprise she didn't slow down a bit. Her built-up momentum and unbridled enthusiasm propelled her nearly all the way to the top where she ran out of gas and faltered. He quickly moved forward to break her fall, fearing that she might tumble awkwardly and hit her head on the rocks, but she hung on, clutching at the last step with her front legs held stiffly out in front of her. With a powerful, super-canine effort that for a moment seemed certain to fail, she managed to complete the ascent and immediately disappeared from view, still in hot pursuit of the rabbit. Jonathan wouldn't be following them as a trailer so he turned around and continued walking along the channel. After a short time Emma popped into view, standing on the rim over his head looking down at him, without the rabbit in her mouth. She was ruffled and had a wild look on her face. The rabbits were driving her crazy today.

For a long time there was no exit from the canyon and Jonathan began to feel trapped. He finally came upon a small side canyon that climbed out of the main channel in a series of irregular steps. He availed himself of this auspicious circumstance and slowly emerged from the subterranean vault and embraced the refreshing openness of the desert. Emma was waiting for him as he arrived, having tracked his progress with her keen sense of smell and her acute sense of hearing. As he began to walk along exploring the infinite details of the desert landscape, he saw that changing his distance to the hill up ahead changed his distance to the arroyo on his right and also his distance to the ridge he had just crossed, which was now receding in the background. This scenario made it seem as if the distances were all connected together. The illusion was created by the fact that the earth was a rigid body. The objects around him were all rooted on the surface of the earth and this relationship fixed the distances between them, so in changing one distance he was thereby changing them all. But this was not a characteristic of space. Rather, it was a characteristic of a rigid body.

A coordinate system was an imaginary rigid body that extended beyond a material object, a continuation of the rigidity of the object into the fluidity of space—or rather the emptiness of space, because fluidity suggested some sort of materiality—a structure that was projected onto space by erecting stiff measuring sticks or locating additional material bodies, or more conveniently using electromagnetic radiation to time signals transmitted between such bodies, and thereby construct the distances through inferences. A coordinate transformation sought to establish the relationship between two rigidities, one moving in regard to the other, neither of which actually existed. The fixedness of mathematical ideas was ideal for representing these kinds of permanent structures. Coordinate systems extrapolated the conditions that were found in rigid bodies out into space—only space was not a rigid body because in space distances were not connected to one another. Using the model of a material body for an understanding of space was inappropriate because they were two entirely different things. A rigid body connected points and distances together. Space did not.

Jonathan could not locate a point in space because space did not have points. A rigid structure, whether it was a material body or a conceptual coordinate system, created the places for him, points he could find and then return to later. But when he returned to the truck after a day's hike, although the truck was still located at the same point on the surface of the earth, parked right where he'd left it, the truck was not at the same point in space, because the earth had been rotating about its axis the whole time and carrying the truck along with it. The earth had also been orbiting the sun. The sun had been turning along with the rest of the galaxy and the Milky Way had been moving relative to the other galaxies. Jonathan couldn't possibly locate the space where he had parked the truck because the space it had originally occupied could not be identified. Essentially he had parked the truck nowhere. Yet he imagined that the space where he had parked the truck was still out there somewhere, untold miles away in some unknown direction, completely lost in the void, but that was just a fantasy. The space had never existed and finding it had always been impossible. The consequences of this were far-reaching. No one seemed to realize that the non-existence of points implied the non-relatedness of distances. The earth changed all that, at least for the people who spent their lives there. The earth provided the necessary rigidity for a determination of place and created a system of objects with stable if not permanent relationships between them. A coordinate system simply replaced this function in an abstract sense and granted the physicist the necessary tools for giving space a structure that it didn't actually possess.

Space could be full or empty, occupied or unoccupied. The problem was that the two examples of spatiality were not the same, nor were they equivalent, because occupied space was structured by the material bodies and unoccupied space was not structured in any way. Mathematicians and physicists wanted to draw pictures of geometrical forms so they chose occupied space to represent space. They needed the paper to hold their fancy shapes and artistic designs for them and without such a stable surface they were completely out of luck. Naturally they said that the material object inherited its spatiality from the space that it occupied, but if space could impart spatiality to an object, then it could impart the same spatiality to itself, which clearly it could not do. Space needed the material object to define the points that then created the structure for the space the object occupied. However, space didn't need objects to exist. Objects were absolutely nonessential to its being. If all the matter in the universe were suddenly taken away, space would still exist—or would it? A universe with nothing in it could not be envisioned. No one could observe a universe without first being inside it, so a completely vacant universe could not be observed either.

Still everyone believed that space came first and then material bodies were placed inside of it, added to the space in an independent move, but these material bodies changed everything. Suddenly there were places, specific places that could be individually labeled and unambiguously identified where previously there had been none. Space did not structure material bodies—material bodies structured space. Material bodies had an ontological priority that enabled them to create the space they occupied. They could also be used to define the space that surrounded them.

If the only thing that existed in the universe was a spaceship, then the astronauts inside could fire the engines but they couldn't go anywhere. They could feel the thrust but they couldn't move. They could not have a velocity. No matter how long they kept it up, they would never be in a different place. They could not say that they were moving through space, because they could only move through something that existed and outside their spaceship nothing existed. They couldn't do any of the things that everyone normally took for granted, namely, reach a destination, achieve a velocity, or know where they were at.

The astronauts could launch a probe and then measure the distance between them and this object. Everyone would say that the probe had been launched into empty space, implying that something was out there to receive the probe. For if there hadn't been space, then the probe could not have achieved a distance from the mothership. If space ended at some point, then the probe could only go so far. The probe would stop moving when it ran out of space. All of this assumed that space was a thing in itself, an object endowed with shape and extension. Living our entire lives surrounded by material bodies had ingrained this way of thinking so deeply into our psyches that we couldn't let go of it. A bigger house meant more space. A larger yard had more room. This made us think that space had size and dimension, a form of its own that existed in addition to the material bodies, but only because we took these material bodies for granted, not fully understanding that our concept of space had been created by them. Outside of these material bodies there was nothing, and nothingness could not come to an end by ceasing to exist. Only something that existed could have boundaries.

Empty space could only end when it ran into something. The closet space ended when it ran into the closet walls. Jonathan could measure the walls of his bedroom closet and then say that in doing so he was measuring the space inside. Just as with the straight line between two creosote leaves, he could use the walls to define a nonexistent geometrical construction, a rectangular cuboid consisting of empty space. Jonathan could tear the closet walls down, but this space would still be there and he could locate it by referencing the bedroom walls. He could say: "This is where the closet once stood and I can still identify the space it once occupied." In fact, Jonathan could tear the whole house down and still find the closet space by measuring out from the curb and up from the ground. This made the space seem like a reality, something that existed in its own right, but now take away the curb and the ground, and then take away every other material body. The space disappeared because it wasn't real in the first place. The space was merely an aspect of the closet walls, one of the many properties of their material existence. Without the walls there was no closet and no space within the closet. The walls created everything and space played no role in any of it. Without the walls, the closet space could not be identified, located, defined, measured or observed, because it didn't exist outside of those walls. Much as the constellation Orion, when Jonathan tried to trace the pure mathematical form of the closet space back to some sort of reality, the form vanished and he had to admit that it was all just an illusion and he was not looking at any kind of object.

Something might in fact be driving the galaxies apart, but that didn't mean that space was expanding. If physicists envisioned the universe as a vessel containing space, then galaxies that were increasingly separated from one another would have nowhere to go. To accommodate the greater distances, the vessel would have to expand. Nothingness on the other hand was unbounded. An unbounded universe had plenty of room and the galaxies could all be moving away from each other without space expanding. Physicists loved nothing more than to picture abstractions as realities. They were already in the habit of seeing mathematical concepts this way, so they could simply add space to a long list of other imaginations, ranging from vector forces and electric fields to gravity wells and worm-holes.

In talking about the expansion of the universe, rather than saying that the galaxies were all moving away from each other, physicists said that these objects were in fact stationary and it was space that was moving, carrying the galaxies along with it. Space changed its dimensions and these objects were now in different places. At first this new conceptual scheme seemed to offer no advantage, basically reformulating the original situation, however, the new way of thinking did something very important. It transferred the action of the corporeal bodies over to abstractions. The situation could now be understood as an interplay of ideas. A visualization of space was performing all sorts of tasks and physical reality was just a pawn manipulated by convenient fictions. Physicists wanted to live in a mathematical fantasyland so they imagined that abstractions were responsible for everything they saw and all the heavy lifting was done by these nonentities, the components of a mathematical universe they had made up in their heads. Physicists cheerfully jumped to the conclusion that they were observing mathematics at work and it was this mathematics that was causing everything to happen. The universe was merely a puppet show where the figures danced as the ideas of mathematics pulled on their strings.

Moving about, reaching destinations and achieving velocities could only exist in a universe full of stuff, a universe occupied by endless entities, both material objects and illuminating radiations. Jonathan could pick a route through this congestion, make his way across the landscape, swim in an ocean of minute particles, but he could never travel through space. Physicists didn't care about any of this. They said that when Jonathan was walking across the desert, he was also moving through space.

When physicists measured the distance between two material bodies, they assumed that they were measuring a third rigid body, an invisible object known as the space between them. This approach created a huge problem. Space wasn't a rigid body and thus it could not possibly have dimensions, so in fact they weren't measuring anything. Once the material bodies were removed, the distance vanished. If space had the structure of a rigid body, then the places that the original material bodies had occupied would still be identifiable and the distance would still be measurable. Jonathan could measure the distance between himself standing on a hilltop and the truck parked in an arroyo, and then walk to the truck and drive away, but the distance between the hilltop and the arroyo remained. Although this still wasn't the reality of the geometrical construction, it was the reality behind the geometrical construction, the rigid body incorporating the distance. Space on the other hand provided no reality for the physicists' geometrical constructions.

This problem could be circumvented in a laboratory by arranging additional material bodies in the area so as to maintain the localities of the original two material bodies, but this had to be done artificially. Space could not do it by itself because space had no structure of its own. Physicists were creating exact structures in material objects and then falsely attributing these structures to space itself. Space had nothing to do with any of this.

Jonathan could place a tape measure over a board and measure its length, but then if he raised the tape measure and held it out in front of him, he wasn't measuring anything. In the first instance he was starting out with the world, the reality of an object that existed. The board meant that he was measuring something. But when physicists tried to measure space, they were starting out with the measuring device. The measuring device did not necessarily mean that something was being measured. Even in cases where an argument could be made that something was being measured, it was often unclear as to exactly what that was. The true meaning of the statement "if you can measure something then it must be real" was that the measurement was a reality in itself, a self-contained act that was able to stand on its own. The instrument, just like the tape measure, could simply be measuring nothing other than itself.

Physicists replaced the indeterminateness of space with a fixed mathematical structure, a structure they had created not just by introducing material bodies, but by introducing precise mathematical objects. The strictness of these objects were what gave space an exact mathematical form. Space could not receive and incorporate this form onto itself, instead shrugging it off. Physicists kept pasting these external designs on space, but with nothing in place to hold them, they did not stick and immediately detached. Thus all further mathematics that physicists derived from such measurements was invalid. They had neglected a fundamental truth that space could not be measured since it had neither shape nor substance.

A rigid body could only structure the space within its boundaries, that is, perform this function locally over a limited region. Since the universe wasn't a rigid body, it could not expand or contract and it didn't have a volume. The dimensions that were posited for the universe were just calculations based on the measured distances between material bodies. Nothing necessarily existed here that was being measured. The techniques of measurement still generated numbers, but these numbers referred to a bunch of geometrical fabrications and they had no meaning outside of them. The metric governing the size and geometry of spacetime involved concepts that were primarily connected together by the mathematics of time and space, and not in any real way.

Gravities and electromagnetic radiations established connections, however, these were only partial and limited, and fundamentally different from the geometries that connected abstract mathematical entities. The connections in the physical world were independent of one another and they were not systematically ordered, that is, consolidated within a single framework. The universe was connected haphazardly and not exhaustively, thus it was both connected and not connected at the same time, that is, connected in some ways but not in other ways, connected at certain times and in certain places, but not connected everywhere all of the time.

With no structure of its own, space became an artistic medium, an empty vessel that physicists filled with their highly creative and original forms, the blank canvas on which they drew their pictures of nonexistent lines and shapes. These pictures told them nothing about the universe because space was not what allowed physicists to draw lines—consciousness was what allowed them to draw lines.

When Jonathan told himself that he could visualize a perfectly straight line, what exactly was he visualizing? Could he really see the line extending to infinity in both directions? Could he really see that there wasn't an imperceptible warp over it's entire length? Jonathan wasn't actually visualizing the line of geometry, but merely the vague semblance of a line, a generic image, a file photo. Jonathan could also picture himself performing a cartwheel. This mental image could take on different forms. At one extreme, he saw himself executing a perfect cartwheel in the manner of a professional gymnast, a vision he could easily entertain because he'd seen these gymnasts performing cartwheels on television. So he imagined that he too was an olympic athlete. By combining these two ideas, he was able to see himself flawlessly pulling off a series of cartwheels to the amazement of his friends. At the other extreme, he saw himself putting his hands on the floor and attempting to swing his feet up in the air when they flew off in the wrong direction and he awkwardly flopped on the floor. This vision was a more accurate description of his true athletic abilities. But when he actually got down on the ground and tried this maneuver, he could not have foreseen what happened. His arms buckled and he slammed his head into the ground and lost consciousness for a while. Jonathan could easily envision all sorts of things and some of them might be real and some of them certainly weren't, but what enabled him to see himself in a certain way, or see a line as being perfectly straight? Performing a flawless cartwheel was not an impossibility, and neither was drawing a straight line. He could practice cartwheels and after a time become very proficient at them. He could also learn to draw straight lines. Sometimes Jonathan pictured himself flying through the air as Superman and sometimes he pictured a straight line extending in both directions to infinity. Was the line any more real than the flight of Superman? Still he could see them both in his mind.

The physicist did not get his idea of a straight line from nature because in nature there were no straight lines. The physicist told Jonathan that the straight lines of a coordinate system represented the actual structure of space, but Jonathan retorted that he could also draw lines, only they were very different lines, the lines of crazy, non-geometrical shapes. The physicist came over and drew his lines over Jonathan's, a grid work of straight, perpendicular lines, and said that Jonathan's lines were really based on his lines, that everything in the universe was based on his lines. Jonathan snapped back: "But they're only lines! They're not superior to my lines—you just like them better. My drawing is as much a picture of the universe as yours, it's just that I have a different idea of what the universe looks like. You only like your lines because they allow you to make calculations, but calculations lead nowhere other than into the emptiness of abstractions, a place where no one in their right mind wants to go. Outside of that, your lines aren't special at all. The idea that the universe was built around straight lines was absurd. If the universe was actually based on lines—an outrageous proposition in itself—it would certainly be based on crooked lines, since everything in the universe was irregular, uneven and askew."

The physicist replied: "It is the straightness of my line that reveals every conceivable deviation and deformation in other lines. I can't tell anything from your line because it's crooked. The structure of my line, by being absolutely uniform, can be compared to the structure of everything else. The problem with your line is that it lacks organization and form. Without structure, your line cannot reveal structure."

Jonathan said: "You're assuming that something lies behind your line that is also straight, that there is a reality in addition to the line. The line isn't straight simply by definition. The line is straight because of something else, so the line isn't just a line. It is already in your mind a representation of some reality, a model for space itself, but you need to prove that. You've leapt from the idea of a line to the reality of a line when they're not the same thing at all. What else is there behind all of your whimsical fantasies?"

Jonathan picked up a stick and haphazardly scratched a line on the ground. The line had absolutely no mathematics. He had not constructed the line using mathematical ideas and he had not even drawn the line precisely. He had simply scrawled it without thinking. Jonathan could take this crooked line and draw a second line underneath it, but in contrast to his first line, he drew this line very carefully, making sure that it was perfectly straight. He then determined the distances between the first line and the second line in an orderly fashion, measuring along additional lines that were all exactly perpendicular to the straight line. If he had drawn the line on a piece of paper using a sharp pencil instead of dragging a stick across the sand, he'd have done this with the utmost precision, as with any geometrical construction. He would have ended up with a collection of numbers, a set of accurate measurements of the distances between the two lines. He could then use these numbers to describe the crooked line. The original line that had been random and unplanned, free-floating and not connected to anything, was now connected to the straight line through an infinity of perpendiculars. In an operation that had been unnatural and unprovoked, he transferred the mathematics of the geometrical construction to the crooked line. The crooked line was now a mathematical object because he had used mathematical ideas to fabricate a mathematics for it.

The physicist sought to identify a structure within the crooked line. The crooked line wasn't simply crooked anymore, but crooked in a very definite manner. This allowed the physicist to see the crooked line as being structured even though it wasn't. The straight line was the reference point for the structure of the cooked line and without the straight line he couldn't show what the structure of the crooked line was. The crooked line, which formerly referred to nothing other than itself, could now be seen as not only having a crooked nature, but as being crooked in a unique way. Thus the straight line created a structure within the crooked line and the physicist now saw the crooked line differently. Now all lines, no matter what they were, had structures, and precise mathematical structures at that. The technique brought about an ideal situation because it was the precision of the geometrical construction that allowed the physicist to mathematize everything, regardless of its form, regardless of the fact that it hadn't been mathematical to begin with.

But if Jonathan denied the physicist's straight line by pointing out that there was no reality behind this line, making the distances he had created arbitrary and pointless, then crookedness remained crookedness and all the irregularities went unspecified. They were insignificant and there was nothing more to say about them. The crooked line was completely unorganized, undefined and meaningless. Without the physicist's imaginary straight line, the mathematics of the crooked line disappeared and everyone saw that the structure of the crooked line was just a figment of the physicist's imagination. The artificial and extraneous introduction of the straight line enabled the physicist to spontaneously create a mathematics for the crooked line. The physicist could then say that the crooked line was actually full of numbers and that the numbers stood in definite relationships to one another.

The physicist's maneuver was a clever trick based on a deception, involving the staging of an illusion, thus making the transformation of a crooked line into a structured entity comparable to the performance of a magician in a circus show. The magician enticed a woman to lay down in a box with just her head and feet sticking out, then made it appear as if he were cutting her in half with a steel saw. Of course, he wasn't actually cutting her in half, but by all appearances, that was exactly what he was doing. In a similar fashion, the physicist made the crooked line appear as if it were a mathematical object even though it wasn't. Physicists would have trouble with this analogy because according to the mathematical way of thinking, if the crooked line appeared to have a mathematical structure, then it actually had a mathematical structure. This was equivalent to saying that if the magician appeared to be cutting a woman in half, then he actually was cutting her in half. By placing the crooked line in an elaborate nest of geometrical lines, the crooked line took on the appearance of its surroundings. Physicists tried to convince everyone, including themselves, that they were witnessing something here that any person who was in full possession of their faculties would clearly see was impossible.

The crooked line didn't have a mathematical structure because Jonathan had not constructed it that way. The crooked line would have to be made up of mathematical ideas to begin with, making the mathematics of the line deliberate. The mathematics that the crooked line exhibited by virtue of the additional geometrical constructions was inadvertent and therefore utterly meaningless, a token mathematics without any real significance.

Whenever Jonathan read textbooks on mathematics, all the mathematics contained within those pages was deliberate because that was what mathematics was: methodical, calculated, and purposive. The relationships in analytic geometry were deliberate. The various aspects of the Cartesian plane were deliberate. Inadvertent mathematics was different. It was not really mathematics at all but only the mere appearance of mathematics. All things in nature had qualities that could be quantified, but that was irrelevant because these objects were not mathematical in nature, that is, they were not deliberately mathematical. The mathematics was added to them in a further series of steps designed to accomplish that end. These secondary and subsequent procedures were what was known as the mathematization of an inherently non-mathematical object.

The ability to change the nature of an object and turn it into something it hadn't been before was the hallmark of the scientific method. The physicist might object that he hadn't transformed the line into a new entity because he hadn't altered the line in any way, that is, changed its shape. From a purely formal point of view, the line was exactly the same as before, but he had changed the existence of the line because he was now thinking about it differently. In essence, it became a different object with a whole new set of properties.

The manner in which a scribbled line was turned into a mathematical object by adding geometrical constructions to it, thereby surrounding it with an artificial mathematical framework which then transferred its own mathematics to the line, was similar to the way that a natural phenomenon was turned into a structured event, taking something that had no mathematics and making it appear as if it did have a mathematics, and specifically a mathematics that was already built into it. The physicist imagined that his carefree introductions of geometrical forms and algebraic relationships had no consequences, no effects on the natural world, when in fact they were responsible for everything he saw. They were what created the illusion of a mathematical universe, since they generated structures that hadn't been present at the beginning of the procedure.

The physicist saw the mathematics of a phenomenon as being real because he had reasoned from the realities of measurements and quantities, but by exactly the same argument, Jonathan could construe these measurements and quantities as being false because the physicist had reasoned from the unreality of geometry, constructions that were based solely on his imagination. If the physicist accepted the first argument, then he must acknowledge that the second argument was equally valid.

The mathematics of the crooked line was thus a mixture of fact and fantasy rather than a simple representation of reality as the physicist had supposed, however, the contributions of the mathematician's imagination and the reality provided by the crooked line were not equal. The facts of the crooked line had been placed in the service of fantasies and these fantasies were given a priority they did not deserve considering that they were not based on any sort of reality. The crooked line was enlisted in a mathematical scheme that was not part of the line and the line played only a passive role in the overall program. Thus the line was being manipulated by the physicist according to his whim, assigned a function to serve in creating distances between the crooked line and what was now an axis of the cartesian plane. The crooked line was merely a pawn in the physicist's plan of action. The entire geometrical construction was an artistic invention, not based on the motif of the crooked line but rather on other patterns, in opposition to the traditional view that the mathematics provided the physicist with the reality of the crooked line. The physicist smothered reality under many layers of artificiality, in effect replacing it with his own thoughts. He plugged the crooked line into his preconceived set of ideas where it took its place among all the artificial lines and carefully constructed angles and assumed the role that had been appointed for it. The crooked line did not bring about its own mathematization because it couldn't. It couldn't because the crooked line wasn't mathematical. All of the mathematics lay entirely outside the line and not within it. The crooked line's structure was based on arbitrary definitions and this meant that the crooked line's structure was completely fabricated.

Jonathan leaned forward, bending deeply at the waist, and slipped through the rusted wires of an old fence, the loose strands hanging limply from a line of blackened posts all pointing in different directions. The fence had been the last barrier and now that he was on the other side, the mountains were immediately accessible, standing directly in front of him. He continued walking and began dropping down into a canyon. The route meandered through a cluster of sharply rounded hills. From the old fence line the hills had looked like a bunch of grapes arrayed along the stem of the arroyo. The wash continued onward and cut into the low peaks up ahead, a meeting of bald heads staring blankly at each other with disjointed faces. Jonathan did not have time to explore this canyon further and hesitated at the entrance. Several terraces of broken rock adorned the vertical walls, each tier having been hollowed out by water cascading down the cliff, then filled in with sand and clay, forming sandstone planter boxes.

The fractured stone reminded Jonathan of the masonry of Anasazi ruins, but the brickwork was just an illusion. The natural horizontal plates had cracked from the stresses inflicted by endless cycles of hot, sunny days and freezing-cold nights, creating irregular vertical fissures. The wall took on the appearance of well-crafted stonework even though its causes had been haphazard and undirected.

Neat rows of miniature prickly pears and agaves stood like soldiers in the stone containers, creating the illusion of a deliberately planted garden. Two desert candles, each with a towering plume, framed the spectacular display. Jonathan decided to take a picture of this strange formation, so he set his pack down to retrieve his camera. He noticed that Emma had discovered a fresh scent and was cautiously pursuing its source. She was acting very strangely, but in his preoccupation with the photo shoot, he paid little attention to her uncharacteristically sheepish behavior.

He circled the two magnificent agaves, sighting through the viewfinder for the best angle to highlight the garden in the background. Distracted by this effort, he failed to notice that he was being observed. As he backed away from the garden toward a line of bushes, his erratic movements spooked a mountain lion that was camouflaged in the shadows and he accidentally flushed it out from its hiding place. The lion had been crouching silently under a dense canopy of leaves and branches waiting for an opportunity to strike when it finally saw its chance. It leapt forward and ran straight toward Jonathan. His muscles tightened and he was frozen in place expecting an onslaught of teeth and claws, but before he could move the lion had run right past him. He spun around trying to keep his eye on the lion and saw that it was heading straight for Emma. In contrast to his stupefied reaction, Emma immediately turned and ran, but she was trapped by the cliffs and could not get away. The lion bounded toward her with long, confident strides. Jonathan knew that his worst fear had finally come true. In a moment of terror he was about to watch helplessly as Emma was viciously torn apart by the razor sharp claws and teeth of a mountain lion.

The mountain lion reached Emma in a flash, but instead of pouncing on her, the cat ran right by her and sprang onto the first terrace, then turned around and glared at Jonathan from this commanding position, a fierce look of indignation on its face, obviously enraged that he and Emma had invaded its territory. Using the ledge as a platform, the animal

111

pivoted and jumped up onto the second terrace, then in a series of magnificent leaps, it scaled the sheer cliff and disappeared from view.

Emma ran up beside Jonathan, her eyes wide with excitement. They both tilted their heads back focusing their attention on the thin, jagged line separating sky and stone, hoping for one last glimpse of the animal, but the lion had seen enough and didn't bother to look back. Emma stomped her feet while standing in place and stared upward, imagining herself to be copying the strenuous efforts of the cat. Although Emma was extremely athletic, she was still no match for this sinewy creature. She could not possibly follow it up the cliff, which was good. Emma didn't fully understand the dangers of wilderness travel, but she had inherited from her ancestors the innate drive to chase after whatever ran away from her.

The wave of adrenaline in Jonathan's body slowly subsided and he began to calm down. He realized that the cougar had merely been threatened by their approach. Feeling cornered by the intruders, it saw the cliff as its only sure means of escape. The lion never had any intention of attacking either of them. When Jonathan got too close, the animal decided to make a run for it, and both of them happened to be standing alongside its chosen exit route.

Impressed by their good fortune and grateful for their narrow escape, Jonathan took a quick picture and put the camera away, then reluctantly turned around and began retracing his steps back to the truck. He didn't want to reconnect with the side road because that would take Emma too close to the interstate, so he decided to cut across the hills seeking a shortcut. He'd been hesitant to halt his outward progress because like Emma he hadn't wanted to go home. Now they were tardy in getting back, so striking out into unknown country at such a late hour was risky. The terrain was rugged and the obstacles ahead uncertain. The steep canyons easily trapped unsuspecting travelers. The canyons were not spectacularly large, but a forty foot drop was still a forty foot drop. Whether a cliff was ten feet or a hundred feet didn't really matter. Jonathan was not an actor in a movie and there was no trampoline or airbag at the bottom waiting to catch him, so one height was as much of an obstacle as the other. A ten foot fall might still kill him, particularly if he hit his head on a rock, and he was unlikely to walk away from such a mishap without a broken bone or at least a severely sprained ankle. Jonathan's ankle was already sore enough and he felt that it could not withstand any more punishment.

Jonathan brushed aside all thoughts of danger and decided to take the chance anyway. He brazenly worked his way down into the first arroyo, carefully picking a course over the uneven ground and sidestepping the numerous throngs of sharp, thorny mesquites. The route was tedious, but overall it looked promising, and not as difficult as he had originally feared. He was still nervous about doing this. He maintained a steady pace, following the arroyo downstream for a while, then cutting a diagonal up the opposite bank until he reached the next ridge. He followed this ridge for a good distance, then struck a diagonal down to the floor of the next arroyo. He crossed the first two arroyos in good time, but the terrain was becoming more irregular. The next arroyo was deeper and more treacherous.

He reached the bottom easily enough where he found the opposing ridge quite formidable when viewed from the perspective of the wash. He forged ahead, but the climb was laborious, requiring a lot more time than he had anticipated. About halfway to the top he despaired and began searching up and down the valley for a way around this obstacle, but the bank was deeply cut at regular intervals by smaller side canyons and each one would be an adventure in itself. He thought about returning to the main wash, but the riverbed kept dropping, meaning that he'd still have a difficult climb to face, so he ended up sticking it out and scaling the height. Once on top, the skies began to darken and he saw that he was out of time. He checked his GPS position. As far as he could tell by the map on the little screen, this was the final ridge and the remainder of his journey would be an easy walk along its even course.

Still he might want a headlamp handy as a precaution. He had read once in a spelunking guide that any crucial item of gear should be carried in triplicate to insure that at least one working version was always available. Since then he'd made it a point to carry three headlamps. Being small and lightweight, they were easy to carry around.

Jonathan unzipped the pocket where he kept the headlamps only to discover that it was empty. His mind labored to answer the question: "How was this possible?" Then he remembered. Last week he'd been working at the house in town rooting around in the basement looking for something and he had left a headlamp on top of the furnace. The low, compact boiler had a rectangular metal cabinet that provided a convenient table for putting things in the summer when the furnace wasn't in use. He could still see the headlamp laying there in his mind, right where he'd left it, sitting next to a box of electrical parts and some coils of wire. The visualization didn't help him at the moment.

Then he remembered that he had used another headlamp a few days ago to check the battery connections on the truck after the engine wouldn't crank. The truck had a winch and the cables to the battery were doubled up, so occasionally he had to fiddle with them. He was supposed to have put the headlamp back in his pack, but he had left it in the truck hanging from the shift lever of the transfer case. The whereabouts of the third headlamp was a mystery. It was probably stuck in one of his coat pockets hanging in the closet at home. This unexpected turn of events made Jonathan's predicament much more dire. He began checking the other compartments in case he had inadvertently put a headlamp into the wrong pocket, but no luck. He zipped everything back up and swung the pack over his shoulder. He was probably close enough to the truck to make it without a headlamp anyway.

Darkness began slowly blotting out Jonathan's vision of the world. The ridge dipped and climbed, not nearly as level as he had imagined, and the distance was also greater than he had anticipated, with still no sign of the truck. Walking became more difficult as the light faded away, slowing him down even more, forcing him to tread carefully so as not to trip over rocks and branches. He finally spotted the truck as he topped a slight rise, its outline barely visible in the scant remains of twilight. To Jonathan's dismay, the vehicle was parked not on the ridge where he was now standing, but on the next ridge. The impassable canyon that he had circumvented at the outset of his journey interposed between him and the safety of the road, the only way out. The contour lines displayed on the GPS screen had been too far apart to reveal the true situation.

The arroyo he had last crossed with much difficulty had since dropped over several falls and it was now impassable. He looked over at the black hole and saw a gaping maw waiting to swallow him up, reminding him of the way that a young bird held its mouth open waiting for its mother to drop a worm down its throat, only this mouth was lined with massive teeth of jagged rocks suggesting the jaws of a reptile or the fangs of a dragon rather than the beak of a bird. Jonathan envisioned a huge desert beast with a voracious appetite for bipeds who inadvertently stumbled into its yawning mouth. He walked over to the cavernous opening and daringly stood at the lower lip of the monster. A ribbon of sand came winding out from the depths of the canyon and in the dim light it uncannily resembled the tongue of a snake.

113

Clearly there was no way to enter the bowels of this demon without simply flinging himself headlong down its throat. Jonathan considered back tracking, but he'd hiked this ridge for some time now thinking the whole time that it was the right ridge, and he recalled no opportunity for an easy descent. Darkness was rapidly engulfing the landscape and he needed to keep moving. Going back would only waste precious time, so he pushed onward in the faint hope of finding a trail to the bottom where the two canyons came together, but it appeared that he was approaching a precipice.

He might be in for some serious rock climbing. Jonathan had done this sort of thing before. Years ago, a hundred miles north of where he now stood, he had similarly come out of the wilderness at the wrong spot. He had taken a short cut and found himself standing high atop an embankment looking down at the truck parked on the level plain below. The sun was rapidly setting and he would have to walk a mile and a half in a circuitous route to reach the vehicle safely. The slope in front of him was steep and treacherous, yet the truck was right there at the bottom, taunting him to go for it.

The sun began sinking below the horizon as he continued to stand on the embankment deliberating his next move, until only one option remained. He no longer had time to go back, so he stepped off the ledge and began crawling down the rocks, scooting on his backside and jumping from one rock to another. He'd made it three quarters of the way down when he landed on an innocuous looking boulder. Looking at the rock from above, he didn't see that it jutted out from the cliff and was suspended over several feet of empty space. The flat stone swung downward as if it had been hinged along the backside, much as the trap door on the gallows snapped opened when the hangman pulled the lever. Jonathan's body jerked straight down, but without the noose around his neck.

He plunged into a gap between the rocks. His left arm had been hanging down by his side and was driven into the stone with the full force of his body weight plus the weight of his pack, dislocating three fingers and shattering the bones in two of them. He painfully lifted himself out of the fissure and onto the adjacent rocks. Jonathan raised his arm and stared in shock and disbelief at the heavy leather glove with the broken hand inside. He was grateful for the glove, because the glove made it appear as if his hand was fine and nothing was wrong. The pain was all inside the glove, while everything on the outside was just as it had been before.

Jonathan clumsily scrambled over the rocks, faltering several times, continuing to work his way down toward the bottom. He felt weak in the knees so he sat down to rest, still somewhat elevated from the desert floor. Jonathan's two dogs at the time paid no attention to him, having already reached the bottom. He observed them running across the level ground thoroughly engrossed in their own lives. After a few minutes he mustered the strength to conclude his descent, stepping carefully this time, fearful that he might trip and not be able to break his fall with just one arm.

Jonathan's fingers were numb with a burning pain. He gently removed the glove with his good hand and then pulled on two of the dislocated fingers, gritting his teeth and popping them back into place, but the bones of the third had slipped completely past each other and they were locked in place. The finger was abnormally large, bulging at the point where the bones crossed. With much difficulty Jonathan lifted the dogs into the back of the pickup and drove to Albuquerque, simultaneously steering the truck and manipulating the shift lever of a manual transmission with just one hand, holding the other hand in a protective position next to his chest the whole time.

The only time slot available for surgery was on Christmas eve. Jonathan ended up being the last person in the recovery room, flanked on either side by straight rows of neatly made beds, his dazed eyes staring peacefully at the walls of the empty chamber, his ears relishing the desert-like quietness of the abandoned hall. Members of the nursing staff peeked in on him occasionally to see if he needed anything. They all had big smiles on their faces, giving Jonathan the impression that they were getting a head start on the holiday festivities, imbibing egg nog and joking amongst themselves in one of the adjacent offices. After a time the drugs wore off and they let him go home where he spent a quiet evening with Sarah and the dogs. Even with metal pins to hold the bone fragments in place, the splintered pieces failed to fuse back together properly and he never regained the full use of his hand. Sadly, he will never grab a handful of popcorn with that hand again.

Jonathan promised himself that he would never do anything like that again, but here he was contemplating a similar maneuver, only this time in the dark of night. The termination of the ridge was just ahead and the drop-off was alarming, the riverbeds barely visible in the dwindling light. He might as well start going back to the point where he first made the ridge and look for an alternate route since he was obviously trapped, but an alternate route might only lead him to another dead end where he'd find himself similarly stranded. His situation at that point would no longer matter because In the complete darkness he wouldn't be able to tell if he was stranded or not.

Suddenly Jonathan was plagued by the irrational urge to stand in defiance on the edge of the cliff and look down at his enemy at least once before he turned around. What he saw startled him. The descent into the secondary arroyo was not as bad as he had thought. He had no idea if the way was clear all the way to the bottom, since such a path often ran into a sheer drop-off along the way, but the closer he got to his objective the better chance he had of reaching it. He might make it most of the way down only to discover that he had to turn around and climb all the way back up, but he didn't really have a choice at this point, so he seized upon the unexpected opportunity and immediately started his journey downward, turning sideways and feeling the uncertain ground with one foot, then after a few steps rotating his body 180 degrees and doing the same with the other foot to balance the strain on his muscles. This strategy also minimized the danger of him snagging his toe and pitching headfirst into the brutal darkness.

The strain on his ankle caused it to start hurting again and he feared that he might not make it to the bottom. The path wasn't easy, but his ankle held out and he was able to deal with all the obstacles along the way. Night was fully established as he reached the bottom and he could barely see anything. He stumbled over disorganized piles of debris, tripping over mysterious, unseen impediments. Whenever he traveled in the wrong direction, the mesquites fought back against his advances, armies of miniature combatants swarming upon him with sharp daggers, devilish hordes of tiny brutes launching coordinated attacks under the cover of darkness, furiously tearing at his clothes and stabbing at his legs, causing him to wince in pain, and curse them vigorously.

Jonathan crossed the junction of the two arroyos and staggered clumsily onto the opposite bank, disheveled and all out of kilter. He was hardly able to distinguish anything in the darkness, yet somehow he sensed that this bank was different. Suddenly he realized where he was at. He was standing at the base of the landfill. He hadn't dared attempt this steep slope coming down, but climbing up was much easier. The formerly bulldozed parcel of land had fewer obstacles so scrambling up to the top, while still difficult and strenuous, was relatively easy in comparison to what he had just accomplished. The darkness gave Jonathan a bold courage and confidence he ordinarily didn't have. He knew that his distance to the river bottom had steadily increased with each step he took, but as far as he could tell he hadn't really gone anywhere and the safety of the canyon floor was still right there behind him, just one step back.

After he'd made it to the top and had the chance to collect himself, Jonathan noticed that Emma was no longer with him. Somewhere along the way he'd lost her. In fact he didn't remember seeing her since he was up on the other ridge. For all he knew she might still be up there, busy exploring the plethora of rabbit burrows, fox holes and badger dens tucked under the bushes. She might have confronted an animal after it had emerged from its quarters and taken off chasing it in another direction, unaware of the path Jonathan had taken. He unlocked the truck and hoisted his pack onto the tailgate, grateful that he had made it back safely. He whistled for Emma but heard nothing. He listened to the silence. He dawdled in the vicinity of the truck with his hands in his pockets, all alone in the darkness. He wondered what had happened to Emma. Tired of waiting, he walked over to the canyon and stared into the coal-black void. He whistled a few times. The feeble sounds faded away into the enormous cavity, the small vibrations utterly defeated by the sluggishness of the thick night air.

Jonathan waited at the edge peering blindly into the blackness. He heard the sound of a rustling off in the distance, approaching from the direction of the interstate, perhaps a breeze coming up the canyon. The volume increased slightly and in a flash he discerned the unmistakable cadence of Emma panting heavily as she raced up the bed of the arroyo. Jonathan was so relieved that she was still with him and that she was apparently unscathed. He could tell that she was running at full speed in response to his calls, so everything was alright and he could start preparing her dinner as usual, on the counter top offered by the tailgate of the truck.

Jonathan patiently waited for Emma, but the sound of her labored breathing suddenly began to diminish. Instead of coming toward him, she was sprinting up the opposite bank. She had picked up his outbound trail from this afternoon and thought that this was the way to find him. He whistled frantically to alert her to her mistake, but his calls echoed off the cliffs and seemed to come from every direction. To make matters worse, Emma was never good at locating sounds. He had always blamed it on her floppy ears blocking her ear canals. Emma knew that Jonathan was calling her, but she didn't know where he was. She began going the wrong way and just kept going, disappearing over the ridge. The darkness swallowed her up and the world was silent once again. Jonathan listened for a faint sound off in the distance. The air was absolutely still. Emma was obviously lost. He feared that she would revisit all the places they had travelled today in the hope of finding him there. She wasn't thinking about the mountain lion that was still out there somewhere, probably hunting nocturnal prey along their previous trail, and she wasn't thinking about the javelinas that were also still out there, hungry and waiting for a hapless victim to fall into their midst.

The danger posed to a domestic dog by a wild animal such as a mountain lion was clear, yet the danger posed by a herd of javelina was even greater. Jonathan had recently talked to a woman at a small music festival up in the mountains who had lost her dog to javelinas. They ripped the dog apart with their tusks and by the time she reached it they had already begun eating it. The surgery required 400 stitches to close the wounds but the dog later died of its injuries. For six years the woman wept every time she thought about the horrible suffering her dog endured following the attack. Well, the dog had been twelve years old at the time and not quick enough to escape. If Emma ever got surrounded by these creatures, she would probably be too nimble and fleet of foot to be taken down. Still if one of them caught her off guard she might falter and that would certainly be the end of her. Jonathan would probably never find what was left of her body

Emma's hearing might be defective, but she was good at making visual contact with him, always turning her head to see which way he was going when they were out on their walks. If he had a light, perhaps he could reveal his whereabouts and help her out. He hurried back to the truck and discovered that the headlamp was in fact hanging from the shift lever of the transfer case. He returned to the edge of the canyon and waved the light back and forth, slowly scanning the terrain, acting as a lighthouse. The beam was powerful enough that he should have been able to spot her, but he didn't see her anywhere. Emma had probably traveled a considerable distance by now and the intervening hills and ridges would block her view of the light. Apparently he was too late. His concern mounted and he considered going after her. He swung the spotlight down to the canyon floor. He was not as agile as Emma and his progress in this terrain would be excruciatingly slow, even with a headlamp. Maybe he should just stay where he was and wait for her.

Then he remembered that he had put a special whistle in one of the pockets of his pack, a high-pitched whistle that was shaped in the form of a trumpet. The shrill note it produced was so loud that he could not blow it in close quarters because the piercing sound it made hurt his ears. Jonathan marched once more back to the truck and retrieved the whistle. There was still no sign of Emma when he reoccupied his lookout. Playing a one-note song with an instrument that was little more than a tiny piccolo in a concert hall of such enormous proportions was so ridiculous that Jonathan laughed out loud when he heard the pathetic sound it made. He doubted that anyone would be able to hear it, unless they were already close enough to see his headlamp.

Still he waved the light and blew the whistle, turning slowly to face the devices in every direction, having no idea where Emma was. He resolved to wait, performing the ritual throughout the night, and at dawn he would go out to look for her. He immediately realized that this was pure nonsense. If Emma didn't show up soon he would strike out into the darkness himself, against his better judgement, to join her, so at least they could both be lost together.

The air temperature plunged now that the sun had disappeared forcing Jonathan to return to the truck to suit up in some heavier clothing. After searching through the clutter in the back of the truck, he finally uncovered a warm coat, mangled and dirty underneath piles of camping gear and other junk. He took a deep breath and turned his eyes toward the heavens. To his surprise, the clouds had all quietly slipped away and the sky was full of stars. He saw Orion standing proudly above him, his cherished hunting dogs running toward him, leaping upward in enthusiastic and heartfelt greetings. Emma was not safe out in the desert all by herself. What if something happened to her? What if she disappeared and he never saw her again? What if he never even found out how she had died? He imagined himself after a long night of waiting and a long day of searching, finally giving up hope and driving home without her.

Jonathan had known such sadness before because he had lost a dog that way once, many years ago. His beloved blue heeler had just turned 14 years old. Two young female dogs had recently been thrust upon him, a black lab mix abandoned near a campground in the Jemez Mountains and a Karelian bear dog found starving on pueblo land. Ralph had been keeping up with everybody just fine, but when Jonathan stopped to water the dogs, he took a close, hard look at Ralph and in the harsh light of midday he saw that Ralph was unbearably old and feeble and realized that Ralph shouldn't be out there with

117

him. Jonathan clasped his hands on either side of Ralph's head and spoke to him softly, saying: "This is your last hike. You've gotten too old to be doing this anymore." Ralph had a sad look in his eyes as if he knew the truth of what Jonathan was telling him.

A mile later the three dogs discovered a fresh scent and raced up a hill single file, following an established game trail. Ralph lagged behind as usual. The dogs vanished one by one over the top of the hill. Jonathan stopped to wait for them and after a few minutes the two girls came back, but without Ralph. Jonathan figured that since Ralph was so old, he had not been able to keep up with the younger dogs, so he continued to wait for him. When Ralph didn't show up, Jonathan trudged up the steep slope, leaning forward, his face close to the ground, his eyes focused on the minute details of the dusty, rubble-strewn trail.

At the top of the hill the narrow world of the trail exploded into an immense landscape. Jonathan could see for many miles in every direction, but the barren hills showed no signs of Ralph. He had expected to find him nearby, busy exploring the desert, forgetting that they were waiting for him at the base of the hill, but everywhere the terrain was empty and silent. He kept thinking that Ralph would pop into view at any moment, but gradually it dawned on him that he wasn't going to find Ralph, because Ralph was gone forever.

Jonathan combed the hills for several hours, staring down into each arroyo from every conceivable angle, retracing his steps repeatedly and calling Ralph's name until his throat was sore. He might have stayed longer, but he knew that there was no point to it. He had covered a large area, eventually just searching for Ralph's dead body, since he must've been killed by a rattlesnake or died from a heart attack, but he found nothing, no clues to his fate, no evidence of his demise. They were five miles from the truck and they needed to get going.

Ralph knew the way back to camp from there. He and Ralph had camped in that same canyon many times before, and Jonathan had often walked in the direction he had taken that day. In the evening he kept expecting Ralph to show up at camp. He stayed up late waiting for him, but finally went to bed. Jonathan woke several times during the night and opened the door of the camper, looking around and calling Ralph's name, shining a light into the nearby bushes. The quiet stillness of the night reinforced his conviction that Ralph would never again come home to him. Jonathan had commitments in the city the following day, but he returned two days later and walked the entire area, hiking 10 miles. He found nothing. Sarah was angry with him and rightfully so. Ralph had been his responsibility and he shouldn't have taken him out that day. The only thing Jonathan could say was that at least Ralph died doing what he loved most in life. That was his favorite place in the whole world and over the years he and Jonathan had taken many good walks there, hiking cross-country through the breaks and badlands. Nobody could ask for more. Over the course of the following year, Jonathan visited the area occasionally and called out to Ralph's spirit, but each time the desert was as serene and motionless as it had been on the day that Ralph had disappeared.

Jonathan stopped breathing and listened to the silence, just as he had done when Ralph had vanished. Emma was undoubtedly gone, perhaps lost forever. She might be miles away by now, alone in the darkness, wondering why Jonathan had abandoned her. She would not find a ranch out there or some lonely outpost where the people would take her inside and keep her safe, nor was she likely to stumble upon any campers. No one ever camped around here other than himself, and he did that only on rare occasions, maybe once or twice a year.

Jonathan walked back to the edge of the canyon and resumed his broadcast, swinging the light and holding the whistle to his lips. Taking a deep breath and blowing the whistle with all his might, he forced it to shriek like a banshee, desperately eliciting a long, drawn-out scream from the tiny instrument. He finally quit the strenuous effort and gasped for breath, then listened for a response. He couldn't say that he actually heard anything, but he discerned a stirring off in the distance, as if some dark figure had begun approaching him. The movements of this unseen entity appeared hushed and deliberate, giving him the impression that it was stalking him. Whatever it was, it certainly wasn't Emma. He began to tremble, suddenly afraid of being alone in the menacing blackness, wondering what evil spirit he had summoned. His first thought was to turn and run, but he held his ground, fearing that there was no place to hide from such a demon or disembodied soul.

The outburst he had just unleashed might have been strong enough to wake the dead. Jonathan wondered how many people had died out here over the centuries with no record of their demise or account of their final ordeal. If some people's spirits continued to haunt the houses where they had once lived, then why wouldn't the souls of the deceased wander the desert at night and haunt the landscape, seeking revenge for the injustices they had suffered during their lifetimes.

The strange entity appeared to be getting closer. He swung the light in the direction of the sound but to his surprise, nothing was there. The sensation immediately stopped. Could he have possibly heard only the suggestion of sound in his brain? Whatever had been out there in the dark had obviously been driven away by the harsh glare of the headlamp. Jonathan reminded himself of a similar incident long ago. Some hunters had erected a platform in the trees at the far corner of an old field near his cabin. One night he was sitting up there under the stars, quiet and alone, lost in thought. He heard footsteps approaching, rustling the autumn leaves in the vicinity of the platform. The steps were carefully measured and unmistakable. They were coming closer, as if this person knew about the platform and was heading straight for it. He sat there without making a sound, waiting to see what would happen next, stunned that anyone would be out wandering the fields and forests in the middle of the night and not at least be wielding a flashlight or a kerosene lamp.

The footfalls reached the base of the tree directly below the platform. The stranger appeared to be ready to start climbing the ladder. A wave of fear swept over Jonathan, not knowing exactly who was there or what he was facing, almost certainly a human being, yet possibly something else: a bear, a deer, or a cougar. Without warning he turned the flashlight on while holding the beam downward, expecting to reveal the identity of the intruder, but to his surprise no one was there. He quickly scanned the area trying to locate the source of the sound, but he saw nothing. The sound immediately stopped. Even a squirrel or a raccoon or a skunk would have been highly visible. Jonathan was left with the hypothesis that a small mouse had hopped along in a series of slow, regular leaps and happened to be under the leaves at the moment he turned his flashlight on where it instantly hid and remained motionless. After many years of walking out in the fields at all times of the day and night, Jonathan had never witnessed such a feat performed by a mouse. If dark spirits did in fact roam the night, how would the physicist possibly know it? He couldn't possibly create the phenomenon in his laboratory, so it didn't exist for him. The experience was not a part of his world, but it had definitely been a part of Jonathan's world on a number of occasions.

Jonathan wondered what other phenomena the physicist hadn't seen. The strangest events might also be the rarest, perhaps even being occurrences that no one had yet observed. Physicists would never know about such things, but Jonathan had put himself in a position where someday he might get lucky and discover something truly unique, a creation the world had produced only once, never to appear again.

Suddenly Jonathan spotted a movement, a brief blur of motion on the other side of the canyon. He held the light steady. It was Emma, still running at the same frantic pace that had marked her departure. He wondered if she'd been racing this fast the whole time and had covered their entire route in the interim.

Emma saw Jonathan's headlamp and now knew where he was. He was standing on the opposite ridge, shining the light across the canyon looking for her. She reached the bottom of the arroyo without much trouble, but then couldn't find a way up to Jonathan's position. She ran feverishly up and down the riverbed looking for an opening, something she could manage. Eventually she discovered a path between the rocks and in a burst of energy quickly gained the rim. Her body was heaving from exhaustion and her eyes had a look of terror in them as she entered the illuminated world of Jonathan's lamp, obviously shaken by her traumatic experience but now comforted by his close presence, and very grateful that she had finally found him.

—5—

Jonathan's first physics textbook when he was a freshman in college had been entitled Mechanics. The opening chapter introduced the reader to projectile motion, which was strange. Shouldn't such a mundane topic be relegated to an appendix at the back of the book? What did the flight of an object propelled through the air have to do with the secrets and mysteries of the universe? Later he discovered that the motion of a particle in a force field would be central to everything that followed, that basically the entire universe would be reduced to this. He was outraged. He went along with it for a while, and then one day after class he approached his professor and presented his criticisms. Jonathan had prepared arguments as to why no one could ever account for the entire universe simply in terms of particles and forces, and braced himself against the emphatic refutation he expected from his professor, anticipating a long and heated debate. Instead the professor calmly explained to Jonathan that physicists liked what they did because they found it interesting and fun, so they weren't interested in asking such questions. They just wanted to get on with the business of doing physics. The professor added that if Jonathan truly wanted to think along those lines, then he should study philosophy. Thus in an easy-going manner, Jonathan's problem was matter-of-factly resolved and their brief discussion ended amicably.

Later Jonathan thought that the professor had merely patronized him in order to get rid of him, but finally concluded that the professor had been sincere. One of the professor's duties as a teacher was to help his students find their ways toward the careers that were most suited to them, to assist them in discovering their true callings in life. If Jonathan was not cut out to be a physicist, then Jonathan should think about doing something else. To this day, Jonathan could still hear the professor's voice echoing down the empty hallways as he lectured in one of the adjacent classrooms. His voice possessed a resonant, melodic quality as it wafted down the corridors, like a draft of fresh air coming in through an open window. Jonathan always found his lectures incredibly peaceful and soothing. He recently discovered that the professor had died a few years back from non-Hodgkins lymphoma. Jonathan, of course, followed his advice and took up philosophy.

When Jonathan was busy studying the humanities, he reviewed papers briefly for an undergraduate journal of philosophy. During that period a woman submitted a paper proposing to take the average distance between two people over an extended period of time as a measure of their love for each other. In his review, Jonathan noted that people often found themselves surrounded by people they didn't really like--neighbors, relatives, co-workers--and due to circumstances beyond their control, they were unable to be with the ones they truly loved. In fact, they might often be separated by great distances from the ones they loved, so she needed to take a different approach, or at least add more factors to the equation: the extent to which the two people communicated with each other, how much one person occupied the other's thoughts, and what kind of thoughts those were. Unlike distances, many of these factors were asymmetrical. Love also took on many different forms. Some people loved money. If a person kept his cash locked in a safe hidden in the bedroom closet, did that person love money more than the person who kept his money in a safe-deposit box at the bank across town? Some people loved Jesus. They said that they felt his presence near to them throughout the day. While this fit the paper's thesis, how did someone go about measuring this distance? What happened if a person was in love with an idea?

A reasonable requirement for someone to mathematize love was that this person first understood what love was, but a definitive answer to that question had eluded poets and philosophers for centuries. When he thought about it, Jonathan found the woman's proposal a bit puzzling and asked himself: "What was the motivation for creating a numerical measure for love in the first place?" The object of the exercise appeared to be twofold: first, to transform a vague concept into a precise one, and second, to take a complex issue and make it simple. Instead of delving into the messy nuances and conflicting interpretations of love, Jonathan could do away with all of that and replace the idea of love with a simple quantity, conveniently bypassing the nasty problem of understanding what everyone meant by love. Few people would say that the essence of love was mathematical in nature, and if that were the case, then quantifying love actually made it more imprecise and took it further away from its true identity.

Perhaps the thesis of the paper wasn't about people who were actually in love, but only about people who acted as if they were in love. The essence of love was about caring, about feelings and sensations, thoughts and emotions that were experienced subjectively, but the proposed formula only dealt with outward behaviors. The approach externalized an internal phenomenon and characterized it solely from the perspective of an impartial observer.

Jonathan envisioned measuring the distances between various people in a city. Let's say one pair of individuals worked at offices that were 30 miles apart, then spent the rest of their time living together in the same house. Another pair of individuals worked in the same office together, but then went home to separate residences that were located 30 miles apart. If each pair divided their times equally between home and work, they would both produce the same number, yet it turns out that one pair were married to each other and the other pair were comparative strangers. Of course, being married didn't mean that the people were in love, and strangers sometimes held secret infatuations for each other, imagining themselves to be in love with a person they hardly even knew.

Geometry was uninformative. It was not just shallow and superficial, but immaterial. Geometry didn't say anything about the actual relationships. It provided no details and cast no light on the specific forms of the relationships. The only thing Jonathan had gained by consolidating the complexities of love into a single scale of magnitude was that he had gotten the mathematics, but the price he had to pay for this mostly meaningless number was that he had lost everything of any consequence. Mathematics did not help him understand the world he lived in, but rather diverted his attention away from what was important and caused him to focus on things that were irrelevant.

The numerical measure was simple and precise only in terms of the mathematical ideas behind it, however this characteristic did not reflect upon how accurately the measure copied the infinite details of the real world. The measure's relationship to love was still messy and complicated due to the fact that the relationship between distance and love was not at all clear, something that first had to be established. Nothing had been gained here other than transferring the problem of defining love to the problem of defining the relationship between love and distance, an equally if not more difficult task. While the distances were easily measured, they ultimately had nothing to do with love. The numerical measure was based on the realities of the world since it was calculated from the actual positions of real people, and this made quantitative love also one of the realities of the world, but none of this implied that the measure captured the essence of love, leaving the concept of love as much up in the air as ever. The investigator had to believe that the mathematical construction was somehow equivalent to the phenomenon of love and here was where the whole procedure fell apart. If love did not have a mathematics of its own, then the quantification of love was divorced from the realities of love, yet closely aligned with the realities of measuring distances.

Repeated attempts to understand everything through the calculation of distances unnecessarily complicated the physicist's picture of the universe and diverted his attention away from what was important. Trying to find a connection between a phenomenon and distance turned out to be a more difficult problem than simply trying to understand the phenomenon on its own terms. Distance was not a fundamental concept of the universe and translating everything into the measurements of distances was only a detour that subverted the task of obtaining a genuine knowledge of physical reality. Distances became obstacles that then had to be overcome with new sets of explanations. A network of distances was simply not a suitable framework for making sense out of natural phenomena. Since distance was a crucial concept in geometry, physicists automatically assumed that it was also a crucial concept in the natural world, but that wasn't necessarily the case.

Love changed many things in a person's life. Love changed one's perceptions, not merely regarding the person who was the object of affection, but also regarding everyone else as well. Feelings and thoughts, actions and behaviors, motivations and goals, were uniformly altered by being in love. The physicist saw temperature and pressure in terms of mathematical ideas, magnitudes that had values at every point in space. He turned temperature and pressure into relatively simple concepts by arranging simplicities in a laboratory, but out in the real world temperature and pressure became wildly complex, influencing many things in diverse ways, becoming highly multifaceted in their effects. Temperature and pressure expressed themselves in innumerable ways, and the simple magnitudes consolidated all of these aspects under simple headings. Could all these subtleties be understood in terms of distances and the concepts that had been built around distances, concepts such as velocity, momentum and energy? Or did the physicist lose his hold on reality as soon as he formulated his mathematics?

Certain people had proposed a similar approach to quantify the experience of solitude, which was, for those who pursued it, a peculiar form of love, the love of being all alone. In contrast to the romantic feelings people felt for one another, solitude increased in direct proportion to one's distance from others. To scientifically determine the most remote place in each state, formulas had been devised to calculate the distances of each point in the state to established roads and population centers. The idea was that the farther a person got from the hubs of human activity, the more that person had solitude. In New Mexico, the computations had identified a particular place in the Gila Wilderness as being the most remote spot, so a group of people, including a local forest ranger who was familiar with the area, hiked to the location. Reaching the point involved a long trek, however the destination was close to a popular and heavily-used trail. Everyone in the group agreed that the spot didn't really feel very remote. The forest service representative said that she knew of a secluded place near the trailhead that hardly anyone ever visited, and that spot felt much more remote than this. Equations just cranked out numbers, numbers that didn't necessarily mean anything.

Jonathan had experienced intense feelings of solitude many times during his wilderness travels, and one thing was certain: no simple rule captured the essence of solitude and no elegant formula solved the problem of finding its location. The only way to move forward was to complicate the theory and take more factors into account. This could be done by adding more terms to the equation and incorporating more variables into the calculations. If people used the improved formula and everyone agreed that the resulting spot felt more remote than anything they'd ever experienced before, did that make the equation a 'Law of Nature'? Or was the whole approach simply misguided and doomed to fail?

Jonathan saw that distances were no more useful in establishing the experience of solitude than they had been in determining the love that people felt toward one another. Remoteness was a complex sensation and it was not subject to quantification. Jonathan had his own list of special places that felt remote to him, and various factors contributed to their status. One factor was the difficulty of access. If the road taking him into an area was rough, with washouts, steep grades, deep rubble or other obstacles, the prize at the end was usually more valuable, but not necessarily so. Better yet, if there was no road or trail at all and he had to bushwhack his way cross-country to get there. Another consideration was the frequency that the place was visited. Long and strenuous trails in the National Parks often received heavy foot traffic so that no one was ever alone for very long. By this criterion, the summit of Mount Everest was not remote even though it was difficult to reach. Other intangible factors entered into the evaluation, strange qualities certain places possessed due to the specifics of the terrain, the unusual lay of the land, the arrangement of the rocks and trees. While distances were easy to measure and simple to calculate, they were of no real value in answering the question of solitude. All the essential factors had to be inspected personally, and that was a tedious and time-consuming process. There were many remote places that Jonathan had never found and would never find. Perhaps the remotest place in the state had never been visited by anyone.

After listening to Jonathan's speech, the physicist laughed and told Jonathan that his accounts of mathematizing love and solitude were some of the best jokes he'd heard in a long time. The physicist complimented Jonathan on his powers of imagination, declaring that Jonathan could tell some really funny stories. Undoubtedly these wildly outrageous scenarios would elicit a hearty round of laughs in almost any social setting, but when Jonathan switched the topic to velocity, the physicist's eyes lit up. Velocity was a very serious matter. Jonathan wondered if that was because velocity was important for understanding the universe, or because velocity was something that could be easily calculated. The desert was often still and quiet, but even then, when Jonathan looked closely at the objects around him, he found subtle motions everywhere: the blades of grass swaying in the breeze and the branches of the bushes bending slightly, the movement of the shadows traveling along the ground, the drift of the clouds across the sky. Jonathan would never waste his time measuring these velocities because they were meaningless. Things moved, but there was nothing special about these movements. Velocity wasn't important at all for Jonathan's understanding of the desert. The physicist had traded concepts of supreme gravity, things like love and solitude, for concepts that were trivial and mundane, like velocity and distance.

The physicist was able to turn all of this around by establishing the importance of velocity in his laboratory, assigning it meanings it ordinarily didn't have, meanings that were defined entirely by the apparatuses. The physicist used mathematical thinking to devise elaborate schemes whereby velocity acquired great significance, because he wanted velocity to have great significance. Velocity was a mathematical ratio that opened the door to further mathematical constructions. Once he had velocity in his hands, he could then introduce the standard fare of mathematical concepts, rates of change and areas under curves, and start playing around with numbers. Velocity was important not because of the role it played in the nature of things, but because of its mathematical significance. The physicist declared: "I'm just going to cast everything in terms of distance and velocity. I'm going to make distance and velocity the keys to understanding the entire universe because distance and velocity are such great concepts." The physicist then set out to accomplish this end by designing experiments where distances and velocities played major roles, but none of this was suggested by nature. The problem was that the physicist didn't start out with nature. Instead he started with the mathematics and found ways to use these preconceived ideas to create artificial phenomena. The resulting theories ended up being more studies in mathematics than responsible attempts to understand what existed.

The spectacular dawn of a new day found Jonathan and Emma flying down the interstate heading directly into the rising sun, their eyes blinded by the heavy covering of bug splotches on the windshield all suddenly glowing in the heavenly radiance of sunlight. Civilization ended after one last failed retirement community, an exit that was the final stop on the way out of town. Exits were hard to come by in this part of the country so Jonathan took advantage of this opportunity to explore the surrounding area. The real estate development was bleak and dingy, a fading grid work of unimproved lanes hastily bulldozed many years ago across the flat desert, then immediately abandoned after no one appeared to buy the surveyed homesites. Weeds had since tried to take advantage of the fresh plots, the highly competitive native vegetation having been unceremoniously removed by huge steel blades scraping the surface, but the weeds struggled to survive in the harsh environment, even with nothing to stand in their way. Street signs stood at each corner, but nobody bothered to utilize the empty thoroughfares since obviously they led nowhere and there was nothing to see at the other end.

Although the envisioned community had never materialized, a few people had bought parcels and set up residency. A large airstrip ran through the middle of the smattering of houses, all politely set a good distance apart from one another. The runway supported a greater number of weeds than the surrounding area, giving Jonathan the impression that no one had ever used it, not even once. The developer had obviously considered the landing strip to be a valuable asset, an attractive selling point to potential buyers. A person could live out here in the middle of nowhere and commute via private airplane, effortlessly zooming over the endless miles of desolate highways, completely oblivious to the ubiquitous potholes and large cracks in the asphalt. Jonathan admitted that the idea was rather intriguing. Similar landing strips were found throughout the desert southwest, including in some of most remote and inaccessible wilderness areas. These makeshift airfields had once been commonplace, the traditional staples of historic ranches and homesteads, most of them now deserted and abandoned. The skies were clear and largely devoid of air traffic, so the landowner would have the world at his fingertips, the entire place to himself. The only problem was, almost no one had a pilot's license—or a plane.

Jonathan was not here to find his dream home but instead to look for a way to access the large expanse of BLM land to the north. Apparently all the old roads had been blocked off when the developer took over the land. The only road he could locate had a locked gate, the authority of which was reinforced by long stretches of heavy duty pipe fencing extending in both directions preventing anyone from easily detouring the obstacle. No trespassing signs were posted on the opposite side of the fence warning travelers not to proceed onto the land where Jonathan was now standing. He guessed that he was not supposed to be here. The situation looked hopeless so he whistled for Emma. She came running and hopped immediately into the truck.

As Jonathan drove slowly back through the community, the eeriness of the place struck him. About two-thirds of the houses were abandoned, the yards unkempt and overgrown with tumbleweeds. The noxious plants had rolled across the open desert for countless miles and then piled up along the walls of the houses. Each year they continued to accumulate, all of them now trapped by perimeter fences and outbuildings, filling the yards with nasty tangles of woody stems and sharp barbs. Jonathan noted that many of the houses were spacious and elaborate, what he would consider to be small mansions, comfortable and even luxurious estates that he might see in more well-to-do parts of the country. The owners had obviously forked out a good deal of money and then either passed away or fled after realizing their horrible mistake. Perhaps the wells all ran dry. The whole area was creepy, a surrealistic juxtaposition of wealth and dilapidation, a ghost town now probably festering with sociopaths and weirdos, characters he didn't particularly want to meet, people driven mad by the isolation and despair, abandoned by the mysterious disappearance of a large number of residents who had gambled everything they'd owned on the success of a misguided venture. A single real estate broker's sign stood by the entrance next to the highway, but obviously no one was ever going to buy property around here again—even assuming that water was still available.

The next exit was 18 miles down the interstate, a straight shot across a barren landscape without a single dwelling in either direction as far as Jonathan could see. The exit had nothing to offer other than a solitary bridge spanning the interstate. The land all around was flat, so huge amounts of dirt must have been excavated from somewhere nearby and piled up on either side of the interstate, then molded into two long ramps. Massive concrete beams connected the two mounds together in the middle, forming a giant arch of earth and concrete straddling the four lanes of traffic. The structure reminded Jonathan of the curved handle of a kitchen cabinet where he'd put his fingers through the opening in the middle and pull the door toward him.

He crossed over the highway and turned onto a paved frontage road with absolutely no shoulders, crowded on either side by imposing stands of tall weeds, the roadway too narrow for vehicles going in opposite directions to pass each other, even though a centerline had been painted down the middle, flanked by narrow half-lanes on either side. The prospect of meeting another vehicle was extremely remote, but of course, it wasn't absolutely impossible. The pavement ended after about a quarter mile anyway. The truck rumbled across a cattle guard and the tires skidded briefly as they adapted to the loose gravel and regained traction. The road suddenly was wide and perfectly straight, vanishing many miles ahead into the distant hills. The lechuguilla and mesquite, standing peacefully in the desert, slowly started to approach him, then they rapidly accelerated, becoming a blur as they passed by the truck. After ten miles of flying recklessly over patches of washboard and deep sand, frequently swerving from side to side and jerking the steering wheel to compensate for the uncontrolled skids, Jonathan reached the side road that he had been looking for. He broke heavily and veered into the sea of bushes onto a narrow track mostly hidden by dense foliage. He looked in the sideview mirror and saw a trail of dust following him down the road. The cloud quickly departed and went its own way. He slowed to a crawl, driving as though he were picking his way through a dense throng of pedestrians in a crowded market.

The road turned to clay and smoothed out as it meandered across a broad, flat valley harboring expansive patches of Tobosa bunchgrass intermingling with tracts of creosote, a place where Apaches had once erected extensive villages of teepees. The mesas on either side had been uplifted, gradually rising then falling off abruptly, mixed occasionally with solitary buttes—overall quite an exotic landscape. The front range was heavily grazed, scores of cows centered around tanks and sacrifice areas where the meager vegetation, an assortment of largely inedible grasses and weeds, had been thoroughly trampled. For hundreds of yards in every direction the ground was pockmarked with hoof prints and now resembled the urethane jacket of a golfball. Just as the degradation from one tank began to thin out, he slowly closed in on the next tank and the situation repeated itself. Jonathan didn't despair because he'd hiked cattle country for years. Somewhere out here there would be ranges that were being rested and sooner or later he would find them.

He kept driving and finally spotted what he was looking for, a fence line crossing the road up ahead. About 25 cows were standing directly in front of a tall gate, a wire mesh stretched across a tubular metal frame and tied to it with short pieces of baling wire. As he approached, the crowd split and backed away just enough to let the truck pass. How did the cows know what Jonathan wanted to do? Was it so obvious that even cows could envision the next step? A maintenance road branched off in either direction and followed the fence line, so how did the cows know that he was not turning left or right, but instead going straight? Apparently the rancher had a different model truck.

Emma gazed silently back at the cows through the passenger side window. She knew what cows were and wasn't threatened by them. She couldn't interact with the cows while she was inside the truck, so there was nothing she could do except passively observe the events taking place on the other side of the glass. Jonathan glanced over at her and noticed that she had a dreamy, pensive look on her face as she returned the cows' curious stares. He wondered what she was thinking. Was she having fond memories of her past encounters with cows? He pulled the truck up as close to the gate as possible and got out. He was surrounded by a multitude of large heads, many of them uncomfortably close, a crowd of giant faces all looming only a few feet away, looking at him and calmly chewing their cud, some peering over the shoulders of others, every eye focused on the new stranger.

126

The chain was way too tight for Jonathan to release the hasp, a ruse by the rancher to prevent anyone from accessing the area. Jonathan looked around at the cows. He could feel their hot breaths on his shoulders and smell the saliva being flicked into the air by the incessant motions of their tongues and mouths. The cows wanted to see if he could do it, if he could find a way to open the gate. They had obvious smirks on their faces, apparently betting against him. Perhaps they had watched others before him, including many who had failed and turned around. He returned his attention to the gate and clumsily fiddled with the chain. He made several more strained attempts to slip the link past the hook of the clasp, but with no luck. At this point one of the cows emitted a gruff, guttural sound, apparently unable to contain its mirth any longer. Another cow erupted into a hearty outburst, a vocalization that definitely sounded like laughter. Jonathan angrily grabbed the gate and jerked it upward. This released the tension just enough to allow him to unhook the snap. He swung the gate forward, thinking that the cows would take this opportunity to escape, but surprisingly none of them broke for freedom or showed any inclination to avail themselves of this fortuitous situation. He hopped back into the cab and drove through. He jumped out as quickly as possible in order to seal the opening, but the cows just stood there with vacant stares on their faces, maintaining their apathetic postures, heads now all turned towards him, mouths churning away, watching him as he returned the gate to its closed position locking them in, somehow knowing that they were not invited to follow him.

The back range showed no signs of recent activity—neither hoof prints nor tire tracks. The road skirted a low embankment and became heavily eroded. He straddled the deep depressions as best he could to keep the truck from leaning too far one way or the other and possibly flipping on its side. The area felt extremely remote, partly because no one had been back here in quite a while, and partly because he was 20 miles from nowhere. If the truck didn't start, he'd be looking at a long walk back to the interstate and the lonely interchange, then another 20 miles to the nearest town. He'd hiked desolate backroads many times before, and into areas much farther away from civilization than this, but rarely had he felt so isolated and alone. He dismissed the experience as nothing more than his unfamiliarity with the region.

He parked in a low-lying area next to a dry wash located in the middle of a broad valley, a place where the dirt was soft and easy on Emma's feet. The desert was generally strewn with various kinds of rubble and much of this debris was coarse and rough, tearing up Emma's feet and abrading her pads as she ran enthusiastically following the numerous scent trails and occasional wild game, causing her to sometimes whack her toes against some of the larger rocks. Jonathan soaked her feet in a solution of antiseptic soap as needed, usually once or twice a week, a ritual she accepted without protest seeing that he fed her a steady stream of small pieces of dog biscuits the entire time, a ploy designed to take her mind off the stinging sensation of the irritated tissue.

Emma wasted no time and ran off exploring the area. Her eager romping and intense tail wagging told Jonathan that she highly approved of this place. He tediously gathered his gear, zipped up his pack, and headed out. The valley meandered for several miles, climbing in elevation the whole way. Jonathan dropped down into the bottomland of the wash and began swimming through the shoulder-high vegetation. The leaves brushed against his arms and legs and the supple branches seemed to flow around his body, offering a gentle resistance to his movements and creating the illusion of being immersed in a fluid. He felt as if he were up to his neck in the imaginary waters of a lake, his head floating just above the surface, drifting along as he treaded water with his legs. He wandered around for a long time, largely hidden in the abundant foliage, with only his head sticking out. Eventually he got tired of plowing through the sea of leaves and branches and set a course for higher ground where he began wading through the shallower pools of plants along the perimeter. He finally crawled out of the sprawling basin by climbing up on a rocky shoreline and for the first time in a while enjoying the freedom of open space. Emma was still submerged somewhere in the vegetation. She had plunged into its depths and disappeared when he first entered the bottomland, and could have gone in any direction. He hadn't seen her in a while.

Jonathan's view of the fanciful lake expanded as he mounted the hillside, extending in two directions as far as he could see. Emma popped into view a hundred yards downstream after noticing Jonathan's change in direction, apparently having caught sight of him through the tight mesh of leaves and branches. She stopped and stared in his direction in order to observe which way he was going next, then climbed the hillside along with him, flanking him to his left.

River valleys in the desert generally had no water in them except during brief summer downpours, yet this precipitation left lasting marks on the landscape by forming perpetual lakes and rivers of green flora. Jonathan topped a rise and entered a sparsely populated forest of juniper and piñon trees that continued onward over the rolling hills. The trees had gathered along the bases of the north-facing slopes and congregated in the declivities of the numerous tributaries to the main wash. He wanted to measure the size of this forest so that he could compare it to the other forests he and Emma visited during their hikes. He immediately realized that this was a strange idea.

Trees had been around for millions of years and during that time they had witnessed many things. They had been able to observe the world around them for countless generations and by now, Jonathan thought, their knowledge of the world must be very great. Therefore Jonathan made it a policy to consult them for advice. He had only been on this earth for a very short period of time and he didn't understand all of its complexities. He often didn't know which way he should turn or what he should do next. Perhaps the trees could offer him some suggestions and guide him along his path in order to help him to reach his destination safely. Jonathan saw trees as repositories of wisdom and ancient knowledge rather than unimportant organisms that existed only on the level of simple biology. Turning toward what was greater than him rather than what was less than him seemed reasonable. He'd often found that other people really didn't know any more than he did. Of course everyone always thought that they knew everything, but they were generally wrong. He'd rather take his chances with the infinity of creation. However, today Jonathan wasn't seeking the wisdom of the trees. Instead, he was seeking the mathematics of the forest.

The trees had all germinated separately at different times and from different seeds, then grew independently of one another, each according to its own blueprint, so what right did he have to tie them all together and create a forest? The disperse collection of different shapes, highly variable in nature, could not be a thing-in-itself, yet he was setting out to measure something called "the forest" that was really nothing more than a concept in his head, an idea that referred to a large number of solitary individuals that had no common origin or any physical connection between them. He felt that he was measuring nothing, a ghostly presence he could not put his finger on, a complete non-entity. Measuring something so unsubstantial, something that didn't exist in its own right, made him feel a bit uneasy, yet he told himself that because the trees existed, he could say that the forest also existed, even though he was adding a mathematical concept to their ranks, something which wasn't out here anywhere. But if it wan't here, then where was it? Where did it come from? The forest did not come from the trees. Instead Jonathan had brought the forest along with him together with his hiking gear, having carried it all the way from his childhood where he first picked it up as a young boy.

Physicists assumed that by defining sets they were studying realities because the concepts they had used to establish the sets referred to real objects and these objects did exist in the world, so through these connections the concepts themselves became real, even though the connection between a tree and a member of a set was simply another idea. As long as physicists began with the physical realities of the world and built their superstructures on top of them, their fanciful notions and elaborate constructions all inherited that reality, as if realities were traits that could be passed from one generation to the next down long lines of ancestries, each new set of concepts receiving this characteristic from its forbears. Reality was shuttled along, transferred from premise to conclusion in much the same manner as the truth values of propositions. In mathematics, the truth of the antecedent was imparted to the consequent, the truth of the axioms led directly to the truth of every statement that could be correctly inferred from those premises, but reality didn't work that way. Reality could not be conferred upon imaginary entities. Nevertheless physicists started tying everything together, drawing lines and sketching diagrams, inventing quantities and linear scales, defining sets and categories, making stuff up and then saying they didn't make it up, convinced that they were simply studying what had been given to them, maintaining with an adamant zeal that everything they had imagined was real. All the free associations and creative license made Jonathan's head spin. He invariably wanted to jump into the conversation and interrupt the flow of artistic inspiration by shouting: "Wait, you can't do that!" The physicist made up reality in his head because he tried to understand reality in terms of concepts and constructions that weren't real. Physicists irrationally added mathematics to nature and then made the startling discovery that nature was mathematical. They concluded that the mathematics had been there all along waiting for them to bring it to light, however, a magician could only pull a rabbit out of his hat if he had already put it there in the first place.

If the forest was going to become a cornerstone of Jonathan's understanding of the world, then the forest had to be more than just a cerebral association. It had to be something tangible. He asked himself: "How could he ever get his hands on the forest?" Maybe he could just intuit the forest. He closed his eyes. He felt the presence of the forest all around him and sensed its reality. He knew that the forest had to be here somewhere, even though he could not put his finger on it. Nonetheless the forest had a size and he could measure it.

The physicist complained that Jonathan was merely harping on trivialities, but if the forest could only be obtained through a series of conceptual maneuvers, this was of the utmost importance to Jonathan. If the forest didn't really exist, if it depended entirely on his way of thinking, then the forest became a tenuous starting point for his analysis of physical reality, a shaky foundation for everything that followed. Essentially Jonathan was basing reality on a figment of his imagination. The size of the forest became a puzzling idea. Many things existed in the forest. In fact, the forest was filled to capacity with actuality: dirt and clay, pine needles, wafers of bark and broken twigs, grains of sand and piles of stones, fragrances and breezes, strong winds and storms. There was a lot of stuff out here in the desert, more stuff than he could possibly fathom, but there was no forest.

Still the forest caused many things to happen. The forest caused the humidity in the atmosphere to increase. The forest caused birds to roost. The forest caused the wind to slow down. The forest caused the mycorrhizae to grow and permeate the soil. But Jonathan asked himself: "How could the forest cause things to happen if the forest was just an idea in his head?" It couldn't. The forest was just a convenient fiction. The forest helped Jonathan conceptualize what he saw around him, to organize ideas in his head, to think. The forest existed only for him. The forest did nothing in the world because the forest did not exist in the world. In the world, there were only details, endless details, mind-numbing, stupefying details. Jonathan's mind was not equipped to deal with such an infinite array of details, details that continually broke down into further details, so he ignored them, all of them, and replaced them with ideas, vague and cursory ideas, inaccurate and incomplete ideas, but that was all he had so he made do with them. He lived in a world of ideas and turned his back on the realities of the world, not because that was the right thing to do, but because he had no choice. He needed to keep thinking. He needed to keep assembling concepts, keep his mind working toward goals, keep formulating systems of thought in his head, keep building imaginary structures in his conceptual universe. His mind needed to keep functioning, somehow, by whatever means necessary, so he fabricated things that didn't exist. He made stuff up because existence was utterly incomprehensible. He could not think what could not be thought, yet he had to think anyway. All his thoughts were fantasies—but so what? He could live in his imagination and the world could be whatever he wanted it to be, as long as he understood that this was not correct, that this was not the way the world actually was, that ideas were just ideas and nothing more.

Jonathan wondered if he should throw the concept of the forest away, given the fact that it did not represent any kind of reality. What obligated him to form the concept of the forest anyway? Was the idea of the forest a necessary assumption? And even if he decided that it was, did such a conclusion thereby prove that the forest existed?

The assumption of the forest was certainly necessary for Jonathan to arrive at a number for its size. In order for the forest to have a size, the forest must not only exist, but it must exist as a mathematical object—unlike the clouds in the sky that were morphing and coalescing, evaporating and condensing, appearing and then vanishing. Turning clouds into mathematical entities was a much more difficult problem and required more sophisticated mathematical techniques. The forest, on the other hand, was a set of discrete, persistent objects and each one could be measured—but that too was just an illusion. The trees certainly didn't change as rapidly as the clouds, but the trees were still born out of nothing only to die and disintegrate into nothing, vanishing back into the soil after assuming a large number of states and configurations. Trees grew, their shapes changed, their limbs died and fell off, reducing their reach. Mathematics, on the other hand, was eternal, the relationships were fixed and permanent. The connections between mathematical ideas would never change, so how did the physicist reconcile the static nature of mathematical theorems and equations with an ever-changing physical reality? Simple. He created a separate mathematics for each place and each moment and then strung them all together with imaginary strings.

The forest was not a collection of trees because in that case the trees were all somehow connected to each other, even though the trees were primarily rooted in the soil, bathed by the atmosphere, and irradiated by the sunlight. They were connected not to each other, but to everything else, all the elements that came in direct contact with them, so the problem was not just a lack of unity between the tress. The problem was that the forest as a set of trees excluded all the connections that really mattered, everything that needed to be understood about the complex matrix of relationships underlying the existence of a tree. The forest as a collection of objects was a misnomer, a mathematical idea that did not help Jonathan understand the true state of affairs, the actual relationships between the innumerable objects of the desert. When he thought about the world in concrete terms, he saw that there was no forest, that this term referred only to a peculiar style of thinking, an aspect of his consciousness. He was imagining something that was not there. The algebra of sets was no more real than the geometry of straight lines.

Physicists granted themselves free rein in the use of their faculties of imagination. As long as they remained within the realm of mathematics, all of their visualizations were real and the conclusions they drew from them were valid. If Jonathan similarly used his imagination to picture a world of conscious beings, he was immediately chided and ridiculed, even though physicists felt no restraint or obligation in this regard. Physicists could do whatever they wanted and still be within their rights, but the moment Jonathan started fantasizing about imaginary entities they jumped on him and howled with contempt. The special status enjoyed by mathematics could only be explained in one way: physicists loved mathematics and they didn't question it. Daydreaming about a fantasyland of inane visions and ludicrous structures was not only sanctioned, but wholeheartedly encouraged, while others were expressly prohibited from enjoying such luxuries.

Obviously the only way to measure the size of the forest was to first measure the size of each tree, so Jonathan decided to start there. How could he determine the size of an individual tree? He could think of several ways of doing this. For convenience, he'd ignore the subterranean part of the tree. The root system was hidden from view and almost impossible to measure. Some plants sent taproots deep into the ground, while others had shallow but very bushy roots. Some roots exhibited frequent branching and formed dense balls while others were linear, thin and sparse. He could excavate the entire root system and weigh it, or he could analyze the root structure and calculate the entire length of the root fibers when added together, but instead he'd just disregard its sizable contribution and focus on the aerial portion of the tree.

Jonathan could represent the tree by two numbers: the height and the diameter of its trunk. The problem was that trees came in all kinds of shapes. Some were squat and fat while others were tall and thin, so even the total linear footage of the trunk together with all of the branches wouldn't do. He decided that the definitive way to measure a tree was to determine the total number of board feet of wood it contained. He could calculate the board footage of a stack of lumber easily enough because the boards were all uniform geometrical shapes. The edges of the boards were straight and mutually perpendicular, but the tree on the other hand, was thoroughly irregular and resisted measurement.

He could pulverize the tree into sawdust and compress the powder into boards of the same density as the tree, then measure the board footage of the resulting pile of lumber. In this way, he turned the tree into a mathematical object and then measured the volume of this mathematical object. It was a lot of trouble to go through just to get a number.

Jonathan thought of another way to measure the volume of a tree. It wasn't easier, but it was different. He could build a giant cylindrical tank and fill it with water. He could immerse the entire tree in the water and measure the rise in the level of the water. The areas and volumes of geometrical figures were easy to calculate—in fact, that was one of the main things he could do with them. The procedure turned an aspect of the cylinder into a property of the tree. Jonathan could not get the mathematics directly from the tree because the tree itself was not mathematical, so he had to fabricate a mathematical object first. The tree could now be understood in terms of the mathematics of the cylinder. This was a clever way of introducing mathematics into a situation that had none of its own, and in the process radically changing the nature of the tree, translating its haphazard structure into the regularity and uniformity of a cylinder.

The volume of the tree could not be constructed out of the other mathematical properties of the tree in the same way that the volume of the cylinder could be arrived at through the other mathematical properties of the cylinder, starting with a mathematical point and a line segment, then using these entities to define the diameter and circumference of a circle, then adding another line segment perpendicular to the plane of this circle, termed the height of the cylinder. The cylinder was comprised of mathematical ideas that were all fitted together, but the tree could not be resolved into such an interplay of ideas.

The tree and the cylinder were so unalike, so contradictory in nature, so different in structure, that when the physicist told Jonathan that he must turn toward the cylinder if he wanted to know the tree, Jonathan was skeptical. Jonathan didn't think he could understand a tree better by studying a cylinder, but the physicist insisted that this was precisely the case. Jonathan wondered how mathematical objects had gotten such powers, strange abilities to reveal the essential truths about the world. The physicist's dictum forced everyone to turn their heads and look at mathematical objects instead, a necessary first step in gaining an understanding of the world. Why couldn't Jonathan just look at the tree?

Although the tree was not a mathematical object when considered as a whole, Jonathan thought that perhaps the tree could be broken down into small enough pieces so that the pieces lost their identities—the complexities, irregularities and variabilities that so characterized the natural world—and they became simplistic enough to be mistaken for geometrical forms. A short segment of twig might be confused with a perfectly uniform cylinder. As the segments of the twig became shorter and shrank into the infinitesimal, the pieces looked more and more like geometrical shapes, until the last vestiges of the detailed configurations and cellular activities disappeared and the twigs were gone—along with the rest of the tree. Now at last the tree could be viewed as a mathematical object, but only after first destroying it.

In fact, the tree was so thoroughly non-mathematical that at no point did it resolve into mathematical elements, but even if it had, what would have been the point? Taking precise and well-defined geometrical shapes or algebraic proportions and then combining them haphazardly neutralized the mathematics and took it all out, thwarted the original impulse to strictly abide by mathematical methods and principles, trampled upon a neat and organized foundation of exact mathematical relationships. The world might as well start out with disorder if that was what the world was going to end up with. If the mathematics was just going to become lost in a chaotic mess, why bother with it in the first place?

The tree was not composed of mathematical building blocks and that was why the volume had to be measured with a cylinder rather than calculated from the spatial properties of the tree, the method that was used to arrive at the volumes of geometrical figures. The tree was fundamentally different from a mathematical object and that was why physicists paid no attention to it. The tree did not interest them because the tree had no mathematics of its own and mathematics was the only thing of any importance to them. Physicists were forced to fabricate the objects they needed in order to execute their plan of creating a mathematical universe, and this could only be done in a laboratory where nothing existed exactly as nature had created it.

A mathematical object demanded mathematics. It entailed a mathematical explanation. A mathematical object required everyone to have a certain level of mathematical knowledge before they could understand what it was. The mathematics was deliberately built into it, an integral part of its being and something that was essential to its nature. This characteristic distinguished it from other objects and set it apart from the natural world. But more importantly, a mathematical object had a form that was clearly visible and not hidden. The mathematics resided on the surface of the object, an obvious part of its outward appearance. Take a graduated cylinder for example. A material such as glass or plastic was externally molded into a precise shape, a rigid tube with equally spaced line segments embossed on one side, the dimensions and markings calculated according to a geometrical formula for the volume of a cylinder.

People who had previously been schooled in mathematics could still ignore the glaring mathematical features of the graduated cylinder and use it in non-mathematical ways, but that meant averting their eyes and ignoring certain aspects of it, essentially turning their backs on the conspicuous mathematical nature of the object. A primitive person who had never been exposed to mathematics could not possibly see the graduated cylinder for what it was and would almost certainly not use it for the purpose for which it had been intended, yet this person could still use it in some other way by investing the object with a personal set of meanings, say by making it a piece of artwork to adorn a costume or a convenient holder for arrow shafts prior to assembly. The object would now adopt roles and perform functions for which it had never been intended. The reverse was also true. A non-mathematical object could be used to perform functions within a mathematical framework, functions for which it had never been intended, and take its place within a system of mathematical ideas where quite honestly it had no proper place. Yet in order to understand a non-mathematical object, a person had to take a non-mathematical approach. The object demanded no mathematics. Its non-mathematical nature was an integral part of its being and something that was essential to its existence. Physicists tried to deny this. They claimed that its non-mathematical nature was just an illusion. They said everyone lived in a thoroughly mathematical universe, meaning that all objects were mathematical in nature, only here the mathematics was hidden, cleverly concealed so that no one could see—other than physicists, of course. Underneath the coarse facade was another world, a world of precise and well-defined objects endowed with mathematical properties. Jonathan replied: "If you believe that, then I have some dirt cheap beachfront property in Florida you might be interested in." Of course, the property was under water.

By sitting on a scale pan, a rock was able to play a role in the mathematics of the scale. A rock wasn't mathematical in and of itself. The rock only became mathematical when it had been placed in the mathematical apparatus. All of the mathematics the physicist saw resided outside of the rock and not inside of it. The rock and the scale were two objects that were fundamentally different from one another—in fact, they couldn't have been more opposed. The rock had been created by nature without the use of mathematics. The scale had been crafted by people who understood mathematics and had deliberately created a mathematical object out of steel and plastic. If scales and rulers and graduated cylinders were everywhere and a major part of nature's repertoire, then Jonathan would have been forced to admit that the physicist was right and nature was indeed mathematical. If natural objects and events were mathematical in themselves, then physicists wouldn't need to build mathematical objects and invent mathematical schemes in addition to them because the objects and events of nature would already be mathematical and physicists could take this mathematics instead of their own mathematics as representative of the true natures of the objects. Instead physicists had to artificially impart mathematics to objects by introducing separate mathematical objects whose sole purpose was to make it appear as if the objects and events of nature were mathematical.

In each case, the physicist called the object in for a cameo. Just as the director of a movie, he told the object to "jump into this pool and submerge yourself" or "sit on this stainless steel pan and don't move." The object obeyed his directions. He then added "That's it, you can go now, I don't need you anymore." The action of the cylinder or the scale would stay and be immortalized in the finished product, the subsequent artistic creation of charts and tables and formulas, just as the footage of cellulose or the digital files in a camera were preserved for posterity in the form of a movie.

In experiments such as these, the object was the star of the show and received top billing whereas the set director, stage hands, and choreographer were not even mentioned, even though they were the ones who had been responsible for everything. Nevertheless the object was given all the credit for the performance, in contrast to the fact that it had played an embarrassingly small role in the overall production.

Rather than visualizing the mathematics as being part of the object, in the sense that the object was exhibiting mathematical behaviors and thus was being the cause of this mathematics, physicists needed to visualize the object as being part of the mathematics, in the sense that it was a passive recipient of mathematical assignments. Physicists plugged the object into a pre-established conceptual framework that they had built independently of the object, making the mathematics extraneous, that is, the mathematics existed completely outside the object. The object inherited the mathematics from its mathematical stage set, ultimately from the people who had built the mathematical backdrops and then directed the object to preform stunts and act out scenes for their personal enjoyment.

Once Jonathan had arrived at a number for the volume of a tree, he could not work his way back from his calculation and return to the tree. All sorts of objects had the same volume so he could not deduce the original object from the numerical value. This wasn't a problem here, because Jonathan had the tree at hand and he already knew what it was, but with something like subatomic particles, no one knew what the numbers referred to. When physicists calculated the volume of an atom, Jonathan wanted to know if this volume was as neat and precise as a pile of lumber or as irregular and elusive as a tree? The physicist said that it didn't matter because the number was the same in each case and the number was his only concern. But what was the reality behind the number? For physicists, the reality of the number wasn't the material object that they had given the name 'atom,' but rather the set of ideas that had been used to generate the number. The reality of the actual atom was another matter altogether. The atom—along with the rest of the world—could never be constructed out of numerical quantities.

Still quantities seemed to be indisputable facts. No one could deny that a large glass held more water as a small glass. A quantity of water was present in each case and the quantity was different with different sized glasses, but the quantity was explicit only because the glasses were mathematical objects, geometrical forms that defined the quantities precisely. Out in the real world the situation was very different. The quantity of water in a lake was much more difficult to define, not just because of its highly irregular shape which alone made accurate calculations impossible, but also because of a number of complicating circumstances. The area occupied by the lake was not pure water, but instead, at least during the summer months, it was a mixture of plants and water. Algae blooms and pondweed, part of an assortment of floating, submerged and emergent species, occupied a significant portion of the lake. These areas would have to be subtracted in order to arrive at the water itself. A thick layer of muck lay at the bottom of the lake, so should calculations of extent be based on the top, bottom, or middle of this amorphous zone? Over many years water from the lake had percolated deep into the ground, and since this water was continuous with the surface water and a direct consequence of its presence, shouldn't this water be considered as part of the lake? Water was constantly being added via underground springs and feeder creeks and disappearing through hidden leakages and surface evaporation, robbing the notion of quantity of much of its significance, particularly as an essential characteristic of the lake. The general belief was that the lake did have an exact quantity of water, only the number was difficult to calculate and ultimately impossible to determine with any precision, but that simply wasn't true. Everyone simply envisioned the lake in the same way that they envisioned the glass, as a vessel that held a precise amount of liquid, even though the lake was nothing at all like the glass. Quantities were intimately bound up with mathematical objects and without them the concepts of quantities disintegrated into rather vague and indeterminate forms, proving that the notion of a simple quantity did not reflect the intricacies and realities of the world. The idea of reality being made up of definite quantities seemed so legitimate in the laboratory, but once outside in the larger world, the concepts lost their footing and slipped away into a morass of incidentals, becoming rather meaningless and inappropriate concepts for understanding anything. The notion of 'quantity' took its primary meaning from mathematics, yet it was then loosely extended to include all sorts of general comparisons whereby it became a fundamental aspect of all things, even though it was frequently ambiguous and confusing. Boundaries were often hazy and imprecise or altogether nonexistent. The atmosphere was a good example. Without clearcut demarcations, quantities were difficult to establish and calculations promptly lapsed into estimations based largely on guesswork and highly questionable assumptions.

The physicist explained that the volume of a tree, unlike the water in a lake, could be defined precisely. The volume was the quantity of space enclosed by the outer surface of the tree. Yet when Jonathan looked closely at the structure of the bark and the leaves, he saw that the epidermal coverings did not form a smooth and continuous surface. The trunk of the tree was a mosaic of ligneous wafers, placed side by side and layered on top of one another. When he pried some of these off, he discovered a loose network of interwoven fibers. There were gaps and openings between all of these distinct parts and some of these openings led to the interior of the tree. The idea of a mathematical surface broke apart and became a mere phantom as Jonathan zoomed in on the reality of the tree. The tree was a conglomerate of subunits that were merely stuck together, a patchwork of independent parts lacking the uniformity and continuity of a mathematical surface. The leaves similarly had various structures affixed to them, minute hairs, nodules and scales, along with stomata, tiny openings that allowed air to enter the interior of the leaf and moisture to exit. These ornaments and tunnels needed to be included in any calculations of surface area.

The tree wasn't a surface enclosing an interior, so it didn't really have a volume—it just looked as if it might—and therefore the mathematical concept of volume did not reveal an essential truth about the tree. Picturing the tree as a geometrical form took Jonathan away from the hard reality of a living organism, a reality that was very different from what he had studied in math class. Lines, forces, sets, spheres and volumes were all mental images and mathematics was all a matter of visualization, of not seeing things as they really were. Nothing in the real world lived up to these idealizations yet strangely this was not considered to be a problem. As long as these imaginations spawned a mathematical universe, they were sanctioned and even lauded. As long as they permitted physicists to spin their fantastic tales of fictitious entities, that was all that was required of them. Physicists didn't actually want to understand the world. They just wanted to create more mathematics.

Jonathan could still immerse the tree in a tank of water and generate a number. He could then assign that number to the volume without worrying about the fact that the tree might not even have a volume—at least not in a rigorous mathematical sense. The tree still displaced water and he could let the technique dictate the results. Physicists could also get numbers for atoms by making similar assignments, but in all cases the realities of these objects were left behind. Was there any sense in which an atom actually had a volume? Probably not.

Envisioning atomic and subatomic particles as concrete objects of great complexity, as objects possessing the infinite particularities that were shared by trees and rocks, rather than as terms in mathematical expressions, represented a radical change in thinking. By applying the ideas of quantification to terrestrial objects, Jonathan could evaluate this program on the basis of what he knew about the world he lived in and perhaps gain new insights into the scientific method. He could find out exactly what was going on with this so-called mathematization of reality.

The desert granted Jonathan a kind of knowledge that was impossible for the atomic physicist. The physicist couldn't compare the results of his mathematization to anything— they were just blind results. The advantage Jonathan had in dealing with terrestrial objects was their availability, and by having them at hand, he could examine them closely and compare their realities to the mathematical forms that he generated, thus he had an exclusive standard he could use to evaluate the constructions and derivations, allowing him to put them into perspective and see what they really were.

This was something that was never done in atomic physics because it was impossible. Physicists couldn't bring the actual objects front and center for inspection; they could not have them testify in person before a court of judges. Physicists could not see the objects directly or know them on a personal level and therefore they had to rely entirely on circumstantial evidence—really just hearsay and secondhand gossip, rumors that could not be traced back to their original sources.

But Jonathan had a unique way to approach the mathematization of the world. By using the same techniques on things that he already knew, the world of his everyday experience, he viewed the hidden realms of atomic physics as direct extensions of the realities of this world, a continuation of this world into smaller dimensions. Instead of supposing a radical transformation in the nature of things and essentially turning the everyday world into its opposite, a world of details into a world of abstractions, Jonathan tried to put the microscopic and macroscopic worlds together and meld them into one. He reasoned from his world to the subatomic world instead of the other way around, where physicists tried to construct the subatomic world out of abstractions and then create a mental path back to where he was now. In Jonathan's view, the macroscopic world was not based on the submicroscopic world, but rather the submicroscopic world was based on the truths of the macroscopic world.

Jonathan had always gotten the impression from physics textbooks that mass was the cause of gravity, but one measured quantity could never be the cause of another measured quantity, just as the mass of a tree could not be the cause of its volume. Certain metabolic processes within the cells of the tree accumulated the substances that had been taken up by the roots, selectively transporting them across cell membranes and storing them in vacuoles, thus adding to the weight of the tree. Certain organelles manufactured proteins and other organic compounds and then assembled these materials into new cells, arranging them as necessary to fulfill specific functions. These additional plant structures took up space, thus increasing the volume of the plant.

The assimilation of substances and the growth of new tissue depended on different sets of cellular functions and the connections between them were confusing or anything but straightforward. In terms of the specifics, the mass and the volume of the tree each had different underlying causes. Jonathan could say that the uptake of mineral salts eventually led to the actions of enzymes, and these actions ultimately led to the formation of cell walls, but the intricate chains of causality all moved in different directions and they passed through many intermediaries that contained plenty of circularities and cross-references. He could bulldoze his way over all the irregularities and bury the ornate structures that nature had created, then say that one quantity caused the other quantity, but if the quantities didn't exist apart from the cellular activities, then what was he talking about? Not being independent entities, they could not be the cause of anything. Mass and gravitational force could also be put together as variables in a single equation, but what were the underlying causes in this case? No one knew. Physicists simply started with the mathematics.

The carnival of subatomic events occurring within material substances were presumed to be simpler than the metabolic processes in trees, but since they were much more difficult to observe, this assessment of affairs might just be wishful thinking. Temperature was typically portrayed as some sort of vibration, but exactly what was vibrating was unclear, and it might be different for different objects under different circumstances. Pressure was usually taken as the kinetic energy of molecular collisions. Even in broad outline, temperature and pressure were at the very least two separate aspects of atomic processes, just as mass and volume were two separate aspects of the metabolic processes of a tree.

A chemist could put a flask over a bunsen burner and alter the reading of a thermometer he had inserted into the fluid, so he could draw a line between the two events and skip over the wealth of particular atomic events, but what he was doing on his workbench was more complicated than he could ever imagine. He attached labels to the unfathomable intricacies of the underlying realities and replaced the specifics of particular atoms interacting with other particular atoms in ways that were completely unknown with vague summations and the simplistic relationships between empty abstractions. His basic action had set in motion a world of activity that he could not comprehend. The exact details of all that occurred could not be summed up in a few words or captured by a string of measurements or covered by a theoretical system of ideas. The derived quantities vaguely pointed to these underlying realities, but the numbers were only fuzzy references to a crystal clear abundance of details. The real causes remained hidden in the tangle of causality that could not be unraveled, the infinitesimal facts of atomic life locked inside the substances. Jonathan could easily imagine atoms bouncing off each other as if they were rubber balls until he imagined electrons spinning as if they were planets orbiting the sun and then he realized that these were just ideas in his head and could not possibly be what was really going on.

So long ago, physicists made the wise decision to take causality out of the physical world where it was a mess, a tangle of incomprehensible connections that could not be traced or even properly catalogued, and place it in the realm of ideas where it was neat and precise, but now they had the strange situation of one idea causing another idea to exist. A change in "temperature" could cause a change in "pressure." Jonathan understood one idea being derived from another idea through a process of logical deduction, or one idea being contained within another idea and implied by it as a corollary, because these were the relationships between the ideas themselves. He grasped the necessity of rational inferences, but causality was something else altogether. The causality of ideas was no longer the causality of the world, but a causality within the mind of the physicist. The physicist could say: "Since I'm just going to play with concepts, I'll forget about the world. All that really mattered was placing one concept after another and asserting that the second was caused by the first, even though this told me little or nothing about actual events."

Jonathan understood the reasoning behind this move. Relationships between clear and well-defined ideas were so much better than the chaotic and complicated interactions of unknowable agents and objects that were well beyond the limits of observation. The truth of the submicroscopic world was mind-boggling, the facts that physicists needed to know were unavailable, the parties responsible for what they saw were inaccessible, detailed knowledge of this tiny world was impossible. A rational person would not hesitate for a moment to trade it in for something better, a nice set of logical propositions and simple connections. So they transformed material causality into something entirely different, a relationship between ideas. That was what everyone studied in books, that was how everyone thought about things in their heads, that was the way the world really worked— wasn't it? Physicists had taken an ugly morass of incidentals and they turned it into an elegant chain of inferences with all the beauty of mathematics. Who could argue with that? Jonathan was sorry to say that he had some bad news. That was not the way the world worked at all. That was just a fantasy, a cumbersome encyclopedia of convenient fictions, a crazy smoke-filled pipe dream. The idea of causality was merely a pleasant narcotic, but Jonathan didn't want to live in the mind-numbing stupor of unrealistic simplicities.

The real problem was that the atoms in books were just ideas in the minds of physicists. Atoms were not like the rocks and clouds and trees of the desert, the concrete items of this world with all their staggering complexities. The transformation of atoms into arrays of numbers glossed over the concrete natures of these objects and made cartoons out of them. The descriptions of atoms in physics textbooks had always made Jonathan laugh out loud, and for him the jokes had never gotten old.

People replaced realities with concepts all the time, so the mistake was perfectly understandable—but not for someone who held accuracy and precision in high regard. Yesterday Jonathan became furious and lost his composure by throwing a rock through the garage window. He could say that "my anger broke the window," but what was he talking about? He must first define anger. Anger was an emotion he experienced in his consciousness. What was the physical reality that corresponded to this emotion? The question was not easily answered. Patterns of chemical and electrical activities in certain regions of his brain—but these too were vague concepts. The truth was that each particular neuron in his brain touched other neurons in ways that only they could know. This contact was not the result of some general wiring diagram of the brain, but a specific reaching out of one neuron to another, a need to touch others and be touched by them in return. The staggering number of electrochemical pathways within each neuron induced

138

cascades of similar electrochemical events in all of its immediate friends and contacts, the individuals with whom it associated intimately. Influences spread from point to point within the living tissues, following avenues of social interactions. Steady rhythms arose in the corresponding electrical potentials. They changed frequency then dissipated. These individual neurons did not have names, their unique personalities were never identified, the exact natures of their encounters with each other remained unknown.

Jonathan's physical outburst had been an equally complex harmony of individual actions, the coordination of independent muscle contractions, an unbelievable number of fibers all moving in unison, each made possible with the help of a huge network of supporting cells. The exact responses were dictated by the variations in the bursts of electrical impulses that were occurring concomitantly within the nerves. Other networks of neurons played their necessary roles. Somehow Jonathan had used the patterns of light and dark occurring on the living tissues of his retinas to identify a rock lying on the ground. He took these flat images and determined the rock's spatial relationship to him, then directed his hand towards it by continuously comparing the sensation of his hand to a visual construction provided by his eyes. He imagined an infinite number of possible trajectories suitable for achieving his goal and selected one out of the many choices, then propelled the rock toward the target, instructing his muscles to act according to his visceral estimations.

Sarah had heard the crash and came running out to the garage to see if Jonathan was alright. She saw the broken window and asked him what happened. He told her the whole story without making anything up. He had nothing to hide. He could say that his account was an accurate description of the events that had transpired, only he never mentioned any of the events that had transpired. He gave her a handful of bland concepts and she was satisfied with that. The description he gave her was based on his understanding of things, a way of thinking about the world that they both shared, and not based on the intricacies of physical reality. His words only made general references to that reality, widely accepted names given to objects that were actually unfathomable collections of much smaller objects all interacting with each other in incredible ways. So how could a description that was so vague be considered accurate? He was saying: "This was not 'sort of' what happened, this was 'exactly' what happened." But where was the precision in his account? He was not just glossing over the particulars, he was completely ignoring them. He was explicitly denying the truth of existence. Yet he could have lied and fabricated a story about a hoodlum and an attempted burglary and lost the few connections to reality that he already had. Or he could have just been honest and told Sarah that he had no idea what happened, that he was utterly baffled by physical reality and the whole thing was a mystery to him.

Untold components had participated in the elaborate production, but the concepts of "anger" and "outburst" were nowhere to be found, nor did these concepts point to the actual microscopic or macroscopic events that had taken place. The truth of the situation did not resemble a connection between ideas. So what did one idea causing another idea mean in terms of the underlying physical reality? The references of the words vanished into the twisted and intertwined clutter of particulars. What kind of existence did the concepts have outside of being thoughts in Jonathan's mind? One thing he knew for certain. They did not have the same kind of existence that physical objects had. Nature existed—but where concepts fit in was anybody's guess.

Jonathan could see the physicist impatiently tapping his foot, wondering when Jonathan was going to move on to more interesting topics. Everybody knew that ideas were generalizations and that they were full of abstractions—so what was the problem? There was no problem, until the physicist tried to create an accurate account of physical reality. He asked the physicist: "How exactly are you going to accomplish this task?" The physicist replied: "With ideas." Jonathan said: " Aha! But if you look carefully at ideation and consider for a moment what ideas are, you can see that ideas by their very nature prevent any kind of accurate rendition of the vast intricacies of the real world." The physicist retorted: "I know. But mathematical ideas are different. I can do it with mathematical ideas."

At this point a light came on in Jonathan's head. Ideas that were plainly obstacles to understanding could easily be turned into vehicles for understanding through a simple ruse. All the physicist had to do was imagine a conceptual universe, thereby making mathematical relationships the ultimate realities of the universe. The physicist didn't have to actually explain anything in the real world. He could just place all of that outside the loop and follow the narrow track of laboratory experiments which were all based on ideas. He could start with mathematical ideas and then create a place for the world to exist within this conceptual framework, carving out a subsidiary role for objects and events to play within his overarching scheme of mathematical definitions. He never had to step outside the world of ideas. Everything was ideas—including reality itself. The entire universe was nothing but ideas—wonderful, glorious, mathematical ideas, all waiting to be thought.

Everyone was so enraptured by their thoughts and ideas that they became easily convinced by their explanations. Jonathan said: "I threw a rock through the window because I was angry." This made perfect sense to him and he felt that he understood exactly what had happened, but when he looked at the event closely, in terms of the actual sequences and the infinite points of contact between the mind-boggling number of participants, he saw that he couldn't possibly comprehend the truth. The way the world accomplished this feat—his thought shattering a sheet of glass on the other side of the room—was fundamentally different from the way that Jonathan's concepts were linked together in his head. His anger was only able to break the window through an unfathomable network of all kinds of interconnections. On the surface, the action had seemed so simple, as if there had been nothing to it, yet when he took it apart in his mind, he saw that the phenomenon was essentially an ornate sculpture made out of sand, dissolving into the constituent grains the moment he touched it. He found that he could not account for each tiny crystal of silica or trace its journey or chronicle its fate. The statement "My anger caused the window to break" didn't really explain anything. He was completely in the dark about what had actually happened.

In everyday life, no one needed to know the truth about such things. Concepts worked perfectly well from a practical standpoint and in fact the truth would only have gotten in the way, causing much confusion and creating endless distractions. But when Jonathan had briefly enlisted in the ranks of physicists, he thought that they were all going for the truth. He thought that they had an agreement that this was their common goal. Instead physicists carried forward the mistakes of the general population and turned their eyes away from the true nature of the world, glossing over this fantastic display of infinite detail with concepts that might be precise in a mathematical sense, but imprecise in the sense of maintaining the unique character of the physical world. Physicists were willing to sacrifice the truth for practicality, always qualifying their statements with "for the purpose at hand"—namely for the purpose of making a calculation—they could substitute one thing for another, a schematic rendition for an actual state of affairs. The fact that they lost the essence of the world in the process didn't seem to bother them.

Jonathan roused himself from his musings and returned to his tree. He realized that there were all sorts of things he could measure about it. He counted the number of leaves and tried to determine the surface area of each one, but here again he ran into a problem defining surface area. He measured the thickness of the bark and traced its decline from the base of the tree toward the outermost shoots. He measured the bark's porosity and density. He invented numerical scales to identify the color and luster of the bark, the shape and texture of the leaves, the flow of sap in the xylem, the molecular composition of the pith. He saw that there was no limit to what he could quantify, and gradually he incorporated more and more aspects of the tree into his collection of facts. He measured the concentrations of hormones in the phloem, the rates of transport of electrolytes across cellular membranes, the action potentials of chloroplasts, the chemical responses of apical meristems to the length of the day.

If the possibilities for measurement were virtually endless, then Jonathan had to admit that the tree could not be measured. Each measurement devoted itself to something else, something that was not the tree but instead a peculiar pattern of cells or a cycle of chemical reactions, an interaction of specific compounds with sunlight, or a quantification of the weight and dimension of entities that bore no resemblance the tree. In measuring the tree, Jonathan was forced to measure other things, things that could be measured, striking off in new directions as he left the tree behind. Since all these measurements depended on definitions, the numbers only had meaning in a world of definitions.

Nevertheless, Jonathan had collected an enormous amount of data. His extensive scribbles and doodles had defaced sheaves of paper, which he then crammed into an old briefcase. He now carried the tree around with him, however, the tree had been radically altered by its conversion into numbers. No one recognized it anymore. Every night Jonathan took the tree home with him and studied it at his leisure, the paperwork spread out on his desk as he sipped cups of tea in the quiet evening hours. The actual tree was many miles away standing on the edge of a broad mesa. Whenever someone came into his library and asked him what those piles of papers on his desk were, he replied: "Oh, this is a magnificent tree that Emma and I found out in the desert on one of our walks." He wouldn't have recognized the tree either if it hadn't been for the fact that he had labeled each page. Jonathan spent a lot of time around trees, but the numbers he generated did not have arboreal qualities to them, and the equations were definitely not tree-like at all. When he looked down at the sheets of paper, he no longer saw the tree. Although he had tried to hold onto it, the tree had mysteriously vanished. He closed his eyes and saw its black silhouette standing against the last light of sunset, its arms waving gently in the breeze as he said goodbye, promising to return at the first convenient opportunity.

Street artists in tourist districts often drew portraits for a fee which were humorous caricatures of the subjects. The interesting thing about these renditions was that anyone familiar with the subjects would immediately recognize them in the portraits even though many of their features had been radically altered—a tribute to the talents of these artists who were able to maintain certain relationships so as to preserve the identities of the originals. Another artist could just as easily have come along and introduced relatively minor discrepancies that would have rendered the faces completely unrecognizable.

Jonathan sat at his desk and drew a picture of a tree using nothing but numbers and geometrical shapes, but if he were a street artist standing on a corner and used a similar technique, he could imagine the response he would get. He closed his eyes for a moment and fantasized about being such an artist. He saw himself hard at work when a tourist spied him from across the street and decided to come over to judge his talents and abilities. The woman faltered as she approached the canvas, obviously confused.

"I'm sorry. I thought you were drawing a portrait of that person sitting across from you in that chair, but this isn't a drawing of that person or any other person, just a mumble jumble of numbers and geometrical shapes."

He immediately replied: "No, you don't understand. These circles represent the person's eyes. I know the eyes have lids and lashes and the openings are more almond-shaped than round, but I have substituted perfect circles for the sake of simplicity, as well as for the sake of mathematical beauty. The nose is over here and it is shaped like a right triangle, not exactly its true form, but one of the shapes of geometry. And all these straight lines represent the strands of hair on the person's head. I made them nice and parallel to each other because I like order and regularity rather than disarray. I used geometrical figures wherever I could, and rearranged all the features according to the demands of my imagination, a personal logic that you're undoubtedly unfamiliar with. I overlaid the face with a matrix of numbers. You can't see the numbers when you look at the face, but I can assure you that they are there nonetheless. I know this with absolute certainty, but I'm the only person who can actually see them. So contrary to all appearances, the face that I've drawn is in fact a perfect likeness of that person sitting over there in that chair."

The tourist stood silently for a moment, carefully eyeing the easel from a safe distance, with a look of disappointment on her face. "Oh," she said, and tactfully walked away, regretting that she had ever come over to meet Jonathan in the first place.

The measurements Jonathan had taken of the tree had created another kind of reality, a metaphorical tree, a mathematical analog made up entirely out of concepts instead of bark and leaves and sap. The charts, graphs and formulas were not the actual parts of a tree, but rather they were pictures of mathematical ideas, ideas he'd learned long ago in far away classrooms, places with no real connection to the tree or the desert where he now found himself. He had summarily deleted the content of the tree, erased all the concrete details in order to arrive at a set of pure mathematical forms, then replaced the actual features of the tree with mathematical ideas, assembling an entirely new kind of tree.

Instead of building skeletons out of abstract concepts, expressionless and impersonal stick figures composed entirely of empty circles and bare lines, nature had built sand castles, vast aggregations of tiny particles, fantastic and enormous sculptures of infinite detail. These structures were beautiful and interesting to look at, but they perpetually crumbled away and the infinite details were lost. Jonathan wanted to bring them back to life, exactly as they had been before, but he could not possibly replace every single crystal of silica and duplicate the position and orientation of each grain of sand, yet that was what he had to do in order to rebuild the original—label and locate all the individual pieces and stick them back together, just as nature had done it in the first place. But how did nature do it in the first place? It was the method that he needed to understand.

Finding another way to produce a similar sculpture didn't count. He could compress the sand into conceptual forms and mathematical molds, externally and artificially imposing shapes on the indeterminate mass, but he could not get the sand to do this all by itself. Nature shaped phenomena out of atoms and molecules by causing them to act on their own. The sand gathered and danced then disbanded and dispersed. If he arrived at the same finished product via a different route, he did not achieve an understanding of nature's methods, but only an understanding of his own methods.

When the physicist observed a thunderhead in the sky, he cast the cloud into mental images—forces and velocities and pressures—and saw the cloud as a system of ideas, still he could not motivate the swarms of molecules to execute this particular dance or assemble themselves into such an astounding entity. The physicist could interfere with nature's work and alter it's course, but he could not stand in an empty field under a clear blue sky and summon the appropriate clouds out of the emptiness, then sculpture these clouds into a thunderstorm of such complexity and sublime beauty, exact in every detail as the one Jonathan had previously witnessed. Yet that was what Jonathan was trying to understand: how did nature go about making the world? The physicist might find a way to trigger certain processes and in doing so cause nature to form a thunderstorm at his behest, and then take credit for the event, but he could not actually do the work himself. The details were overwhelming and so many things were happening at once, his meager intellect could never untangle the events and put them all together in the appropriate patterns.

Emma came wandering over to Jonathan looking for a drink of water. He studied her mannerisms and thought to himself that if he could make something like Emma—just take a chunk of carbon and a bucket of mineral water and turn it into such a wondrous creature —he would truly be an artist. Not to trick nature into doing the work for him, but to do it all by himself. Not to meticulously sculpt every hair on her body out of clay or some other material, but to assemble each one from scratch, carefully placing each protein in its proper location, locking the lipids together as if they were the blocks of a Lego set, but instead of a model of the Millennium Falcon or some other ridiculous cultural icon, he ended up with a living being, then he could honestly claim to have created something worthwhile.

Jonathan didn't understand the world and all the mathematics in the world wasn't going to help him. He saw no hope for the future of the scientific method. The absurd strategy had been doomed from the start and the early investigators should have known better and seen what was going on, or perhaps thought about it a little more. His only advice to physicists at this point was to put the chalk down and step away from the blackboard. He wanted to tell them: "Go home and for god's sake stop all that goddamn scribbling. Look around you. The world isn't mathematical, so what do you think you are doing with all those symbols and equations? Can you make yourself a canine companion—or even understand what that is? Will the abstract relationships help you understand the truth when you look into Emma's eyes and peer into the infinite depths of her being? Can they possibly unravel the mysteries of her unfathomable existence? Let's get real about all of this. Mathematics has nothing to do with anything. You would need to inject the same level of creativity into your equations as nature had originally put into the phenomena and that was a tall order, completely out of the range of possibility, and not only by a small margin, but by a degree comparable to an infinite abyss."

Nature's modus operandi was truly unique and it formed the essence of all that existed. The molecules assembled themselves into events in ways that needed to be understood, yet the processes were mysterious. Molecules got pushed around in all directions. They collided with one another. They became agitated. They boiled in an atomic soup containing a staggering number of ingredients, but then the broth rose up and performed stunts, turned itself into an artistic display of sublime beauty, rapidly evolved and multiplied into entities of vengeance that wreaked havoc on human habitations, or sat around and did nothing. Why did molecules get together and make a sunrise? What motivated the molecules to arrange themselves into a fearsome thunderstorm when they could just as well have done something else? Why did they steal Jonathan's hat? Was he supposed to believe that the molecules did all of this without having any idea of what they were doing? Was such a belief a rational supposition? The molecules could not possibly have been working together in harmony, coordinating their efforts toward common goals, but wasn't that what it took to make this world what it was?

How the world accomplished the infinite tasks that Jonathan witnessed every day was an enigma, an unsolved problem. The infinite particularity was impenetrable. Sometimes Jonathan thought that the universe had been created specifically to stifle any attempt to comprehend it, that it had been deliberately set up to make any kind of rational understanding impossible. The problem was that the universe had existed long before human understanding had any meaning, when the very prospect of a living being staring out at the universe through two eyes was nowhere on the horizon of possibility.

Measuring a tree turned out to be a bigger problem than Jonathan had realized, a problem for which he had no practical answer, but perhaps he could replace the trees with ones and then proceed to count them. That would be a lot easier, but it raised new problems of its own. The forest didn't just end—it gradually dwindled away. He would somehow have to establish a boundary where he could say that the forest stopped, or in other words, the density of trees would have to drop below a certain value for him to say that it was gone. He would have to construct a density map and then select a minimum number to qualify as a forest.

A generic tree was just a convenient fiction and the label "one" could refer to all sorts of plants from the tiniest seedlings to the tallest pines. Of course, trees seemed much more straightforward while sitting in an armchair merely contemplating trees, but out in the field it was a different story. Botanists considered cacti to be trees, given that they had wooden trunks hidden beneath their thick fleshy exteriors, and in fact Jonathan had walked through many magnificent cholla forests in his day. The spruce trees growing at high altitudes were concealed by heavy growths of moss that were draped from every branch. These trees were little more than bare racks that existed simply to hold up the mosses. Dead ponderosas became bird houses, apartment complexes where various avian species commonly took advantage of the soft wood to hollow out enclosures for themselves, still they were examples of standing trees. Trees could be many different things, both living or dead, or often a mixture of both, certain parts having died and decayed while other parts continuing to thrive. Jonathan couldn't imagine a tree in his mind if he was only given the number one. A tree became an "x" in an equation, a name without a face, an empty reference. The density map was more a picture of the mathematics than it was a picture of the forest.

In mapping the forest, Jonathan had not transformed it, but replaced it with something else, substituting an object of the intellect for an object of the senses. The forest was not the forest that Jonathan knew, the one he walked through and experienced. The sights and smells that were so familiar to him were put aside and instead he gazed into a pattern of numbers and geometrical forms. Ordinarily Jonathan viewed the forest from many different angles, sometimes lying on his back and looking up at the canopy of leaves and needles, sometimes standing on a ridge and looking down at the treetops from above, sometimes sitting in the shade with his back against a solitary trunk. He became familiar with all the moods of the forest: the glistening of the sunlight on the leaves during the cool and tranquil mornings, the swaying and creaking of the branches when raging thunderstorms swept down from the mountains, the oppressive heat during dry and sunny summer afternoons, the whisper of the winds in the evening. He put all of these perspectives together in his mind and developed an idea of the forest. The physicist did something completely different, arriving at his own idea of the forest, but for Jonathan mathematics was merely a distraction. At some point he had to make a decision: he could spend all day sitting at his desk making calculations, or he could go walking through the woods.

Just as geometrical sample areas could be used to study the distribution of trees in the forest, the geometrical lines drawn on a Monopoly board could be used to study the rolls of the dice, only instead of counting trees within a circle drawn on the ground, Jonathan was counting dots on the die faces. The play of the game was determined primarily by the relationships of the board layout and the definition of each rectangle, but the dice were introduced at certain points in the play for token appearances. Similarly, the results of the density map were determined mostly by the setup of the grid, but as the dice decided the result of each move in Monopoly, trees decided the number at each point on the map. Both games were primarily mathematical configurations that included a connection to the outside world as a minor element in an overarching design.

The dice in Monopoly existed within an established framework of ideas. This magnified their behavior by making the consequences of their behavior more complex. The contestants saw the dice through the veil of individual plays on the board and the dice became the kernel of a larger entity. No one could discount the role the board played in creating the game, and no one could discount the role mathematics played in the physicist's understanding of the forest.

The Monopoly game had been intended to mimic the capitalistic acquisition of wealth and property, events occurring everyday on the floors of stock exchanges and the plush offices of towering skyscrapers in every major city, transactions enacted by real people in real-life situations involving real money. The elaborate trappings of big business were for the most part well beyond the ken of economics professors and market analysts, and far more complex than the simple play of a board game, just as the forest was incomparably more intricate than any mathematical rendition of it.

Jonathan saw that Emma was overheated by the hot sun and all the exertion. They had wandered out of the forest some time ago and now the only shade within sight was a stock tank sitting on the crown of a large hill, so Jonathan set his course for this prominent landmark. As he got closer, he noticed that the tank was historic, a relic from the heyday of early 20th century cattle grazing, constructed from heavy steel sections riveted together with large, mushroom-headed pins, the curved surface adorned with faded patches of yellow and blue paint worn thin by years of intense ultraviolet radiation, now sharing the exterior with encroaching spans of brown rust.

The gentle sound of trickling water told Jonathan that the tank was still in working order. Emma gladly took refuge in the ample shade provided by the large structure and laid down alongside its concrete foundation. Jonathan broke out her dish and supplied it with cold water. A metal ladder leaned up against the giant cylinder and was firmly affixed at the top. Jonathan decided to climb it to get a look at the interior.

The rungs were extremely narrow rods, way too small to adequately serve as steps. The ladder had obviously been fabricated out of materials that had been laying around the ranch and not suited to the purpose for which they were now being used. Placing all his weight on these thin bars was uncomfortable, even with shoes, but they seemed sturdy enough, so he climbed the ten feet to the top.

His head rose above the lip of the container and he was astonished to find that the tank was filled to the brim. His nose was right next to the sparkling water and with a childlike curiosity he swung his arm up over his head and touched the liquid. He found the water to be surprisingly clean and clear. Most stock tanks were foul, clogged with slimy green algae or disgusting clumps of yellowish yuck, typically having a filthy layer of muck on the bottom, but this water looked inviting. The afternoon breeze rippled the surface as if it were a tiny lake. The water was cold, much too cold to consider going for a swim. Jonathan was not really a water person anyway, and there was no platform to access the pool, only the sharp edge of the rim, yet somehow the thought of a refreshing dip in this unorthodox body of water was strangely appealing.

The sound of distant barking captured Jonathan's attention and ended his aquatic fantasies. He looked down at the ground, calculating his descent back to earth. He hadn't been paying attention to Emma and saw that she had disappeared. She apparently cooled off and resumed her hunting and tracking, but he hadn't noticed in which direction she had gone. The barks seemed to originate from somewhere along the base of the hill. Jonathan hurriedly climbed down the stairs, grabbed his pack and Emma's water bowl, then walked in the direction of the commotion, the noise muffled by the intervening hillside. He reached the edge of the embankment and looked down. Emma was busy confronting a herd of eighteen elk. She had apparently intercepted them as they were traversing the valley floor.

The elk were all standing in place, calmly lounging in small groups facing each other, reminding Jonathan of theater goers milling about in the foyer during intermission. The sizable crowd consisted of all ages, from the smallest youngsters to the most imposing adults. Emma was strutting between them looking up at their faces with her head cranked back, marching from one huddle to another, much as a waiter serving glasses of wine and mixed drinks to idle patrons. She stopped in front of a large party and looked at each one in turn, acting out the role of taking their orders, but the elk looked the other way and ignored her, evidently bemused by her friendly demeanor. Emma was frustrated, doing her best to get their attention, but failing completely.

Jonathan whistled. The elk paid no more attention to his call than they had to Emma's behests to bolt, but Emma snapped her head and ran toward him. She welcomed the invitation to withdraw because the command saved her from the embarrassing situation of being politely ignored by mammoth creatures who were supposed to panic and take flight at the first indication of her approach.

Emma disappeared from view as she reached the foot of the hillside and for a long time Jonathan saw no sign of her. She should have easily made it back by now. Jonathan knew what she was doing. She was watching the herd from halfway up the slope and thinking about returning once more and attempting to induce them to run so she could chase them. Jonathan shouted her name hoping to dissuade her from this further action. Emma finally appeared, gradually coming into view as she gained the top of the hill. He turned around and started to follow the course of a narrow saddle leading up to an area of higher ground. From this commanding position he cast his gaze back toward the place where the elk had been standing, but they had all mysteriously disappeared. He looked around searching for them. This was open country, yet he didn't see them anywhere. He wondered how they had gotten away so quickly.

Looking down on the broad expanse of the valley reminded Jonathan of another place he used to frequent, the vast bed of an ancient lake now known as the Plains of San Augustine. The last time he'd been there was several years ago on the day of the summer solstice. He'd set up a secluded camp in the piñon and juniper forest along the southern edge of the plains. The land had a magical quality to it. He would spend hours sitting on the rocks looking down at the grassland below, absorbed in thoughts and meditations, as a deep serenity swept over him, coming from all directions. The tranquility was so intense and powerful that it became hypnotic, and he could still feel it to this day.

The morning had been getting late when he finally left camp, embarking on a long trek toward Mount Withington to the east. Strange things often happened on the solstice when he was alone in a strange and remote place. First the clear blue sky suddenly and inexplicably became populated with a collection of white puffs, widely dispersed over the region. The small clouds all began rising vertically and as they did so each one was replaced by another cloud, identical to the first, each new generation materializing out of the air at the same spots as their predecessors. Each cloud gradually expanded as it rose upward. Jonathan stood in one place and twirled about. He was surrounded by numerous columns of clouds all methodically streaming upward, reminding him of the bubbles rising in a mug of cold beer. He appeared to be standing in the center of a broad expanse of evanescent liquid, the atmosphere everywhere fizzing and frothing with fermentative action. The clouds reached a certain height and then spread out horizontally, rapidly merging together and filling the entire sky, becoming the foamy head in the huge bowl of heavenly beer. He'd lost track of the time, but in a rather short span, by this mysterious process, the sky had changed from perfectly clear to completely cloudy. He had never seen anything like it before, nor had he since.

Turning away from the events taking place above his head, Jonathan continued walking toward his intended destination. He came to a barbed wire fence where the trees thinned out and the land was relatively open. Here the sky seemed incredibly close at hand. In fact, he felt as if he could reach up and touch it. He stopped for a moment and marveled at the sensation, but then pulled himself together and marched off toward the distant mountains. The experience of being up in the sky and mingling with the clouds quickly faded, even though the ground continued climbing as he walked. Fascinated by this strange phenomenon, he turned around to retrace his steps thinking that it must've been an illusion, but when he returned to the fence, there it was again. The effect was uncanny. A strange storm had materialized nearby and he had an excellent view of it.

Raindrops began tapping the ground, but he was transfixed by what he saw. Two clouds had come together, interacting with one another in ways that he could not even begin to understand. One was a spherical orb with the outlines of something inside of it, a crystal ball of immense proportions depicting an indistinct scene inside, some vague structure of intersecting lines. The orb was merging into a tall vertical column, tearing at the translucent cylinder and pulling material away from it. He got the impression that the clouds were plugging into each other and completing some type of circuit, channeling some unknown form of energy from a remote region of outer space. He felt as if he were peering into another dimension, looking at an alternate universe through a portal.

The clouds were engaged in an activity that was plainly beyond the scope of human understanding. Jonathan squinted his eyes and stared keenly at the spectacle taking place before him, but the details were hazy, blurred by the mist and the rain. Important things were happening which few people ever had the privilege to witness, and he could not look away. Forces were at play that he could not fathom, processes that were utterly beyond his comprehension. After peering into this cosmic realm for what seemed like an eternity, he finally gave up trying to understand it. The storm was coming his way. He'd wasted too much time trying to figure out what was happening up in the sky, so he decided to abandon his plans and simply head back to camp. He observed a group of deer along the way. The deer didn't run from him, but stared back in mutual fascination.

The rain steadily increased. Water droplets fanned out from the clouds and plunged into the heavily laden atmosphere, dispersed far and wide by fickle agencies only to coalesce once again into puddles on the ground. The sequences of events were easy to trace and the connections between them were clear. The vapor in the sky turned into water droplets, the droplets fell to earth, the water they contained soaked the ground, the soil became saturated and the excess water accumulated, the puddles formed. The relationship of the puddles to the clouds was rather straightforward.

Despite the illusion of the clouds being so close at hand, Jonathan could not possibly reach the clouds. He had no means of accessing the sky and entering the realm of the clouds, but the puddles were right there at his feet. He could study the puddles all he wanted. Instead of vainly dreaming of floating in the air and consorting with the clouds in order to get to know them better, he could turn away from these enigmatic beings and focus his attention on the puddles. After all, the puddles stemmed directly from the clouds and they were a direct result of the actions taking place within the clouds, therefore the puddles could be a substitute for the clouds. He could make them his objects of study. He could base all of his observations on the puddles and say that he was actually studying the clouds. He could say that he was viewing the clouds through the puddles, just as the physicist could say that he was viewing the infinitesimal and unobservable realities of the submicroscopic would through the macroscopic phenomena of his laboratory apparatuses, phenomena that were directly caused by the atomic and subatomic interactions.

Even though the puddles came squarely from the clouds, Jonathan could not possibly know the clouds by studying the puddles. Puddles were just puddles. They did not reveal the complex structures of the clouds or divulge the intricate interactions occurring within them. Concrete events could not be studied by analyzing other concrete events that were simply the consequences of those events.

Physicists had no access to the ultimate realities of the world, so they studied other things, the complex activities of their instrumentation. These activities produced all sorts of phenomena: the patterns imprinted on a photographic plate, the numbers displayed on a digital readout, the positions of a dial pointer, the tracks in a cloud chamber. They told themselves that they could know the mysterious entities causing these phenomena by analyzing the structures of the phenomena themselves.

Physicists pictured in their minds one idea implying another idea and this viewpoint validated the method of studying causes through their effects, but in terms of the real world, one concrete reality could only produce another concrete reality, so what physicists were actually doing was trying to study clouds by studying puddles. They had fabricated a conceptual universe and based everything on the belief that the world was built out of ideas, however, concrete realities weren't ideas and they could not be treated in the same way. Ideas allowed physicists to go back and forth between them because ideas were connected in ways that concrete realties could never be—through other ideas. The ways that puddles were connected to clouds illustrated a central problem with the scientific method. The reasoning that was so critical to the success of the method was lost in the infinite particularities of the actual physical processes. One concrete reality could not be studied through another concrete reality in the same way that one system of ideas could be deduced from another system of ideas. The procedures were so radically different that neither copied the other and therefore no parallel could be established between them. Mathematical thinking only functioned in a world of mathematical objects and that was not the world that Jonathan observed everyday, nor was it the world of submicroscopic events. It was only the world of the laboratory.

Jonathan's life had been a long chain of events, each one leading up to the next. He had changed much over the years and now when he looked back upon his former selves, they were all strangers to him. Beliefs, routines, and feelings that he had once considered essential to his identity had subsequently died away. Those aspects disappeared and were now long forgotten, while other things had been born to take their places, attributes that became the new Jonathan. No one could deduce who he had been by analyzing who he was now, because many days had passed in the interim and he could not separate himself from what he had done everyday. If he had adopted a different lifestyle, pursued different dreams and embraced other passions, he would have become a different person. No one could have predicted when he was young how things were going to turn out, yet his childhood had been the foundation for everything that followed. Gaining a knowledge of something through its effects and consequences was impossible. Physicists could have figured this out for themselves, but they were too busy making calculations to be concerned with such things, playing games with numbers and the mathematics produced by instrumentation.

The thing about transformations was that a person didn't end up with what he had started out with, yet the mathematical way of thinking suggested that the original conditions were still present in the final product, that not everything had been transformed. Everyone was told not to confuse transformation with the radical nature of creation and annihilation. Transformation was just the alteration of something—the same entity adopting a new form. The physical world, on the other hand, was inundated with births and deaths. When something died, it was gone forever, and when something was born, it was created out of nothing. Of course, some people tried to see birth and death not as beginnings and ends but as transformations in an effort to establish a sense of continuity, in the vain attempt to connect a disconnected universe. Fear of death did not permit these people to accept the notion that transformations could be so radical as to leave nothing untouched, that they could leap across the abyss between all and nothing. Yet apparently that was what happened.

When Jonathan thought about his own death, he did not find comfort in the fact that his mass and energy would persist, that both would be unchanged in magnitude and that in this sense he would continue to exist as before. Was he really supposed to console himself with the notion that these empty abstractions were eternal and therefore he would live forever as the values in a mathematical equation? Given the particulars of his concrete existence, nothing was conserved and no aspect of his being remained unchanged after his death. Where he had once stood on promontories overlooking spectacular river valleys, no one stood anymore. Nothing was the same.

Transformations were what mathematicians did with equations. They canceled terms, replaced expressions with new symbols, combined factors, but they still had the same equality that they had started out with. The equal sign preserved a fundamental relationship, maintained a constancy that all the manipulations of the symbols did not abrogate. Of course, a mathematician could always pick up the eraser and wipe the blackboard clean, then start over with a different equality. The world, on the other hand, did not undergo transformations, although people often talked as if it did. For example, the farm that Jonathan had once known—his grandparent's farm—was long gone. New people whom he'd never met had taken over the place, razed the house and presumably all the barns and outbuildings, and erected their own structures.

149

He could summarize these changes by saying that the old farm had been transformed into what it was today. The farm lost its former character and a new ambiance was gradually developed. He'd never gone back there so he had no idea what the place looked like now, but that didn't matter. The farm steadily morphed into something else. He could say that the parcel of land was the one common denominator in the transformation, the one thing that connected the old with the new, the central factor corresponding to the equals sign in the mathematician's equation, however, this was not an accurate description of what had happened. What actually happened was that the old farm ceased to exist. It died, disappeared, vanished forever. It did not somehow turn into the current homestead. Instead, the present dwellings, the atmospheres, the meanings of each day, were all created out of nothing. They were not fabricated out of the old materials, based upon the original designs, or built in the spirit of his grandparent's initial endeavor. The old farm passed away and a new place was born to take its place. The property that tied all of this together was not the equal sign in the mathematician's equation, but the blackboard on which the mathematician had written his equation. The new owners had merely picked up the eraser and wiped the slate clean. But even if they had only remodeled the old farmhouse, upgraded certain features and put additions onto what had previously existed, the old farm that Jonathan had known would still be utterly gone. The traits that he had so loved and cherished were now only memories. Many of the people with whom he had shared his summers were now dead and the rest had moved to places he did not know. They had all changed and they were not the same people he had once known.

They all used to play baseball games in the yard, sit around campfires in the evening telling stories, orchestrate barbecues and congregate at long tables adorned with elaborate banquets, ride their bicycles to the lake to lounge on the sandy beach and swim in the water, jog together down the road to the railroad tracks a mile away to watch the freight trains rumble along. He didn't know how the new people lived or what they did with their time, but the activities of Jonathan's relatives during those wonderful summer months were not transformed into the behaviors of these new people. The world did not undergo transformations. This way of looking at things was only a product of mathematical thinking. In the real world, nothing equaled anything else because everything was based on its own existence.

Thinking in terms of outlines—bare patterns and blank forms—engendered the concept of transformation. One design could be followed by another design and the two designs could be seen as being connected, however, thinking in terms of concrete realities, in the terms of the fullness of being and the infinite particularity of the world, showed that existence could only be created and destroyed. Existence could not be transformed. Since empty forms had no existence, they could flow into one another, morph and mold into new shapes, become a plastic reality. The only reason why physicists could say that energy was not created or destroyed was because it didn't exist. Energy was just a calculation, a pattern, an imaginary structure. The fact that the same energy could take on different forms proved that it didn't exist. Material reality, on the other hand, could only exist or not exist. There were no other options, no intermediaries, no shades or degrees, but it was these qualities that were absolutely necessary for transformations to occur. The very nature of existence did not allow for transformations. Existence was an all-or-nothing affair. Existence could not slowly fade away or gradually come into being. In the real world, everything was created and destroyed and done so constantly and irrevocably.

Jonathan's life on the farm had consisted of many things, each of which once existed. These things could have been taken away one at a time and thus his former life could have dissolved piecemeal. In general outline at least, this change represented a gradual transformation because his life was not a solid block or indivisible unity but a conglomeration of individual elements each of which existed in its own right. One person or several people could have dropped out of the summer scene but the rest would have carried on without them. Instead, what actually happened was that the whole scene fell apart all of a sudden because many elements withdrew in a short span of time and the rest disappeared directly thereafter as a result.

Existence wasn't homogeneous, but infinitely nuanced and detailed. The details were connected only by the fact that they coexisted together. Details could appear to be related because the ideas people used to view these details were interconnected, that is, people related the details through their ideas and not through any actual connections in the world. The transformations they observed occurred through the appearances of things. What was really happening was that details ceased to exist and others came into existence and these were the replacements of nothingness with being and being with nothingness. This wasn't the transformation of one into the other or the substitution of one for the other, because substitution implied a fixed role that different entities took turns fulfilling. One detail was succeeded by a completely different detail over the course of time. Since the details were not intrinsically connected to one another, they could appear and disappear independently of each other and this created the appearance of a gradual change or in other words a transformation, but in fact everyone was dealing with the all-or-nothing character of existence operating through an infinite particularity. When the world was viewed merely as appearances—and this was the world that was presented to everyone by mathematical thinking—change was transformational and rather superficial, but when the world was viewed existentially, change was monumental and rather abrupt.

Jonathan remembered the view from the second story bedroom window of the farm, looking out across the roof of the back porch and down onto a small backyard full of trees and flowers, the narrow hallway leading downstairs then abruptly turning left when it ran into an exterior wall, the worn rubber footpads crudely tacked to each step, the linoleum on the floors slightly curled around the edges where it met the walls, his mother and grandmother talking softly in the kitchen speaking the language of the old county that he did not understand, the smell of a freshly killed chicken being plucked by pouring boiling water over the carcass, the kitchen table centered in front of a tall window and covered with a red and white checkered cotton tablecloth that matched the curtains hanging directly behind it. These were the sensory patterns he remembered, the sights, smells and sounds of the farm. He could not sum up the farm in a few words. He could not write a book where he could even begin to describe it—an encyclopedia would barely be a start— and there were many things he didn't even know about the farm. He could never catalogue all the wormholes in the apple trees or every termite embedded in the logs of the milk barn. There was so much to note about the farm and he had no way to encompass it all.

The farm had not been the result of natural forces. It had not come into existence spontaneously of its own accord. The farm was a huge collection of deliberately fabricated objects, each one designed to complement the others and work together in harmony. Jonathan's grandfather had assembled the birdhouses in his workshop, dug the postholes for the fences, cut the cedar posts and strung the wires, built each outbuilding with a definite purpose in mind, purchased tack for the horses then hung the halters and harnesses from nails in the barn, hired a neighbor to mow the fields and bale the hay then stacked the bundles in the loft directly above the cattle stanchions. The farm was based on ideas that Jonathan's grandfather had entertained in his head. Where he had gotten these ideas was a long story, dating all the way back to his father and all the other prominent figures in his childhood. Perhaps he had gotten some of his ideas from books, but the ideas came first and material reality was subsequently made to conform to them. Jonathan could say that mental patterns had caused the objects to exist, but in nature the relationship was reversed. The mental patterns came last. Material reality allowed people to see things, create patterns in their minds. The patterns didn't exist on their own and thus they could not be the cause of anything. Nature didn't even see the patterns—only people saw the patterns. In mathematical thinking, it was the patterns that did things, it was the patterns that dominated the situation. The details were only the constituent parts of these overarching forms and thus they played subsidiary roles defined and determined by the patterns themselves. In nature, a pattern was merely a postscript, an aftereffect resulting from the interaction of the details. The details came first and they were the only realities. The pattern wasn't an essential part of anything and it formed the basis of nothing. The pattern itself had no means of accomplishing anything. The world functioned solely through the plethora of details, each detail causing other details to exist or not exist. A pattern was nothing more than the fleeting appearance of an infinite and incomprehensible swirl in an endless swarm of pinpoints, the hypnotizing reflection of a primordial existence that was not based on any pattern.

Physicists saw patterns that were different from Jonathan what saw, the shapes and forms of mathematical relationships rather than clusters of sensory impressions. Still these patterns were mere phantoms when compared to the substance of brute existence. This was no more clearly evidenced than in Jonathan's death. Physicists could preserve only two aspects of Jonathan's passing away, two ideas they could maintain throughout the transformation, two flimsy skeletons to which they could cling, two bare patterns they could match together. Everything else was gone. What they had here was trivial, yet they took these meager forms and elevated them, turned them into events of the utmost importance. Why? Because the equations were the mathematical laws that issued directly from the supreme authority that wielded dominion over the entire universe. Physicists claimed that they weren't acting in accordance with their own exaggerated ideas of self-importance. No, physicists were acting on behalf of something they saw as being greater than themselves, something that was not of this world. They were merely agents of a higher power, emissaries of a new way of thinking. They had devoted themselves to a preeminent deity who ruled the universe, and based on their unshakable faith and conviction, they sought to proselytize the entire world. Much as Christian missionaries firmly entrenched in their beliefs and perspectives, they intended to drag the ignorant and superstitious masses out of the darkness and into the light of reason so that all people could behold the divine glory of mathematics for themselves. Was this a realistic appraisal of the true nature and power of mathematics? The physicists certainly thought so, so off they went on their path of discovery, their holy crusade for the acquisition of truth. But if they had only mathematical truths, truths that were empty and hollow and not the substantive truths of this world, then they were imposing doctrines without just cause. They were inducing people to believe in things that might be mathematically correct, but irrelevant in the big picture, things that were utterly meaningless when placed alongside the fullness of being.

Nevertheless many people considered conservation laws to rank among the most sparkling jewels of modern physics. Jonathan never understood why these laws were so important given the nature of physical existence. Physicists ignored the bulk of substantive details in order to latch onto minuscule tidbits of precious mathematical reasonings. Could the trifling statements put forward by conservation laws overshadow the realities of what was actually happening in the world around them? Could these empty abstractions deny the truths of this world? In their minds, a smidgen of logical formalism was far superior to an infinite mass of actual existence. In the manner of many religious thinkers, physicists sought to transcend corruption and death and find solace in the fact that the most crucial aspects of the universe were eternal and indestructible.

Mass and energy would never die, but unfortunately, these mental fabrications didn't exist. They were mere calculations, formal relationships harbored in the minds of certain individuals. Physicists imagined to themselves that the mass and energy of Jonathan's body would still be out there somewhere after his death, broken apart and intermingling with the fragmented masses and energies of countless other bodies, a number of objects so large that it could not be given a name. The ideas behind these quantities were monolithic and incorruptible, nothing at all like the objects and events themselves, things that sprang into being then vanished without a trace. According to physicists, after Jonathan's demise the materials and agencies that had once been him would be scattered far and wide. In doing so they would become unidentifiable and undetectable and this was how Jonathan would live on—as nothing. Mathematical thinking created weird perspectives, truths that were untruths, truths that were obtained by ignoring what existed and then pretending that all sorts of imaginary constructions existed. The truths of this world were very different. The characteristics that had once been Jonathan would not be scattered. They would simply disappear.

Everything in the universe was constantly being created and destroyed and the only things that were able to escape this fate were abstractions. Magnitudes did not capture the intricate realities of the world, but introduced something else, an artificial reality made up of mathematical definitions. Quantities required mathematical objects to define them and Jonathan was anything but a mathematical object. How the calculations of mass and energy should proceed in his case weren't clear. Should the water in his stomach and the urine in his bladder be considered part of his mass? Did the hair in his brush this morning, the dead skin scraped away by the loofah, and the nail clippings that were flushed down the toilet, all figure into the calculation of his mass and therefore need to be subtracted from it? Was Jonathan's energy the sum total of the energies of the individual molecular bonds in his tissues, the energies of the electrons orbiting the nuclei, the heat generated by a wide array of metabolic chemical reactions, or the gain in potential energy as he climbed a hill? All sorts of calculations were possible. In truth Jonathan didn't really have mass and energy.

At first glance, the calculations of mass and energy came out much better when applied to mathematical objects. Newton's Cradle seemed to be absolute proof that the conservation of momentum and energy were realities of the universe, but when the details of the motions were examined more closely, the action of this device was rather complicated and the mathematics a lot messier. There were slight motions that might not be noticed at first sight. Numerous factors were responsible for these aberrations: the elasticity of the balls allowing them to compress, the surface areas at the points of contact, the precise spacings between the balls since they couldn't actually touch one another, the pressure waves being transmitted and reflected from one end to the other, the resistance of the air and the stresses placed on the wires, the heat generated by mechanical deformations. Many of these aspects were complicated subjects in themselves. Despite the fact that simple and precise mathematical forms had been carefully imparted to every feature of the design, the phenomenon still failed to exhibit simple and precise motions. Nice try, but what happened when the well ordered and highly accurate mathematical forms of the setup itself were removed? Pandemonium and chaos resulted, and all was lost. How could anyone say that the natural world obeyed the laws of physics? Even the behaviors of mathematical objects were difficult to control and hard to predict, as Jonathan had learned many years ago in grade school. Now he had to plow through several pages of explanations just to grasp some of the processes taking place in the rocking motion of desktop toy. Why didn't more people see that physics was a sham? The physicists' attempts to create a false world of precise mathematical relationships always fell short of perfection, and quite often failed altogether.

Jonathan saw that an important conservation law was missing from physics and it desperately needed to be added to those that had already been established, but it was a law of a different sort. What exactly was conserved under the transformation of physical realities into mathematical forms? Were the preserved qualities insignificant aspects of the world in the same way that mass and energy were insignificant aspects of Jonathan's death? Jonathan knew that the physicist was willing to sacrifice everything for the sake of a beautiful formula, but he could not so easily throw the world away like that. In physics, the move from the real world to a system of mathematical formulations wasn't really a transformation at all, but an existential leap where the real world effectively ceased to exist and a mathematical world emerged out of the dark recesses of the human imagination to take its place.

Night fell as Jonathan and Emma lingered up on a hilltop watching the sky. Two massive storm cells lumbered across the western horizon, the second accompanying the first on its slow trek to the north. The thunderheads were each many miles across. The leading storm released a collection of lightning bolts originating from diverse points, the initial bolt apparently triggering the others, always repeating the same specific pattern and sequence. Jonathan thought that the first discharge should have relieved the tension in the clouds and reduced the static buildup of opposite charges, thus making subsequent discharges less likely, at least until the potential differences were restored, but the opposite held true. The first lightning bolt caused more to follow in short order, not together at one point where a connection between them might be plausible, but widely distributed across the dark and inscrutable mass. Several vertical shots, two near the trailing edge, one in the middle, and one along the leading edge, drew a fixed design of brilliant arcs in the sky, followed by a long horizontal streamer which hooked downward at the end, finishing its long flight with a belated hit to the ground. This exact pattern recurred with rhythmic periodicity, its location transposed as the storm marched along the horizon. Jonathan always thought of clouds as nothing more than shifting, indeterminate blobs, but the arrangement and timing of these discharges suggested some kind of stable inner structure, an architecture that was unique to this particular cloud.

The trailing storm on the other hand was highly variable. Some bolts ignited and extinguished quickly like the instantaneous bursts from a camera flashbulb. Other bolts were long and drawn out, pulsing with surges of energy punctuated by hesitant pauses, as if the cloud were emptying a reservoir and pumping the contents out in spurts. Jonathan thought that a pattern might still exist here, a longer sequence of events spread out over time, but he hadn't picked up on it yet.

In studying clouds, physicists broke the clouds down into columns of air. This was done because of the demands of the mathematics and not because the model was suggested by the clouds themselves. Jonathan saw that clouds did in fact have structures, but these structures weren't bundles of vertical columns. Starting with mathematical forms that had nothing to do with the clouds was a strange way of analyzing clouds, but physicists didn't care how the method looked. They had complete faith that once they plugged in the numbers, the structures of the clouds would magically appear.

—6—

The seasons gradually changed and the day finally arrived when Jonathan and Emma had to forsake their winter haunts in order to escape the increasing temperatures of the approaching summer. Jonathan packed up his camper and headed north seeking refuge in what was known in New Mexico as the high desert, the foothills of the Sangre de Cristo Mountains. The land was a transition zone and not a true desert, the sand and gravel yielding to clay soils which turned to muck at the slightest rainfall. The Taos mud was legendary in these parts, able to suck the shoes right off your feet or bury your four wheel drive truck up to its axles. Although the creosote bushes, ocotillos, and mesquites were gone, chamisas, prickly pear cactuses, and walking stick chollas were broadcast across the hills, maintaining the exotic desert atmosphere, but then dwindling in numbers as Jonathan and Emma continued to climb in elevation, until the last flickers of the desert were finally extinguished and they found themselves in what most people would consider to be a true forest. The rocks were mostly buried under a thick carpet of conifer needles causing the earth underfoot to compress and expand with each step as if Jonathan were walking on a mattress. The extensive openness of the desert was also gone, and Jonathan was no longer able to strike off in any direction he chose. Confined to the trails and roads that had been blazed by others, he helplessly followed their lead and wound up wherever they took him.

The damp smells of soggy earth and rotting wood permeated the air as Jonathan strolled down an old logging road gouged into the side of a mountain at 10,000 feet, the former thoroughfare now masked by an assortment of lush foliage. Emma was eager to explore this unfamiliar territory and went running off down the road. The road curved up ahead copying the contours of the mountainside, causing Emma to disappear from view, but she quickly reappeared once again because she had oddly stopped. She stood at the edge of the road looking down at something hidden below the grade, growling at it. She was making a rather peculiar sound, a vocalization that was only vaguely menacing. The noise reminded Jonathan of his bass fishing days in northern Wisconsin where he'd hoist the anchor—a coffee can filled with cement—by dragging the chain over the gunwale, the clunks of the heavy links reverberating off the hull of the boat and producing a sustained drum roll.

Jonathan looked at Emma fondly and thought that she was being especially cute, but as he came up behind her he realized that she was gazing directly into the eyes of a very large black bear. They were almost nose to nose, each one transfixed by the piercing glare of its newfound adversary. Not thinking, he lost his composure and yelled out in a most authoritative voice, "Emma, No!" His bellow resonated in the hollow created by the road, echoing up and down the canopy of trees. He stared up at the leaves and branches in amazement, listening to them repeat his words over and over again.

The bear had been so captivated by Emma's arrival that it hadn't noticed Jonathan's subsequent approach. Taken by surprise at the sound of Jonathan's jarring shout, the bear instinctively recoiled. It jumped back and fell away from the embankment, then whirled about in mid-air, immediately tearing off down the side of the mountain with Emma in hot pursuit. In an instant, both of them disappeared into the maze of trees and bushes below. Jonathan stood at the lip of the escarpment dumbfounded, suddenly finding himself alone in the forest. He eventually regained his composure and broke the silence by grumbling to himself, "Alright Emma, go ahead and meet the bear!"

He peered down through the snarls of branches and tall stands of trees looking for a sign of his lost companion, wondering what he should do. Finally he decided that he must go after her. Examining the terrain with his eyes as it disappeared into the depths, he figured that he could strike a diagonal and then switchback, climbing over the numerous tangles of fallen deadwood, and by proceeding in this fashion make his way down the treacherous slope. He belatedly began his journey, stepping carefully onto the steep incline. The long pine needles were woven into a dense mat and each strand was encased in a waxy coating. Together they formed a deceptively slick surface. The heel of his shoe took off sledding down the slope as if it were a toboggan on frozen snow. He failed to keep up with its rapid acceleration and fell to the ground with a bone-jarring thud, continuing the slide several more feet on his backside. Accustomed to the dry sand and gravel of the low desert, he hadn't anticipated this. He tentatively stood up and swept the duff from his sleeves and pant legs with his gloves, then took a few more faltering steps. He felt as if he were walking on ice. This wasn't going to work. Making his way down to the bottom of the mountain would take all afternoon.

The truck was parked only a short distance away. The mountainside dropped for a considerable distance then leveled out into a broad valley. From this vantage point, Jonathan could see only small patches of this valley through narrow openings in the trees, but he had crossed this valley on the way in so he was familiar with it. If the bear kept running and Emma kept chasing it, she would end up in this area sooner or later. He could drive back down the road and look for Emma down there. As he stood awkwardly on the slope, paralyzed by indecision and allowing time to pass quietly on its own, maintaining his shaky foothold on the slippery ground while simultaneously weighing the merits of his new rescue plan, he noticed something moving out in the valley below, a tiny white dot slowly crossing a pale grating of hazy lines and markings, an almost indiscernible speck drifting lazily against a mottled background of yellows and greens. He realized that he was watching Emma crossing the valley floor, racing at full speed, only she was not running away from him, but coming back toward him, hastily returning from the dense forest located far away on the opposite side of the valley. He peered at the seemingly insignificant speck so as not to lose track of it within the immense confusion of objects spread out before him. He saw no black spot following her so she must have lost the bear. Emma disappeared from view as she reached the base of the mountain, but after a while she reemerged in the trees below, running excitedly. Jonathan monitored her progress as she wove back and forth, her image gradually enlarging. She arrived in due course, panting heavily and completely out of breath, and stood a short distance away. He remained silent, inwardly happy to have her back. He visually examined her body, searching for signs of injuries or wounds, but she appeared to be alright. Finally with a sarcastic tone of voice he said, "Well, Emma. What happened to the bear?" She ignored his query, staring blankly at the trees, stoically putting up with his taunts. Jonathan turned around and climbed the short distance back up to the road, walking on his hands and feet hunched over like an ancestral ape, finally standing upright on the road. Emma immediately joined him and they advanced together along the picturesque track, the narrow trail in many places completely overgrown with saplings and brush, happily following its winding course for several more miles. The path abruptly ended for no apparent reason leaving Jonathan stranded overlooking a remote valley full of steep and inhospitable terrain.

Finding themselves back at the truck, they both climbed into the cab and Jonathan continued driving along the side of the mountain. He encountered a long stretch of road where the roadbed was notched into the side of a nearly vertical slope, the narrow thoroughfare barely wide enough to allow the passage of the camper. The appearance of a second vehicle would mean that one of them would have to back up along this tight and treacherous route all the way to the previous turnout. The last turnout Jonathan had noticed was now several hundred yards behind him around a very sharp bend in the road. He didn't relish the idea of driving in reverse over such a long distance, particularly with a precipice looming so close and the camper sticking out and reducing his visibility to the rear. On a whim, he stepped on the brakes and jammed the column shift into park, then quickly exited the truck and peered over the edge. Emma waited for him on the seat even though the door was open, watching him with mild curiosity, somehow knowing that he wasn't really going anywhere. Just as he had thought, the ground continued falling well beyond his ability to see through the maze of trees.

The drop-off was dizzying, but probably not all that dangerous. The large number of trees would arrest the truck's fall before the vehicle had gained enough momentum to begin toppling them, and many of the trees were too big to be uprooted anyway, even by a three-quarter ton truck with a heavy camper. The trees all insisted on standing vertically rather than growing away from the embankment, so if Jonathan could stand perpendicular to the side of the mountain, they would all appear to be laying down. Still the trees were there, ready to catch them should the wheels inadvertently slip over the edge. The imagined safety might just be an illusion and the truck would easily find a way through the labyrinth, bouncing like a pinball and accelerating rapidly, achieving enough velocity in the first few moments to cause a deadly crash. If the truck got turned sideways and the wheels got caught on tangles of downed wood, the truck would roll and that would not be good for them. Jonathan hopped back into the truck and started driving. The answers to all these questions became unimportant as he reached the next spur and drove onto a broad carpet of grass filled with beautiful flowers and luscious weeds. The opportunities for spending the night here were numerous. A charming aspen grove provided the most attractive spot, so he pulled over at the far end of it to begin the well-practiced routine of setting up camp. Soon everything was in its proper place.

Jonathan had plenty of time before dark to relax and enjoy his new camp, so he decided to take a walk. The evening was beautiful and absolutely still. Emma explored the more inaccessible regions of the aspen grove, poking her nose into piles of fallen deadwood and examining the exposed roots of the standing trees, while Jonathan strolled up and down the quiet and deserted dirt road under a canopy of leaves. The dark-colored leaves intermingling with the light-colored patches of sky created the illusion of an arched stained glass window, a mosaic of green and clear panes all supported by an irregular array of marble columns, the pillars of smooth white bark that comprised the tall and mighty aspens. He sauntered from one end of the large hall to the other, then back again, climbing and descending the gently sloping road, until the heavenly light began to fade and the windows grew dark.

Without warning, a strong wind rose out of the east and buffeted the branches of the trees destroying the illusion of the grand cathedral dome overhead. The trees began swaying madly and the branches started knocking into one another, producing a cacophony of hollow-sounding clunks and jarring raps. The leaves started growling and hissing and emitting all sorts of menacing sounds, adding their individual voices to the awful din. Jonathan made haste and returned to camp where he had a better view of the cold front sweeping over the mountains. An eerie light pervaded everything, the glow of the sunset having been radically transformed by the bizarre atmosphere. A yellowish-green luminescence came from every direction and enveloped the landscape, turning the trees and grasses into the inhabitants of a planet located in a faraway galaxy, a place where everything was very strange.

Jonathan found a narrow path between the pines and came out in a clearing after which the ground fell away sharply, plummeting down to the river valley below. He sat on the lip of the steep incline and observed the approaching storm play out before his eyes. Emma appeared out of nowhere and sat beside him, her attention now also captivated by the spectacle taking place before them.

A lake of silver clouds welled up on the ridge over their heads, steadily enlarging in depth and diameter until it flooded the highlands, the fog being driven against the barrier of trees and rocks by the strong winds. The storm continued to push the clouds against the natural dam until the condensed moisture overflowed its bounds and spilled over the rim in a vast river of clouds streaming down into the canyon below, just as if someone had tilted a giant pitcher of fog and started pouring its contents into a glass. The narrow stream of dense vapor churned and tumbled as it swiftly and silently raced past them, the chutes and rapids and eddies all clearly visible, the unstoppable surge bulldozing its way through the complacent atmosphere, cascading down the side of the mountain and disappearing into the dark depths below where another reservoir must certainly be filling up with mist and fog. The size and magnitude of the torrent was awe-inspiring, its sheer power and massiveness was terrifying, a giant river suspended above the tops of the trees, strangely contained by invisible banks forcing it to follow a definite course.

The temperature dropped quickly indicating that a snowstorm was heading their way. By morning Jonathan feared that he and Emma would be stranded here, the road buried under several feet of snow. He considered fleeing but the storm was already upon them and he had no time to pack up and run for it. The highway was only ten miles away and although the road was winding and rutted, the grade was downhill all the way. Jonathan could always walk out no matter how much it snowed that night, and he didn't want to miss this spectacular event. He'd always loved storms since he was a child and he had weathered more than a few in his camper without undue hardship.

To Jonathan's surprise, he awoke at daybreak to find a peaceful landscape showing no signs of snow. He hiked up to the ridge where the silver lake had been, but the lake was gone and the area was not what he had pictured it to be in his mind. The storm had transformed the ridge into a mystical realm, but he saw no evidence of any of the previous night's activities. In fact, the forest was rather bland and ordinary—dingy to be honest. He followed the old logging roads and game trails eventually finding himself completely on the other side of the mountain with Emma scouting up ahead. Even though he was at the highest point in the entire region, he had absolutely no views of anything, the forest effectively blocking his vision in all directions.

As the morning heated up, ominous clouds developed and rapidly turned into severe thunderstorms that began sweeping across the higher elevations, forcing him to return to camp and pack up, then retreat down the mountain to the friendly confines of the high desert. He established a pleasant camp situated among the pinions and junipers, a rarely visited spot where he'd camped several times before, located a safe distance away from the violence of the storms. The site did not appear to have been used by anyone in his absence, even though it lay at the edge of an open valley and offered a grand view of the faraway mountains, now with the angry thunderheads lording over them.

Jonathan and Emma spent the rest of the afternoon enjoying their new surroundings. He played games with her, throwing the ball and trying to entice her to chase after it. She invariably showed no interest. Despite his unflagging efforts, she had never displayed the slightest inclination to engage in this kind of activity, not even when she was an enthusiastic young puppy. He'd owned a couple of labs before her but never one that wouldn't chase after a ball. Emma adamantly refused to chase cars, skateboarders, bicyclists, or joggers, but hunting was her one passion in life. She came from a long line of hunting dogs in Missouri where the people took hunting seriously. The dogs were kept in kennels and let out for the sole purpose of running down game in which they found great sport. When Emma was younger, she would chase after every kind of wildlife, coyotes and foxes, quail and wild turkeys, elk and antelope, but now that she was older, she generally chased only rabbits and geckoes—and of course bears. She'd chased her first bear when she was only five months old.

Some dogs chased bouncing tennis balls because their motions resembled the motions of rabbits, the balls apparently hopping up and down and scurrying away, but Emma knew what a ball was and she knew that a ball was not a rabbit. She felt that no dog in its right mind would ever hunt tennis balls. Hunting rabbits was another story altogether. A mere pattern of behavior was not enough for Emma. She wanted the real thing. But how did she know what a rabbit was? The physicist said: "She knew the rabbit by its properties." Emma inhaled the scent of a rabbit and saw the image of a rabbit and put the two together. Breaking down events into schemata and comparing bare outlines was the essence of the mathematical way of thinking, but Emma was not thinking about pure forms. So how did she intuit the rabbit if she didn't construct the rabbit out of its properties? Jonathan wondered where she had gotten her knowledge of rabbits, a knowledge she had apparently possessed long before she ever encountered one.

When Jonathan threw the ball for Emma, the physicist told him that the path was a parabola--that is, if the wind wasn't blowing. But in the desert the wind often blew. Sometimes the motion of the air was only a gentle breeze, but frequently it was a stiff wind, so he needed to add another term to the equation of motion. There was also a coefficient of drag, which was the resistance to the motion of the ball from the air itself— without the wind. If he threw a baseball, which was relatively heavy, it was much less affected by wind and air resistance than say a tennis ball. But what about a whiffle ball? The trajectory of a whiffle ball was hard to predict. The object was carried away by even the slightest breeze. Still it was nothing compared to the path of a leaf in the wind--the paradigm of unpredictability. Equations seemed to leave Jonathan completely in the dark here. So what happened to the original parabola, the perfect curve that the physicist had handed him? Physicists abandoned it in the realm of abstract ideas when they came back down to earth. There were no parabolas in the desert. The wind made sure of that.

Leave it to the wind to be the one to steal the physicist's parabola and spoil his pleasant day outdoors. Jonathan had learned from experience that the wind was an impish prankster and responsible for all kinds of mischief. It loved to sneak up behind him and knock the hat off his head. The trick had merely been annoying at first, but the routine became enraging after a while. The wind enjoyed nothing more than seeing Jonathan throw himself into a tantrum. When Jonathan loaded up with packages from the truck and headed toward the house, the wind sensed that he was helpless, that his arms were tied up and that he was unable to defend himself, so here it came, rapidly approaching out of nowhere, but there was nothing he could do about it, and with an invisible hand the wind knocked the hat off his head and sent it sailing away. Now he had to put everything down and chase after his hat, because if he didn't, there was a good chance he wouldn't find it when he came back later.

Jonathan had once made the mistake of putting his hat on the patio chair next to him in the evening as he was relaxing after a long hike. He forgot about it and left it there. The wind came up during the night and took the hat away. The next morning it was gone. Jonathan scoured the surrounding fields over the next few weeks—over the course of the entire summer in fact—but he never found his hat. It had strangely disappeared and where it had gone he did not know. He had worn that hat for years and had grown quite fond of it.

Another time, Jonathan had hiked to an old quartz and mica mine located in a remote part of the wilderness, a secluded canyon which fed into the Rio Embudo. He took his hat off and set it alongside his pack while he descended into the circular depression. He found nothing of interest there, metal detecting and poking around the entire afternoon. The day was getting late so he hastily secured his belongings and started walking back up the trail. About halfway to the truck he realized that he'd left his hat at the mine. The sky was growing dark and there was no time to recover it. This wasn't a problem since he could simply return the next day and retrieve it.

He did return the next day, but the hat wasn't there. He looked everywhere for it. He stood motionless before every bush and tree, slowly and carefully scanning its branches, thinking that the hat might have gotten lodged somewhere. He meticulously combed every inch of ground, examining every detail for a considerable distance in all directions. As summer wore on, the hunt for his favorite hat became an obsession with him because he was sorely vexed by its loss. He returned to the mine on numerous occasions, gradually expanding his search area even further with each visit, until winter came and he abandoned the cause.

The following spring Jonathan once again visited the mine, but no longer with the thought of looking for his hat. He'd long since given up on it and gotten himself a new one. To his shock and dismay, the hat was laying right there in plain view within a short distance of the mine. While he was happy to have his old hat back, he was angry that the wind had played such a cruel joke on him. The wind must have gotten a good laugh at the amount of time Jonathan wasted on his many foolish and futile searches.

Did this kind of behavior by the wind require an explanation? The physicist said that the wind was random and haphazard. First it blew one way and then another. Eventually the hat ended up where it had started because after a period of time all the variables cancelled each other out.

Instead Jonathan imagined his hat going on a long journey over the winter, taking it miles away into unknown country. He saw the hat skipping and sailing along, leaping over trees and deftly sidestepping outcrops of rock, eventually tumbling home exhausted after its long trek where it waited for Jonathan to return. Or did the wind maliciously steal his hat and then whimsically give it back to him six months later? He imagined his patio hat similarly off exploring the world without him, taken on a grand tour by his persistent antagonist. He wondered if someday the wind would give it back to him and he'd find it lying there on the patio next to his chair.

Emma had an abbreviated attention span so Jonathan attempted to rekindle her interest in playing catch with him by switching from the ball to the frisbee. He told Emma to stay, then walked away about twenty paces and, after getting her attention, threw it towards her. She made a half-hearted effort to open her mouth and grab it. He had practiced making good throws many times, releasing the frisbee at precisely the right angle, letting the object go at exactly the right point in his arm motion just as a pitcher in baseball. He aimed the frisbee directly at Emma's mouth in order to encourage her to catch it. He had one of those frisbees that was made out of nylon cloth and after being wedged behind the seat of the truck, left out in the rain, stretched and twisted in tug of war games with Emma, the disk was no longer even remotely flat, so it flew funny. A piece of vinyl tubing had been sewn into the perimeter which was now off-center and not quite circular. The disc was weighted on one side so that it not only wobbled but described epicycles. Mathematics had generic equations which Jonathan could use for each of these types of subsidiary motions, so he could simply add them together into a patchwork quilt to get the resultant trajectory--well, more or less. If the math didn't come out right, he could always add another term as a correction, a fudge factor. The frisbee slipped out of his hand as he threw it and went careening off in the wrong direction, tilting and slicing into the Apache plumes, still somehow bound by Newton's Laws of Motion. Emma stood there watching it, then stared back at him with a comical look on her face. She seemed to be saying: "You didn't seriously expect me to go after that one did you?" He gave her the command telling her that the game was over: "That's it! Emma, that's it!" He wasn't interested in the idealized conditions of a laboratory. He wanted to calculate the course of his tattered frisbee on a blustery day, but that was a problem he could not solve with mathematics.

Jonathan had to resign himself to the fact that the winds of the desert did not follow simple rules. The erratic and unpredictable behavior of the frisbee merely exemplified this intractable nature and added its own capriciousness to it. Newton's Law vanished into the unfathomable complexity of the real world and became lost—as lost as anything could possibly be lost. During his daily excursions hiking long distances over inhospitable terrain, these formulas had proven utterly useless to him. The notion that somehow the laws were still out there somewhere, buried deep within the infinite morass of incidentals yet still controlling the outcome of events, was insane. The desert winds had swept all of these simplistic idealizations away and they were nowhere to be found.

The frisbee acted as if it had a mind of its own, bound to no laws whatsoever, but rather than throwing Newton's Law away as the situation seems to demand, Jonathan decided instead to try to salvage it. Coming up with these laws was a lot of work and physicists had already invested a great deal of time and energy getting their hands on these precious ratios, then devising laboratory experiments to confirm their veracity. Perhaps he could find a way to hang onto it. He had it! He could build the flight of the frisbee out of Newton's Law by replicating the law over and over again at many different points. Since these mathematical relationships were imaginary, he could put them wherever he wanted them, so he could just arrange them so that they added up to the intact phenomenon and then he could not only say that they existed, but that they were moreover the cause of what he saw. This approach was so ingenious that it superseded the inspiration of the original law itself. He could build the phenomenon out of whatever ideas he chose, so why not choose the ideas of Newton's Law? This was absolutely brilliant! Instead of starting with the intact phenomenon which showed no signs of Newton's Law whatsoever, he could turn everything around and start with Newton's Law instead, then reconstruct the phenomenon out of that. Newton's Law was still not visible anywhere within the phenomenon, but Jonathan could now say that it was there anyway, working behind the scenes to guide the frisbee to its destination. He could say that Newton's Law was an essential part of the frisbee's motion.

The only way to argue against Newton's Law at this point was to refute the proofs of mathematical theorems or challenge the outcome of laboratory experiments. Checkmate. Critics were trapped in an inescapable web—unless they questioned the right of physicists to construct physical realities out of imaginary building blocks. Now it was the physicist who had to defend himself. Where did the physicist get this "Newton's Law" anyway? Not from anywhere around here, and certainly not from the frisbee.

Jonathan could not deny that a heavy, streamlined object dropped in a still room with no obstructions had a motion that closely followed a mathematical formula—but so what? What did that have to do with anything? Objects in nature didn't behave that way, so why did physicists think that this formula was so important? It was certainly not important for understanding nature.

The physicist, of course, was outraged that Jonathan had the audacity to bring up the name of that unkempt vagabond called Nature while they were discussing these topics in the elegant sitting room of formality and propriety, a place where everyone acted accordingly. The physicist claimed that he had found the law to be extremely useful in his experiments—but that was Jonathan's point. The law was important because it represented a giant step toward generating a false world of idealizations and abstractions in the laboratory, a world that was destined to replace the vast desert—at least in minds of physicists.

In an experiment, phenomena were shaped by the mathematical designs of devices, similar to the way that potter's wheels were used to craft pots. Both the apparatus in an experiment and the potter's wheel in a pottery workshop were loaded with mathematical precision. The wheel was a round disk mounted at the end of a straight steel rod, incarnations of the mathematical ideas of a circle and a straight line. In order for the potter's wheel to work properly, the line must intersect the exact center of the circle. The precise arrangement of the mathematical parts was a necessary introduction of further mathematics. The simple machine became a material object endowed with the mathematical property of radial symmetry. The mathematical ideas came together in a harmony which then could be used to create clay pots with similar attributes. The shape of the finished pot hadn't been a part of the clay and it didn't represent some law of nature. The configuration of the potter's wheel was just an idea in the head of the potter. He then used the idea to make the clay conform to his idea of a pot. He then went on to fashion all sorts of plates and bowls and cups, all imbued with the symmetry of the potter's wheel. The physicist similarly went around putting mathematics into everything, then turned around and exclaimed: "Isn't that amazing! Mathematics is everywhere!"

The shape of the finished pot was not a property of the substrate, just as the trajectory of the frisbee was not a characteristic of the frisbee. Jonathan was the one who made the frisbee fly. He was the one who gave it a trajectory. Physicists turned the tables around and said that Jonathan was studying the frisbee, as if it were acting on its own. The object was the recipient of the motions Jonathan had imparted to it, much as the clay was the recipient of the forms the potter molded into it. If conditions were right and he made his throws precise, then the object behaved in a precise manner, but only because he had caused the object to do so. The object didn't fly by itself and it didn't possess inherent motions of its own. Jonathan artificially created these motions and then claimed that he

162

was studying the nature of the object itself when in truth he was only studying its responses to his influences, which were essentially just the influences themselves realized through the material object. But this was true only if Jonathan was able to accurately control the frisbee.

A frisbee in the wind, on the other hand, took off in any direction the moment he let it go. His arm motion was rather irrelevant and he had little say in its flight. He could not dictate the results and therefore there was no mathematics. In order to make the flight precise and unambiguous, Jonathan had to control everything: every nuance of wind, every detail of aerodynamic shape and size, every aspect of the initial impetus. But if he controlled every feature of the environment along with the aerodynamics of the frisbee, then he was also controlling the result, and to control the result was to create it.

The frisbee had no equation of motion of its own because the flight of the frisbee was unique each time Jonathan threw it, but he could solve this problem easily enough. He could build a machine to throw the frisbee, a machine whose action was exactly reproducible, a device that he had instilled with mathematical precision. The device then released the frisbee with the same tilt, the same linear velocity, the same angular momentum, so that if he performed the experiment in a room with absolutely no air movement, then he could get the same trajectory with each trial. The reproducibility of the machine action caused the reproducibility of the frisbee's motion, artificially transferring this character to the frisbee. Now the frisbee had an equation of motion and he could specify what that was. He had employed mathematical precision to create mathematical behavior in the object. He had imparted mathematics to the frisbee where previously it had none. He had fabricated what he had wanted all along: exactitude and a strict conformation to rigid forms, the vaunted hallmarks of mathematics itself.

Precision was important in physics because precision provided the physicist with a means for creating mathematics. Jonathan's cloth frisbee was not perfectly rigid and that would affect its trajectory even with the machine is in control, but he could correct that problem by replacing it with another frisbee, one having a precise, mathematical form. The problem had just been a lack of precision in a critical area that resulted in unwanted deviations in the outcomes.

But what had Jonathan done to the frisbee? He had imprisoned it and taken its freedom away. Now he was concerned for its welfare. He was concerned not only for the freedom of the frisbee, but for the freedom of the wind as well. The frisbee was a part of something greater—as all things in this world were. Breaking those connections deprived the frisbee of essential aspects of its nature. Material objects were inherently social beings: they mingled, they worked together in harmony, they fought against each other. Taking all that away prevented the frisbee from expressing itself. In a controlled environment, the frisbee acted properly for the sake of its audience, marching across the room in step with the music that the physicists had provided for it. Jonathan, on the other hand, wanted to let the frisbee go. He wanted it to fly away and play with the wind. The laboratory-controlled frisbee was boring and held no surprises for him. He imagined the experimenter turning to him with a grin on his face, saying that he had done it. He had mastered the frisbee. Jonathan just sighed. That was not the frisbee he knew.

The experimenter hadn't exposed the underlying mathematics of the frisbee simply because he made the frisbee act in a strange way. He had only subjugated the frisbee to his will, much as a strict school teacher broke the will of his students. The frisbee obeyed his commands. He could make it follow many different paths of his own choosing, but now he was only studying the lessons he had taught it. The frisbee played the part perfectly well, performing exactly on cue, rehearsing the well-choreographed routine and hitting all the marks in perfect time. No one needed to wonder about any of this. Everyone was watching a stage show, just as if they had been sitting in the theater watching actors mouthing a scripted play. The audience could memorize the lines and predict the next utterance in the verbal exchange with uncanny accuracy because they weren't watching a spontaneous conversation, a free discourse that mimicked the way nature spontaneously created the flight of the frisbee. The performance instead resembled an experiment in a laboratory.

The method had almost universal applicability. To prove the point, the physicist drew parallel lines on a sheet of paper, separated by exactly the length of a sewing needing. If he dropped the pin on the paper haphazardly, he found that the pin came to rest on a line 64% of the time. None of this was particularly interesting, except for what followed, the physicist's interpretation of events. The behavior of the pin, he said, was mathematical in nature because the pin was acting on its own. He wasn't controlling it. In a similar way, all sorts of things were mathematical in nature. The pin could take its place alongside a huge number of other objects that also displayed mathematical behaviors, proving beyond all doubt that everyone lived in a mathematical universe.

But the only thing the physicist had done was invent a mathematical game. He selected a mathematical object, a piece of metal cast into the form of a line segment, and then observed plays on a mathematical game board, a planar surface with precise geometrical constructions painted on it. None of this existed naturally in the world. The physicist introduced the mathematical ideas by drawing parallel lines with equal spacing on a flat surface, then in a strange twist of reasoning concluded that it was the behavior of the pin that was mathematical. The behavior of the pin wasn't mathematical at all—it was only the structure of the game that was mathematical and this caused the pin to appear as if it were the source of the mathematics, that it was governed by some law of nature. The problem was, of course, that nature didn't play those kinds of games. Nature never fabricated mathematical objects and it never created game boards. Nature didn't know what mathematics was so it couldn't produce the mathematical setups that were necessary for objects to exhibit mathematical behaviors.

Paul Dirac, a famous quantum physicist who had an equation named after him, once wrote that "mathematics was a game where mathematicians made up the rules, and physics was a game where nature made up the rules." He was wrong about about the second part. Nature never made up the rules to any games, because nature never played games. Only physicists played games. Dirac then went on to say that when the physicist was writing the equations of nature, he should strive to achieve mathematical beauty, presumably because that's what nature had done in the first place. Of all the conceivable mathematical schemes, nature had chosen a particular set of mathematical forms to adorn itself and make itself presentable, just as someone might peruse a department store inventory and select a suit of clothes off one of the racks. Acting in the manner of its human counterpart, nature cloaked itself with the most beautiful outfit it could find—limited, of course, by the choices at hand—ostensibly with the idea of courting mathematicians. But no mathematicians had existed back when the universe was created, so the cosmos must have looked into the future and astutely anticipated their arrival. But why didn't the cosmos choose ugly mathematics instead, cluttered and awkward equations with no elegant conceptual framework, a hodgepodge of ill-fitting fragments with uncoordinated patterns? Well, perhaps it did. Given the real life situations outside the laboratory, the equations of physics rapidly got complicated and messy. So how then was ugly mathematics different from no mathematics at all?

If theorems were established through chains of reasoning and mathematics was a complex interplay of definitions and deductions forming a network of interconnected thoughts, all envisioned, formulated and comprehended by the human mind, then how did mathematics ever get into nature? How could mathematics possibly exist as something other than ideas in a person's head? If physical reality embodied mathematics, not just in a token, superficial way, but in the meaningful sense of deliberate mathematics, then the material world must have been made according to the thoughts of some intellect. Jonathan was forced to assume that the creator of the universe had adored mathematics so much that he built an entire universe simply to showcase these ideas. In this grand project, the creator employed mathematical reasoning as a guide in the same way that a geometer used mathematics to make a protractor, or a physicist used mathematics to make a cloud chamber.

For a mathematical universe to exist, physicists not only had to assume that a supreme intelligence existed, but that this deity was also a talented mathematician. Since God couldn't write anything down or scribble symbols on a blackboard—the crutches of mere mortals—he must've had the mathematics already figured out in his head, much as a virtuoso pianist could play a lengthy piece without ever referring to the sheet music. How the existence of deliberate mathematics in nature did not imply some sort of intelligent design had never been clear to Jonathan.

Back in his student days, Jonathan had been introduced to the notion that science was a thoroughly cultural affair. New theories gained widespread acceptance through the assertive personalities of dominant figures rather than from the resolution of vital issues. The opposing point of view simply faded away with its criticisms unanswered, drowned out by the enthusiastic cheers of the proponents of the new way of thinking. Physicists rarely subscribed to this view and generally decided that they had won the debate because they held themselves to higher standards than everyone else. They relied exclusively on the empirical verification of their theories—a premise that was much more complex than they realized and not an accurate description of the scientific method at all—but if that were true, then where was there an account of the origin of mathematics in nature? Where was there a justification for the legitimacy of mathematics in dealing with the infinite particularity of the world? If mathematics was just a tool to organize data, then it didn't really exist in nature, but only in the minds of the organizers.

On the other hand, if the whole of existence had actually been founded on mathematical equations, then these equations must have been devised first and then subsequently invoked, the universe exercising the laws to create itself. Existing laws were required in order to guide the formation of stars and then to organize them into galaxies. It was these equations that condensed the glowing gases into colorful nebula and collected the widely distributed asteroids into aggregates which later became planets. It was these laws of nature that populated the planets with life forms and then organized them into complex societies. Either the final result had all been worked out in advance, or the laws of nature had been created without knowing the consequences. What if the universal consciousness had made an error in its calculations and the gamut of creation unfolded only to arrive at a contradiction? The supreme mathematician would embarrassingly pick up the cosmic blackboard eraser and go back to review his work, obliterating the offending sections and installing the essential alterations to his monumental equations. He would then have to reset inflation and the big bang and start all over again.

If Jonathan dispensed with the idea of an ultimate mind, how was it possible that the motley collection of astronomical objects comprehended the beauty of abstract thoughts? How was it that the universe shared the same vision as earthly mathematicians and possessed the same aesthetic values as they did? This all seemed a bit self-centered and presupposing to Jonathan—and smacked of religion. Scientists had eagerly proclaimed their renunciation of such naive anthropomorphisms, yet their sun still revolved around them and the world was still created in their image.

Some form of primitive matter might have existed prior to the laws of physics, utterly undirected and disordered, and then this amorphous muck formulated mathematics as it went along, creating a coherent and sophisticated worldview out of its youthful experiences, the physical and mathematical sides of its personality growing together as one. Or perhaps the mathematics of the universe evolved from lower and simpler forms such as Newton's Law to higher and more complex forms such as Relativity Theory, in the same manner that Darwinian evolution created sophisticated forms of life out of the elementary ones. The earliest equations of the infant universe must have been on the level of high school algebra, but over time they mutated randomly and by a process of blind trial and error they became graduate level tensor analysis. The advanced physical laws plainly out competed the more basic laws and over countless eons of time these laws became universally true. Just as primitive forms of life still existed in the world today, Euclidean geometry and linear algebra stuck around alongside their more inscrutable evolutionary cousins as relics of a distant past.

Relinquishing the idea of a supreme being or a cosmic consciousness left the physicist in an awkward position, his job of mathematizing the universe having already been done by inanimate matter, that is, by something with no intelligence. Physicists could escape this demeaning characterization if they maintained that subatomic particles didn't behave as they did because of mathematics, but were motivated by other impulses, urges that physicists didn't understand. The particles pursued unknown goals and hidden agendas, objectives completely unrelated to the forces and energies ascribed to them by physicists. In the process of going about their unfathomable business, their motions and interactions followed certain patterns when viewed in the laboratory situations orchestrated by physicists, but these were incidental to the actual purposes which animated the particles. Something else entirely was going on here and no one had the slightest clue as to what that was, but someday, if by chance physicists ever found out, they would all be shocked. Everyone would hear the refrain from a chorus of voices chiming that once again "we had no idea." The theories that physicists concocted to explain phenomena came from the minds of physicists and not from the tiny realities they called particles. They could not get inside these entities to see what was really going on and they could not know them on a personal level. Their perspectives were limited to merely external views from great distances due to the vast discrepancies in size.

Without knowing the truth about the world, physicists would still get to have their mathematics, but only as a secondary effect. The regularity of events would allow a mathematical interpretation by default. The universe would assent to mathematization through its silence, but ultimately mathematics had nothing to do with anything and the formulas were all mere appearances. None of the mathematical structures actually existed in nature. They were just convenient fictions.

The universe had no mathematics of its own. Mathematical forms were added to the world by simply drawing lines over everything, superimposing bizarre images on top of realities that were vastly different. In this way objects and events could be decorated with all sorts of mathematical designs, stenciled with distances and lines, cloaked in abstract formal concepts, but these coverings were extraneous and not implied by the objects. Physicists could dress realties in all sorts of costumes and create all sorts of looks, come up with stylish appearances and unique fashions, and in the minds of physicists the age-old adage that "the clothes made the man" rang true. By and large, the clothes that people wore determined their identities, and everyone judged their characters and inferred their private lives by the way they presented themselves in public. The superficiality of mere appearance became the only thing that mattered. This was the essence of mathematical thinking.

Since physicists could not look past their mathematical constructions and grab hold of the underlying realities, the wealth of details that formed the actual substance and content of the world, the mathematics concealed the truth of that world much in the way that a suit of clothes concealed not only a person's anatomy, but a person's thoughts and feelings as well. Clothes distracted the viewer's attention away from what was most important about the individual and caused the viewer to focus instead on a rich conceptual world of aesthetic values, an often intricate interplay of colors, patterns and forms, even though the clothes were removable, changeable, and nonessential to the existence of the person wearing them. The clothes were an independent reality that presented its own concerns and issues, and the person wearing the clothes adopted a role not unlike a mannequin in a store window, displaying the clothes in public for an admiring audience of fashion-minded individuals. The person became a blank outline that existed only to hold up the garments. The clothes became the main point and they received all the attention, while the real person faded into the background and essentially disappeared from view.

In daily life, many people assumed identities by donning uniforms. Society was populated in no small part by policemen and firefighters, cooks and waiters, plumbers and repairmen, doctors and nurses, athletes and referees, postmen and deliverymen, white-collar businessmen and fast-food workers, each with a peculiar set of clothes. Real life criminals and Hollywood movie characters frequently took advantage of this fact and pulled off various rackets and heists by masquerading around as people they weren't, and those around them always fell for the ruse because everyone judged others by their outward appearances. By cloaking everything in mathematical forms, physicists could disguise the true nature of physical existence and pass it off as something else, a mathematical system, but they never seemed to ask themselves about this entity that was sporting the equations or ponder the question of what remained once the facade had been stripped away. Physicists were thinking incorrectly about mathematics, and the metaphor of a fashion ensemble was a much more accurate description of the actual roles that mathematics played in their understanding of the world.

The heat of the afternoon subsided and it was time for Jonathan and Emma to embark on their customary evening walk. Jonathan struck out from camp once again seeking new territory to explore. He walked about a mile over flat terrain and then crossed onto the next range by climbing over a tight, well-maintained barbed-wire fence, about four feet tall, using one of the metal stakes as a handhold and the wires as steps. The strands were separated by only a few inches. Emma waited patiently, watching Jonathan struggle to balance himself on the shaky tightropes. Once on the other side, he barely found enough slack in the wires to create an opening for her. He stepped firmly on the second wire with his shoe and pulled up as hard as he could on the third wire, stretching the wires apart just far enough to let her squeeze through.

The next range was pristine and not currently grazed, but the sun was setting and they were out of time. Jonathan decided that they should turn around and head back to camp. He gave Emma the command: "Let's go home!" She immediately took charge with a burst of energy and flew past him in order to get a head start on the return journey, while he trudged along some distance behind her.

A rabbit materialized out of thin air, appearing directly in front of Emma's nose, popping into existence already running with extreme vigor. Emma did not hesitate and immediately started chasing it. The intense pursuit unfolded just as Jonathan gained a slight rise granting him a perfect view of the events. The rabbit wasted no time and set its sights on the fence line partitioning the two ranges, apparently knowing that Emma could not possibly get through the fence without Jonathan's help. It shot between the wires like a bullet, imagining that it had thereby acquired safety.

Emma was not far behind, flying at breakneck speed. Jonathan muttered an admonition under his breath: "Pull up, Emma, pull up!" but it was no use. She was so focused on her prey that she had forgotten about the fence. Emma was a blur of motion, yet Jonathan was able to study the pattern of her movements. Her front legs reached out grabbing as much turf as possible, while at the same time her hind legs shot in the opposite direction, propelled by each powerful thrust of her hindquarters. For a brief moment Emma was stretched horizontally to the maximum extent that her body would allow. Then her feet bunched together under her as she began another cycle of this scissor-like motion.

Emma was about to hit the fence at full speed. Jonathan feared that the force of the impact would break her neck or the sharp barbs would cut her chest open. He wanted to look away but instead kept his eyes fixed on her, expecting the worst. Emma froze for the briefest second just as she was about to hit the fence, holding her body in the extended position a little bit longer than usual. Then suddenly she was on the other side of the fence. Jonathan was confused, not sure what had just happened. She couldn't have possibly gotten through the fence. Finally he realized what happened. She had leapt completely over the top of the fence. Without missing a stride or faltering in any way, and without giving any outward sign of her deft maneuver, she had created the illusion that somehow she had magically gone right through the fence, seamlessly preserving her pursuit of the wildly fleeing animal. Jonathan's initial apprehension quickly turned to devout admiration. At an all-out sprint Emma had cleared a four foot fence without stumbling—and without losing an inch of ground to the rabbit.

Despite Emma's heroics, the rabbit successfully eluded Emma's attempts to overtake it. She loped through the tall grass searching for evidence of her lost prey, tracing out large figure eights, then finally circled back. Jonathan rendezvoused with her at a point along the fence. He reprimanded her verbally: "Emma, don't you ever do that again. You could have killed yourself." She was still out of breath and had a dazed look on her face. She didn't understand what Jonathan was saying to her, but he still felt obligated to express his objections in an upbraiding tone of voice. Secretly he was proud of her athletic abilities and dazzled by her amazing performance, compelling him to soften his words and tell her in a more soothing tone: "In the future, Emma, just be more careful."

The return trek was uneventful and they arrived back at camp earlier than Jonathan had expected. The camp was an excellent place to watch the full moon rise so he unfolded his reclining chair and oriented it toward the eastern horizon. He was inspired to perform an experiment to pass the time until dark. He rummaged through the items residing in the back of his truck and located three objects: a book about desert plants, a block of wood that he used as a jack stand, and the defunct CD player he had removed from the dashboard a week ago. He cradled the miscellany in his arms and scrambled up a neighboring outcrop of large rocks. Standing on the crown of this untidy assemblage, he threw the three objects successively out into the air with the same arm motion. Emma turned her back on him and quietly moved away pretending to sniff the ground. Apparently she thought that he was trying to dragoon her into a game of catch. He observed the three objects land at the same spot on the desert floor. The trajectories were exactly the same, yet when he examined these items directly, he saw that they weren't even remotely the same. They all described the same parabola due to the influence of the earth's gravity, but that didn't mean that they were the same object, and in fact, their trajectories didn't tell Jonathan anything about the objects. The objects could be simple or intricate, fabricated or natural, useful or pointless, symmetrical or lopsided. Gaining a knowledge of something by studying its motion was utterly impossible.

Jonathan stepped carefully down the chain of lopsided footholds and arrived safely on the ground below. He scrutinized the book as he gathered his belongings and returned the items to the clutter of the truck bed. He thought about expanding his study of books. He realized that the movements of books were quite complicated. The motions of books were linked to the motions of backpacks and purses, and these in turn were functions of the movements of people. The motions of books were also related to the motions of vehicles. He brought to light the strong force of attraction between books and bookshelves and translated this into an equally strong attraction to libraries. He noted that there was also a weak force of attraction between books and classrooms, and wondered if he could relate this force to the strong force of libraries. Just as atoms were often the carriers of electric charges, people and vehicles were often the carriers of books. Books were routinely transported great distances and passed between homes, offices, restaurants and hotel rooms, carrying mass and energy with them. Sometimes they moved in unison with buses and cars and Jonathan wondered about the nature of this bond. He calculated the velocities and accelerations of books under all these forces and wondered if larger and heavier books traveled shorter distances compared to smaller and lighter books, due to their greater inertias.

Eventually Jonathan became an expert on books. He had studied their motions in great detail. He had worked out the statistical distributions for every size book and every location. He had categorized the interactions of books with all other kids of objects. He finally came to the conclusion that he knew everything there was to know about books, but unfortunately he had forgotten one thing. The mechanics of books was incidental to their natures. If he didn't actually know what a book was, then he really didn't know anything about books. The empty forms of mathematical relationships didn't tell him anything important about books. They never gave him a feel for their place and purpose in the universe or clue him in on the details of their existence. That was something he would need to know if he was going to have a genuine knowledge of books. The mathematics of motion stayed outside the objects. It only told him how something behaved in response to the external forces acting on it.

A book could not be understood by following its trajectory through space or by studying its mechanical interactions with other objects. Jonathan could not possibly know what a book was by how it acted under the influence of forces. So why was it, Jonathan asked himself, that electrons and protons could be understood this way? The answer was that their motions were the only things physicists could discover about them. These objects were way too small to be observed directly, but their motions took them over much greater distances and these distances could be more easily measured. Physicists couldn't look at electrons and protons in the same ways that Jonathan could look at books. This lack of direct personal experience allowed physicists to replace electrons and protons with abstract ideas and transform them into mathematical entities since there was no means for objecting to this maneuver. The same ploy when used on macroscopic objects was clearly erroneous.

The failure of mechanics to reveal the nature of books forced Jonathan to reexamine his approach and look at books more closely. He put together a list of some of the properties of books, other than size and weight. A book always had a cover and inside this cover it had pages: pieces of paper glued together along one edge. Symbols were printed on the faces of each page. All books were the same up to this point and they were indistinguishable from one another. Jonathan could just as well pull one book off the shelf as the next one if this was all he knew about books. Just as with electrons and protons, they were identical and interchangeable. But what if Jonathan learned to read. Then, for the first time he saw that each book was exceptional. It was a concrete individual of immense complexity, and not an abstract entity identified by a symbol and a table of numerical values.

When Jonathan turned twenty-one, he built a log cabin at the back corner of his grandparent's farm, long after his grandparents had died and left the farm to his parents. The farm was located in a heavily wooded region of Forest county in northern Wisconsin, not far from the upper peninsula of Michigan. The cabin was meant to be a practice cabin, since Jonathan fully intended to drive to Alaska and build a second cabin where he would live out the rest of his life, but he felt that he needed to first see what was involved.

He began the work after everyone had left for the summer. The farm was a popular place during the summer months, cousins and uncles and aunts visiting from the city and enjoying a bit of country life, but suddenly the place was quiet and deserted. The morning that he began his project was misty and cloudy and the air was absolutely still. The trees had not yet turned color since it was only the first week of September. He decided to plant his cabin next to a large marshy area, one of the few substantial patches of forest left standing on the property. He stepped through the bushes along the margin of the swamp and chose a fine spruce of the appropriate diameter. He proceeded to cut the tree down with an antique chainsaw he had purchased for fifty bucks from a neighbor who lived a few miles down the road. He then went around cutting more trees. He carefully selected each tree in such a way that after he felled it and hauled it off to his construction site by dragging it with a heavy chain slung over his shoulder, he could not tell that anything had been removed from the forest. He meticulously collected the slash and carried it away so that the area looked just as he had found it. Of course, this forced him to take fewer trees from each location and thus he had to heave the timber over much longer distances, but since this was going to be his front yard, preserving the aesthetic quality of the forest was of the utmost importance to him.

Trees were simple things to him back then. They were not complicated because his interaction with them was not complicated. He was only concerned with diameter and length and the basic operations of cutting, limbing, and peeling off the bark. He saw trees as nothing more than tiers in a cabin wall. Later in life he studied trees as living organisms and discovered the truth about them. They were unfathomably complex.

The particles of physics were really just the tiers or building blocks of a mathematical framework, much like the trees that Jonathan had used to build his cabin. They formed the essential parts of a theory of matter, but unlike the logs stacked to form the walls of

Jonathan's cabin, the requisite features of particles were not limited to fixed dimensions and static shapes, but also included motions and dynamic equilibriums. Still the only point of any concern was the way these concepts could be fitted together to form a harmonious structure. The individual pieces had to mesh, to interlock and match up against one another, to comprise an overall design. In building a cabin, Jonathan never asked the question "What is a tree?" because his project didn't require an answer to that question. He didn't need an explanation of a tree before he could proceed. All he needed to know was how the tree could be made to serve his purposes and in this context only a few basic facts were relevant. The tree could be reduced to a cursory outline of its spatial extent without sacrificing anything in terms of the goals Jonathan had set for himself. In physics, physicists were similarly constructing an edifice for themselves. By employing mathematics as the means of construction, they had decided ahead of time that the only relevant aspects of particles were going to be mechanical in nature. They needn't answer the question "What is a particle?" in order for them to erect a beautiful structure out of ideas. A particle could simply be an "x" with mass and spin just as a tree could simply be an "x" with length and diameter. The concepts could all be cut to fit, just as Jonathan had trimmed each log to a specific length so that it could occupy a certain space. Perhaps someday physicists would gain an appreciation for what particles really were, in the same way that Jonathan had later found out the truth about trees.

By mid-October the cabin was finished, except for plastering the oakum chinking in a few places and putting a small roof over the front porch. Jonathan lazily walked the abandoned fields of yellow timothy and basked in his sense of accomplishment. He hauled firewood each day to the small opening in the trees and stacked each load alongside the others behind the back wall of the cabin. Wood smoke gathered in the tops of the trees as autumn advanced and the elms and maples began shedding their leaves. The days grew unnaturally warm and Jonathan was convinced that the beautiful Indian summer would never end, but then one night a foot and a half of snow fell. Winter arrived in the dark of night just as if someone had snapped their fingers and summoned it. The harsh reality of the brutal northland snuck up on Jonathan while he was asleep in his new cabin. By Christmas the snow was four feet deep. Wearing snowshoes each day, he was able to establish hard, encrusted trails that criss-crossed through the swamp and over the fields, but if he unbuckled the wooden frames and stepped off the trail in his boots, he sank up to his chest in the snow.

Sometimes instead of spending the night in his cabin, he camped in the woods in a tent. He often went out at night, under the stars in the pitch black of a midnight sky or bathed in the bright glow of a full moon where he proudly took his rightful place among all the other mysterious shadows laid out across the drifts of snow. He sat by a lonely campfire or walked for miles blazing a fresh trail to parts of the forest where he'd never been before. Once, when he was returning to his cabin in the early morning hours, the howl of a timber wolf erupted and resonated in the cold air, a haunting sound in such a desolate environment. He soon discovered its tracks. The wolf had been standing on his front porch when it had let out its mournful cry. Jonathan stood on his porch and let out a similar cry, mimicking the wolf's pitiful serenade. Loneliness had crept into his soul and the interminable isolation of a long winter had taken its toll on him. He now understood the existence of the wolf all too well.

Jonathan did not have the mental fortitude for such a lifestyle and he went crazy. He started talking out loud to himself, raving about things that made no sense, flinging himself into thickets of young saplings and falling helplessly into the deep snow that surrounded them. He was a wreck and he couldn't even drive himself out to the safety of civilization, the society that he'd left behind months ago, now hundreds of miles away. The old farmhouse was only a quarter mile away and it still had a phone, so he called his parents. The next day his father drove up from Chicago to rescue him. His father had consulted a psychiatrist before he left town and advised him of the situation. The psychiatrist had prescribed sedatives to give to Jonathan. He and his father stayed at a motel the first night. Jonathan remembered slipping into a hot bath as the drugs started taking effect. He was so grateful for the company of others. The long ordeal was finally over and after that he only stayed at his cabin for short periods of time. Very quickly his dream of a wilderness life had been shattered once and for all.

Everything could be reduced to simple outlines and considered in terms of basic mechanics. Jonathan remembered reading a book where a chemist triumphantly announced that the energy relationships of organic molecules followed the same rules as inorganic molecules so there was no vital principle involved in the chemistry of life. Jonathan knew that the complex behaviors of these compounds suggested otherwise. The chemist had arrived at his conclusion only by looking at the forces of molecular bonding and the rates of chemical reactions and as far as these went, there was nothing unusual in the realm of living organisms.

But organic molecules when placed in the workshops of living cells were capable of amazing feats. Either atoms had latent potentials that were not realized in the inorganic world, or some hidden factors were added to the atoms when they became incorporated into living bodies that made them come alive, perhaps factors that were passed along from one organism to the next. The inexplicable behaviors of these molecules still required some sort of vital principle, only this principle did not reside in the mechanics of the chemical reactions. Atoms had much more to them than the chemist realized. Something more was needed to account for the bewildering acrobatics of living tissues.

If physicists concerned themselves only with momentums and energies, their knowledge of their subject matter would be correspondingly limited. Their view of things depended on their level of interaction with them. Since they couldn't get inside atoms and molecules and see what they were all about, they had to settle for a merely external perspective. They continued to deal with these things without ever learning the truth about them.

Jonathan looked up and noticed that the sky had grown dark. He cast his gaze over to the eastern sky and searched for evidence of the full moon. A wall of black clouds had risen above the horizon, towering over the lowly mountains of rock and stone, extrapolating the designs of these earthly denizens into the heavens and projecting their majesty on a much larger scale. The clouds appeared to be just as rigid as their terrestrial counterparts, but if he stared at them intently, he caught them drifting and morphing into new shapes. Without warning the various patches of grays and blacks assembled themselves into a human face. The tousled hair, the protruding nose, the cheek line, the ear lobe and lower jaw, became suddenly all clearly visible, molded into the facade of the dark clouds, the head facing to Jonathan's left, presenting him with a partial profile.

After a few moments a small rent appeared, adding a vacant eye to the mix of other facial features, the eye staring at nothing in particular, much as the eye of a statue. A few moments later the full moon slipped up behind the aperture and supplied the key ingredient. The face abruptly came alive just as if someone had plugged it into a wall socket. The face shifted its gaze towards Jonathan, staring directly at him with a raised eyebrow and an incredulous expression, apparently astonished to find a man and his dog

172

standing on the rim of a broad valley, all alone in the gathering darkness, lost in the middle of nowhere, occupying a place that no one ever occupied. Jonathan adopted a similar look on his face. The celestial stranger's unexpected intrusion into his private domain had made him feel uncomfortable. To his dismay, the cloud's quizzical expression morphed into a look of disapproval. The cloud had apparently discovered who he was.

The eye abruptly closed and the moon withdrew from view. The parts of the face broke apart and Jonathan was left once again with an amorphous mass of black clouds. The moon vanished and he found no sign of its whereabouts. It sulked behind closed doors for a while, but then coyly peeked out through a rift. Jonathan waited patiently, hoping for a second appearance. After a long delay the flirtatious deity made a grand entrance into the night sky, soaring triumphantly above the leaden bank, its muted light washing over the incredible range of billowy peaks.

Jonathan glanced over at Emma and noticed that she was the color of moonlight. She stood at attention by his side, full of expectation, stimulated by the appearance of this new lantern in the sky. She appeared to glow in the dark, much in the manner of the moon itself. The kinship was now strikingly apparent. He had always known that Emma was born on the night of the full moon, but in all this time he hadn't understood what that meant. Emma was one of the moon's daughters, and had inherited the moon's lovely aura. Jonathan returned his eyes to the heavens, exhilarated by his new insight, and gave thanks to the heavenly maiden for the precious gift he received, the little bundle of fur he'd been handed on that wonderful March night four years ago.

—7—

They arrived at the adobe farmhouse after dark, thirty years ago, during a raging blizzard in mid-October. The renters had finally left, and Jonathan and Sarah were eager to reclaim her old house. They went back to their rental property on Llano San Gabriel where they'd been staying since they had gotten kicked out of La Bolsa, but Jonathan was unable to climb the snow-packed driveway in his old Honda Civic. The storm was gathering momentum and night had already fallen. Jonathan parked at the bottom of the long driveway and they both carried their belongings down on foot, peering helplessly into the impenetrable mass of snowflakes. They then drove through the height of the storm, blinded by the reflection of their headlights off the incredible profusion of flakes. They were the only car on the road, all alone in the blackness of night, swerving uncontrollably on the sloppy dirt road.

Eventually they arrived at a quiet and deserted asphalt highway and after twelve more painfully slow miles creeping over the buried pavement, Jonathan swung the car onto another dirt road, sliding once again on the slick ice and sodden clay, desperately trying to keep a treacherous course between the overly deep drainage ditches along each side of the road. The trenches were now invisible and he could easily imagine that they were no longer there. Every sign of the danger had been obliterated, but if a tire accidentally slid into one of those ruts, they'd be leaving their belongings behind and walking the last few miles to the house on foot.

The car finally topped the ridge and slipped over the edge and they began the long descent down to the river, skidding and veering from side to side. Many times Jonathan stopped the car completely, then crawled slowly forward, repeating the process over and over again as they inched their way down the steep incline, the tires barely holding onto the icy surface. The valley below was utterly pitch black. As far as Jonathan could tell the valley was completely uninhabited, with no faint light or sign of life anywhere. Of course, the electricity had gone off. The electricity always went off whenever there was weather, even if it was nothing more than a powerful gust of wind.

The house was dark and bleak as they approached the sagging wooden gate. Flakes were still falling and the snow was too deep to open the gate, so Jonathan just parked the car directly in front of it. He and Sarah circled around on foot through a part of the adjacent field carrying what they could, then walked down the remainder of the driveway stumbling and teetering on the collapsing crusts of snow. The rooms of the house were cold and empty, a kitchen table and a couple of chairs, a dresser under a window, not much else. Jonathan sighed deeply as he toured the dismal interior. They were stuck there. The car would never climb the steep grade out of the valley in these conditions, and they had nowhere else to go anyway.

A bank of solar panels stood in the upper field near the house. In the past Sarah had used these panels to power an irrigation pump during the summer months. Jonathan wasn't sure if the system still worked, but he felt that it was worth a try. He found a box of long extension cords in the shed and ran a line out from the house to the field, trudging through the snow in his arctic pacs. A full moon began rising over the hills as he plugged the cords together, distracted by the sparkles of moonlight dancing all around him on the snow. To his surprise the batteries still held a good charge and he was able to use the inverter to power a light bulb and a radio, which gave the grim, lifeless house at least a touch of civilization.

Sarah's ex-husband had fashioned a **banco** in one corner of the living room, a mud bench he had molded into the walls, and it was here that Jonathan decided to put their bed. In the poor light of a naked bulb, he nailed together a wooden pedestal matching the height of the **banco**, filling in the void and changing the L-shaped seat into a rectangle capable of supporting the platform of the bed. Sarah had started a fire in the cooking stove in the kitchen and by this time the heat had begun to infiltrate into the living room.

In this manner Jonathan inherited an historic adobe house dating back at least to the 18th century when the area was first settled by Spanish-speaking peoples, but almost certainly occupied even earlier by native Americans. The house was constructed in the style of a traditional pueblo, each room added to the existing constellation of rooms as an independent cell with its own four walls, not relying on any of the previous walls for support. Curiously the house was halfway buried in the ground on two sides. Jonathan didn't realize it at the time, but this subterranean feature was the result of other rooms that had once existed adjacent to the house. The roofs had been dismantled long ago and the lumber removed, whereupon the remaining mud walls dissolved from exposure to the rain and snow, eventually settling into a relatively uniform surface elevated several feet above the original ground. He had since excavated the soil in a few places and discovered stone foundations radiating out from the existing rooms.

Jonathan's first years in the normally dry desert were exceptionally rainy. Areas around the state became flooded. The ground surrounding the house gradually became saturated and the constant moisture rotted the adobes of the half-submerged walls causing one of them to collapse inward. He and Sarah awoke one morning to find sunshine streaming in through a large and steadily widening gap near the top of the wall. Jonathan immediately abandoned his plans for that day and drove down to Santa Fe to purchase a number of construction jacks and heavy duty timbers. He suspended a wooden beam from the ceiling using screw hooks and jute ropes. He then laid another beam on the floor directly opposite the ceiling beam and placed the jacks between them.

The flat roofs had originally been covered with only a thick layer of dirt, which was left exposed to the elements, but the modern invention of tarpaper was later added to the dirt and spread out on top of it. The paper invariably deteriorated over time and developed leaks. The solution was always to add more layers of new tarpaper rather than remove the old worn-out layers. As the years progressed, the roofs steadily grew in magnitude and eventually evolved into monstrous weights, yet the rainwater still managed to find ways to seep in and soak the dirt underneath. Once inside, the moisture was now trapped by the multiple sheets of tarpaper and unable to escape.

The wet dirt became extremely heavy, so the jacks had to be strong enough to support this load. Once a wall had been relieved of its burden, it could then be broken apart and the debris hauled away. As Jonathan worked down to the base of the wall with pry bar and shovel, he managed to pull out some intact adobes. He found the bricks to be curiously small, only two to three inches in height. The bond beams atop the walls were also miniature items, nothing more than slender saplings. The doors to the house were unusually small, barely five feet tall. As Jonathan worked, he thought about the families of tiny people who had done all of this, trying to picture them in his mind.

The dirt walls were porous and after a while dust, soot, and cooking grease impregnated the surfaces. There was really no way to wipe the walls down or effectively clean out the filth. The only solution was to cover the grime over with a fresh coat of mud plaster. This had been done a number of times in the past, and if a previous application of mud had been troweled smooth, then the next plaster coat did not adhere to it very well. The new layer could be peeled away in rather large, intact sheets, revealing the old surface beneath it. This gave Jonathan a restored view of the wall that everyone had once enjoyed, granting him a rare glimpse of a past now long forgotten, the occupants having perished many years ago. The rooms had gone through many transformations over the years, times had changed, new generations had taken over the property from their parents, and perhaps new relatives had moved in to join the household.

Jonathan was musing over the long history of the house as he pulled away the layers of plaster along one wall, working toward the bare adobe blocks underneath, when he noticed that the original plaster had been blackened by soot. Further excavation revealed that a fireplace had once existed at this spot, but then it had been subsequently removed. Digging into the base of the wall produced numerous pieces of wood charcoal and some animal bones. Not being knowledgable about vertebrate anatomies, Jonathan speculated that the bones might be dog bones, being approximately the right size and shape. He was aware that times had often been tough and it had been a common practice for people to eat their dogs. He had read that both mountain men and Indians considered dog meat to be a delicacy, and apparently engaged in such feasts regularly.

He went over to his dresser in the next room and took out a clean handkerchief. He spread the white cloth on the floor and placed the bones on top of it, folding the corners into a neat bundle. He grabbed a shovel on his way out the door and carried the items into the desert. The dogs followed him thinking that they were all going for a walk, but when he stopped to perform his duties, the dogs quickly took off running to investigate the area up ahead, playing and chasing each other with great zest. Jonathan located a nice spot at the base of a piñon, dug a small hole, then buried the bones, just in case they were actually dog bones.

He wondered what he would have done had he been starving in the middle of a long winter, stuck in a bleak world of cold and snow with nowhere to turn. He imagined walking up to Emma and petting her gently, then cutting her throat open, watching the look of betrayal in her eyes as she died on the floor, her smiling eyes forever banished from his life. What kind of existence would he have had following such a heinous act? Was the grim fate of both of them starving to death really any better? Certainly it was. Jonathan had decided that he and Emma would share everything together.

As Jonathan was thoroughly engrossed in the project of building a new kitchen wall, a wall in the back room also collapsed. He now had another major construction project on his hands, but he was getting good at this. One day after the old wall in the back room had been broken apart and taken away, Jonathan was laying the cement foundation for a replacement wall when a thunderstorm swept into the valley. With a heavy layer of dirt above his head, he felt safe in his earthen cave, so he ignored the inclement weather and happily kept working. The missing wall gave him a wonderful view of the fields outside. He paused often to watch the neighbors' horses peacefully grazing off in the distance. As he was laboring away, a bolt of lightning came down from the heavens and hit the ground about fifty yards away. The bolt was broken into segments, a series of incandescent strips each several feet long separated by vacant spaces. It hissed and sizzled, smoke issuing from the white-hot ribbons. Since the air was rather humid, the vapor was probably steam.

Jonathan was quite impressed by this event, yet he still felt immune to the danger, so he continued working as before. After a few minutes a second bolt struck in about the same place, only this time the bolt made an abrupt u-turn approximately twenty feet above the ground. The lightning continued onward, rising and falling like an archer's shot, soaring over the fences and trees and flying above the heads of the horses standing out in the fields, finally slamming into the rocks at the mouth of the valley about 300 yards away, the incident accompanied by a fierce crack of thunder. Jonathan was shocked. The house sat on a terrace slightly elevated from the floor of the valley, granting him a perfect view of the whole affair, a privileged front row seat to the spectacle. In all other respects the second bolt was identical to the first, broken into segments and sizzling and smoking. He instantly realized that he wasn't safe standing under a dirt roof. If lightning could travel horizontally, then it could easily come into the room through the opening provided by the absent wall. In fact, he was lucky that the lightning had gone the other way, because if it had come toward him, he probably wouldn't be standing here today.

Jonathan had been led to believe that this kind of behavior was impossible. If lightning was nothing more than the discharge of static electricity seeking a ground, it would never have turned around like that, being already so close to its intended target. The long arc over the valley could not have possibly been the path of least resistance—300 yards compared to the remaining 20 feet. If the lightning bolt had actually been attracted to the earth, the earth was right there in front of it, only a short distance away. None of this made any sense according to what he had been taught in class and what he had read in books. Since physics textbooks didn't normally have chapters devoted to lightning, he must've read it in some other publication, yet these were the kinds of stories that everyone was made to believe, blatant lies, ridiculous fabrications proffered by people who really had no idea of what they were talking about.

Many times Jonathan had observed lightning traveling in a crooked but roughly linear path, only the path was canted from vertical at about 45 degrees. He'd considered the possibility that this angle was aligned with the earth's magnetic field or the earth's rotation, but neither of these possibilities seemed to be the case. On several occasions he'd watched a number of successive bolts, all emanating from the same cloud, all possessing the same diagonal trajectory—basically repetitions of the same bolt linearly transposed across the landscape—but he saw no reason for this strange behavior. No discernible force was pulling the lighting to one side, certainly not the wind or gravity or the separation of charges, and the path was about forty percent longer than a straight line to earth, thus correspondingly increasing the resistance of the air.

A few years ago Jonathan was camping near Mount Withington and on the final day he was caught in a spectacular lightning storm. The next morning he drove up to the summit where the forest service had built a fire lookout. A woman came out of the cabin eager for some conversation. The cabin was perched high atop the highest outcrop of rocks in order to provide the occupant with a panoramic view in all directions. Given that the cabin was a prime target for lightning, a fine opportunity for lightning to easily reach ground, Jonathan asked the woman how many times the cabin had been hit by lightning, figuring that it must've been a lot. She replied that, to her knowledge, it had never been hit by lightning. Puzzled by this, he inquired about yesterday's fantastic storm. She said that yesterday the lightning repeatedly passed by the cabin, each bolt canted at a sharp angle, but then struck a group of rocks about eight hundred feet below the cabin on the face of the mountain.

Jonathan had read something recently that might at least provide a partial explanation. The article proposed a strange idea: the cabin, by being made entirely of wood, was invisible to the lightning. A similar story was currently being passed around on Airstream forums. Since he'd been spending the last few summers living in an Airstream that was parked in the middle of an open field, the topic was of interest to him. The theory had been put forward that since the trailer was mounted on rubber tires, if he placed wooden blocks at least 5 inches high under the stabilizing jacks and unplugged the ground power, the trailer was completely isolated from ground. The lightning then did not see the trailer as a target, that is, as a convenient way of reaching ground. It might still hit the trailer, but it would be a lucky shot due to blind chance. On the other hand, a grounded metal structure, it was said, was known to attract lightning.

Jonathan was skeptical of these explanations. A metal storage building stood near the Airstream. It was built on a wooden frame so he didn't know how well it was grounded, but it had never shown signs of ever having been hit by lightning. Since a tall tree was often the target of a lightning bolt, the underlying premise of the story seemed patently false, unless it was the column of sap within the tree that conducted the electricity down to the roots, but in that case, lightning would never hit a dead tree, only living trees with large amounts of sap in them. He had seen dead trees with blackened scars that were undoubtedly due to lightning strikes, but he didn't know if the trees had been hit before or after they had died. The evidence seemed rather patchy. He wondered about the old adobe house. Being made entirely of wood and dirt, it would be indistinguishable from the surrounding terrain.

The years went by, the summer monsoons came and went, and neither the Airstream nor the house was hit by lightning, yet the leaking roof became a huge problem. One day Jonathan got the brilliant idea of removing the tarpaper and placing a two inch layer of cement over the dirt. The cement, he thought, would form a waterproof barrier and provide a level surface and for the first time he would have a proper roof over his house. He climbed up on the roof and cut the thick sandwich of multiple tarpaper layers into three foot squares using a portable reciprocating saw, but even these small sections were so heavy that he could barely drag them across the roof and fling them over the side. The dirt underneath was soaking wet so he let the exposed roof dry in the sunlight for a few days, but this did little other than air out the top surface. Many weeks of hot sunny weather would be required to thoroughly dry all the dirt, and up in the mountains during the middle of the summer that was not just a rarity, but an impossibility.

He framed a section at the far end of the roof to serve as a concrete form, mixed the cement on the ground, then hoisted buckets onto a scaffolding and onto the roof. The work was hard but it went well and at the end of a long day the section was finished. The slab looked beautiful, just as he had imagined it in his mind. He celebrated his accomplishment that evening and toasted to the glorious future of the house while playing albums and dancing around the kitchen. His dogs eyed him suspiciously from the adjacent bedroom desperately trying to get a head start on a long and restful night's sleep.

Jonathan began to have his doubts the next morning after sleeping on the idea overnight. The cement was very heavy. Over time it would crack and deteriorate and start leaking. A friend came over and brought up the fact that cement was by no means impervious to moisture. The incorporated moisture would freeze in the winter and expand, fracturing and splitting the concrete even further. Jonathan became convinced that the whole scheme had been a bad idea and while the cement was still green, he broke it apart and threw the crumbly, jagged pieces off the roof and back onto the ground from whence they came.

He had another idea. He would remove all the dirt and frame a flat roof out of lumber in its place. He backed his pickup truck against one of the walls and shoveled the dirt directly into the bed of the truck, then hauled it away. When he had collected the last of the dirt and swept the surface clean, he discovered that the ceiling boards had rotted through in many places. The ceiling looked fine from inside of the house, but from above he could see that the damage was worse than anything he could have imagined. Thoroughly committed to a new roof at this point, he tore off all of the ceiling boards only to find that several of the **vigas** had also rotted in the middle. They looked like hollowed out canoes. He had no idea that any of this had happened. From below, the ceiling appeared to be fine.

He didn't really need to replace the **vigas** because the framing alone would easily support the roof so he left them in place. He decided he would cement cap the tops of the walls, halting the progress of the rats and squirrels and mice as they burrowed upward inside the dirt walls. He would nail hardware cloth around the perimeter, bending the metal mesh into L-shaped sections extending to the height of the framing and then well out onto the ceiling boards, generously overlapping the sections and securely fastening them in place with roofing nails. The rodents would have no way to gain access to the interior of the new roof.

The plan had been well thought out with a clear understanding of the problems involved and the reasons why the original roof had failed. The scheme appeared to be foolproof and he was certain that his worries were over. To Jonathan's dismay, the rodents didn't take long to figure out ways around his trustworthy barriers. They somehow managed to get into the roof and once inside, they built nests and carved out networks of tunnels through the batts of fiberglass insulation, apparently in order to socialize with each other. They raised families and multiplied freely, happily urinating and defecating over everything the whole time. They choked on strands of fiberglass and died from asphyxiation while others succumbed to virulent rodent diseases. The carcasses began to slowly rot away in the dark recesses of the elaborately framed roof. Maggots fell from the spaces between the boards and landed on the bed where Jonathan slept. Sometimes he found droves of them squirming and wiggling underneath his pillow when he turned the covers back at night. The roof began to leak and the fetid seepage thoroughly wetted the foul chambers of the famed roof. Black mold quickly spread through the moist environment and mushrooms sprang out from the ceiling boards, then up from floorboards where the water repeatedly dripped. The room began to stink in a way that was indescribable.

Jonathan's ideas had fit together so well that he couldn't understand where he had gone wrong. The reasoning had been absolutely ironclad yet his conclusions had been utterly false. He had merely created a fantasy room out of his ideas, just as the physicist had created a fantasy world out of his mathematical relationships, but this room in his head was not the same room as the one in the house. He could start over with a fresh set of ideas, but he had no guarantee that things would work out better this time. Ideas were just ideas. Reality frequently mocked his most brilliant brainstorms and he often struggled to maintain his sense of humor in the face of such abuse. Reality apparently didn't care what he thought.

His calculations had been straightforward. He added the idea of a cement cap to the idea of a sturdy wire mesh and put them on one side of an equation. On the other side of the equation he added together the intelligence of rodents and their physical abilities to dig and gnaw through materials. In the middle, he inserted a "greater than" symbol to represent the formal relationship between the two expressions. The strength of the barrier was greater than the rodent's capabilities to break into new spaces. The logic was impeccable, so he took the equation to be an irrefutable law of nature, much in the manner of Newton's Law of motion. He should have known better. Nature had easily overthrown Newtonian mechanics, just as it had overthrown his Theory of Rodentivity.

The roof experiment had been an unintentional study in rat behavior. Even though Jonathan's ideas had failed miserably, the result of the experiment taught him something about rats he hadn't known before. He understood rats better now. The reality he fabricated exposed the rats' natural inclinations and allowed him to better gauge their capabilities.

Jonathan had provided the rats with an interesting new playground, a complex of rooms, hallways and living quarters, an elaborate maze of obstacles and materials, but he was primarily studying the effects of his house on the rats, seeing firsthand the consequences of his imaginative ideas, the often unforeseen implications of the many arrays of wire mesh and unlikely assemblies of concrete, adobe and fiberglass, the bewildering structures of wooden frames. His laborious constructions had influenced the basic facts of the rats' existence and changed their lives forever. He'd given the rats opportunities to adopt new roles and create new patterns within the parameters of the game he had set up for them. Since he wasn't dictating the behaviors of the rats directly, he could say that he was observing the rats and not just studying the impact of his work, yet what happened up there in the roof was all Jonathan's doing. He alone was responsible for the progressive deterioration and eventual dissolution of the house and the rats only played a subservient role in the overall scheme. He should never have given them the chance to move in because they easily won the game. They also paved the way for all sorts of other organisms to take over. Jonathan's actions had so many consequences which he had not anticipated, and his original impetus quickly carried events into previously unknown territories. Similarly, the imaginative ideas and fabricated realities of laboratory setups carried with them sets of hidden consequences, setting in motion chains of events not intended by the experimenters. Measurements did not simply influence the realities of the world—measurements were the only realities that physicists could observe. Measurements were assemblages of mathematical forms in which the objects of nature made themselves at home. The objects invented themselves within the confines of novel situations and strange things happened, new realities emerged out of the fusion of disparate elements, and a dialectic was born which then assumed a life of its own, taking the participants far beyond the control of the instigators.

The collective activities of the rats became just another aspect of Jonathan's roof. Jonathan could say that the roof provided him with a means for studying rats, but he could also say that in studying the rats' behaviors, he was actually studying the roof. The rats provided him with another picture of the roof, taken from a novel perspective. However, there was one thing Jonathan could never say. He could not say that in studying the roof he was studying the nature of the universe.

Jonathan could build as many roofs as he wanted but this would never lead him to a truth about what existed, just as the physicist could build all sorts of devices and none of them told him anything about the world outside. The physicist could claim that he was studying the laboratory phenomena through his instrumentation, and Jonathan could counter that the physicist was really just studying the instrumentation through the laboratory phenomena, but none of this went beyond the circle of fabricated realities and artificial constructions that the physicist had built for himself, the labyrinthine environment where new behaviors emerged as light and matter adapted to new constraints, and atoms and molecules discovered things about themselves that they hadn't known before.

The leaking roof and rotting walls were forced to take their places alongside the many other problems of the old adobe house, some of which went well beyond mere structural insufficiencies, involving things that were not so easily dealt with. Once, long ago, Jonathan had fallen asleep at night with the outside door left wide open. The season was early spring and patches of snow still lingered in the fields. The cold, damp air filtered into the dirt rooms and threw him into a deep sleep. The heavy scent of rotting leaves and decomposing grass settled into the rooms, but otherwise he was alone in the house, having no canine companions that night.

Jonathan's eyes sprang open and he awoke with a start during the darkest part of the night. Something was in the room with him. He wanted to sit up and search the darkness with his eyes, fearing that a wild animal had crept into the bedroom while he was asleep, but strangely he couldn't move. He couldn't even turn his head. He stared helplessly up at the old wooden boards above his head, gripped with terror. He could feel the creature approaching him from the side, slowly nearing his head, much as the reptilian monster in the movie "Alien." His body was paralyzed and he was powerless to fight back.

The demon was about to whisper something in his ear. He could sense the strange presence next to his head, when a surge of energy fueled by a sudden panic broke the spell and he lurched forward into an upright sitting position. At that exact moment he heard a blood curdling scream coming from the vicinity of the open door, or perhaps from the orchard immediately next to the house. He had been released from the clutches of whatever it was that had held him and the demon bellowed in anguish, furious at its failure to accomplish its goal, apparently departing into the night with a wail of sorrow, or the articulation of a farewell curse. Jonathan's mobility gradually returned to him and as soon as he could muster enough strength he walked through the blackness of the empty house and peered out the open door at the dreary night. The trees loomed ominously outside, supporting spiderwebs of bare branches under a gray, overcast sky, the stark outlines standing quietly in the vague mists, much as the trees in a graveyard lording over the headstones of the dead. He saw nothing moving in the orchard. He closed the door and bolted it, then went back to bed.

This was the first time he met the dreaded Curse of the Ancient Ones, the ones who had come long ago during bad times and in their destitution and depravity they raped their daughters out of blind lust, beat their wives in fits of drunken rage, brutally clubbed their neighbors over petty disputes, stabbed their enemies with steel knives then watched with satisfaction as their victims bled to death on the dirt floors. Years later the Curse came out of the ground at those places where the blood had been spilled, oozing from the splattered walls that were still outraged by the crimes they had witnessed. The dead cried out for vengeance and the tortured souls demanded that their sufferings be redressed. Many bad things had happened over the years and justice had not been done.

Numerous other visitations of the Curse were to follow. On several occasions the Curse tried to kill Jonathan, casting a spell over him and getting him to do things he ordinarily would never have done. The Curse affected others in the valley in much the same way. Once the Curse got his neighbor to run a circular saw across his thigh, tearing it open. The guy had just moved into the valley and he was confused. He kept muttering: "I don't understand how this happened. I know better." Jonathan didn't say a word. This guy would have to figure it out for himself. He'd have to come up with an explanation of his own, one that suited him. For Jonathan, the Curse was very real and it was not something to be trifled with.

In the 30 years that he lived in the valley, he never again heard a scream such as the one he heard that night. The sound might have had no connection to the events inside his bedroom, but what were the chances that the gruesome howl occurred randomly at the precise moment that he sat up? Since Jonathan only resided in the valley half the year and there were 473 million seconds in 15 years, the odds of the scream happening at that very second were 473 million to one. Was it rational to suppose, against such odds, that there was no link between the two events?

The encounter hadn't been a dream. Jonathan had left the door open in the evening and he had gotten up in the middle of the night to close it. He could not honestly say that nothing happened. His experiences were connected to the events in his world much as the experiences of the physicist were connected to the events in his laboratory. The general outlines were the same in both cases, but trying to understand the Curse in terms of physics was absurd. According to physics, whatever was in the room with him was an assemblage of molecules and forces, but how those forces held the molecules together or how the molecules acted upon the tissues of his brain was a mystery. A phenomenon such as this fell outside the province of physics, along with all the other experiences Jonathan had witnessed while living in the desert. Physics was played on the game board of the laboratory and nowhere else. So unless Jonathan built a laboratory, he was excluded from the activities of physics. He might as well try to play solitaire without a deck of cards. But setting up a laboratory would only be a distraction for him, diverting his attention away from the world he so desperately sought to understand. Jonathan found the narrowness of physics stifling.

Physicists created the situations where the methods of physics could be applied. The whole thing was a setup. The visitation of a demonic spirit was not reproducible so the phenomenon could not be investigated, but more than that, according to the physicist, the phenomenon wasn't real. If physicists could only understand what could be furnished on demand, then the bulk of natural occurrences must be summarily dismissed. But in neglecting this wealth of phenomena, they were omitting the most essential pieces of the puzzle, and thus they could never form a complete picture of the universe in their minds.

Sarah had bought the property 45 years ago during the hippie days of the sixties. She had learned gardening from her immediate neighbors, a traditional hispanic couple still living the old way of life. Maria made a meager living growing green chile, a difficult task given the short growing season at such a high altitude. Alfred was a sheepherder who was gone much of the time tending his flocks up in the mountains of Wyoming. One year Alfred got very sick and was dying of cancer. Sarah would see him out in the garden all night, hoeing and cultivating by the light of a kerosene lantern hung from the handle of a shovel stuck in the ground, desperately dealing with the pain and unable to sleep.

Two days before Christmas Alfred hung himself. The piñons in the area were stunted from the extremes in temperature, so Alfred tied the rope around a sturdy branch that was a bit too low for the purpose. He was forced to grab his ankles to hold himself off the ground while he swung from the noose, but he didn't let go until the deed was done. The neighbors found him the next morning and cut him down, leaving a portion of the rope tied around the branch.

Jonathan visited the site regularly. Someone had placed a nice hand-carved cross under the tree to mark the spot. During a cold and lonesome winter, Jonathan became depressed living in the valley and obsessed with the idea of following in Alfred's footsteps. One night he got very drunk and decided to kill himself. In his wasted and befuddled state, he could not locate a rope, even though there were several around to choose from, so he grabbed an extension cord from the shed and lurched out into the darkness. He had no trouble finding the wooden cross on the cloudy, moonless night, even in his condition. He fumbled with the cord, working feverishly to tie a good knot, then passed out cold under the tree. He awoke some time later, probably around midnight, lying flat on his back, staring up at the short piece of rope dangling from the limb directly above his head. He laid there for the longest time, just staring up at that piece of rope. He carefully studied the stark network of branches radiating out from the main trunk, then shifted his gaze to the ominous, gray clouds in the grim winter sky. The night had gotten very cold and he was freezing. He finally roused himself and stood up. He untied the rope from the branch and flung it as far as he could, then stumbled home and went to bed, leaving the extension cord strewn on the ground. He went back to retrieve it several days later, but the rope was nowhere to be found. He wanted to put it back on the tree where it belonged, as a symbol of Alfred's passing.

Jonathan had no reason to kill himself. He was still young and strong. He had many things he could do to relieve his unhappiness, many directions to move his life, but he was under the spell of the Curse. The Curse wanted to get rid of Jonathan in order to get him out of the house, the house that rightfully belonged to the Curse. The Curse wanted Jonathan dead, but it didn't have the means to accomplish this task, so it had to find a way to entice him to perform the deed himself. The Curse acted the only way it could, by corrupting Jonathan's thoughts. The Curse had never given Jonathan permission to take over the house and turn it into a place of his own. Who did Jonathan think he was anyway, to just march in there and wrest control away from its previous owner?

Jonathan didn't know where the Curse was today or what it was doing right now. The old adobe farmhouse had been uninhabitable for six years now, overrun with pack rats, field mice, ground squirrels, black widows and brown recluses, molds and mushrooms, and families of snakes. The Curse had won the war for control of the house and Jonathan had long since admitted defeat and fled to the safer and friendlier environs of the desert wilderness. He continued to spend parts of his summers at the opposite end of the property, as far away from the old house as possible, where Sara sometimes joined him to help with the gardening. He refused to sleep in the rotting, infested, and crumbling old dwelling with its dank atmosphere and large populations of small creatures creeping across the floors and crawling up the walls. He always wore a full respirator when he first entered the house and immediately opened all the doors and windows to force the unpleasant odors to evacuate the premises before he dared inhale the remaining whiffs of toxic gases and foul smells. He much preferred the fresh air of open spaces.

Emma was with him today participating in a venerable New Mexico tradition. He had opened the gate on the *acequia madre* and let the water out. The water gushed and streamed onto the field and cascaded over the broken ground, collecting the fragmented reflections of sunlight into a marathon of little swimmers all bobbing up and down on the surface as if they were enjoying themselves in a miniature water park. Emma stared at them in fascination, straddling the rippling currents with her head stuck in the most peculiar position, her legs stiff with her nose pointing straight down. She watched them closely, suspecting that they might suddenly take off running across the fields hopping up and down like rabbits. She wanted to make sure that she caught them before they got away.

The irrigation required only Jonathan's intermittent attention and occasional intervention. While flooding the final section of land along the lower fence line, he was practically standing in the front yard of his closest neighbor. The neighbor had hired a friend of his, whom Jonathan happened to know, to construct a flat roof over a new adobe room he had recently added to his house. The guy was up there today doing the entire job with nothing more than a hammer, a piece of sting, and a chainsaw. Jonathan watched him work during his frequent spare moments. The guy approached the space where he wanted to put the next board, stretched the string out from one end of the opening to the other, and pinched it with his fingers. He then walked over to the pile of boards and stretched the string across a fresh piece of lumber. Keeping his eye on the spot where his finger had been, he picked up the chainsaw, pulled the cord, and started the engine. In one quick move he lopped the end of the board off, then placed the board in the opening and nailed it in place. He proceeded in this fashion, board by board, until the entire roof was finished. The result wasn't pretty. The chainsaw tore out small chunks of wood and the cuts weren't even remotely square, but since the whole thing was going to be covered up anyway and no one would ever see it, why bother to do a nice, neat job? Otherwise the measurement technique worked quite well. He accomplished the whole task without the slightest recourse to numbers and calculations.

This guy was simply following an unwritten rule in northern New Mexico that a person should never exert more effort than was absolutely necessary to accomplish a task. The traditional method of measurement involved a lot more work and required a tape measure, a rather expensive item that had to be purchased at a hardware store. Few people had that kind of money to throw around. A tape measure had units of length that were repeated over and over again along its entire length. Numbers were printed on the side which counted these units, yielding a numerical figure that represented the length of the board. Instead, the neighbor's friend copied the whole length directly in one move. The only problem was, this route did not lead to mathematics, since at no point did he convert the world into numbers.

Each year Jonathan measured the effect of various compost teas on the growth of the plants in his vegetable gardens. For a time he kept a log of the heights of the plants at regular intervals of time, and at the end of the growing season he plotted them against each other on a graph. This approach involved a lot of work, not only getting out there with his tape measure and moving around in the dirt on his knees, but also writing down all the figures in his notebook, but it was very scientific. Later Jonathan learned to simply plant two rows of the same cultivar alongside each other and apply one fertilizer to one row and a different fertilizer to the second row. Then he could walk out there each day and see the difference with his own eyes, but he could never publish the results in a scientific journal because he had not taken the crucial step of quantifying his results. No one could do a mathematical analysis of his findings, yet in every way the conclusions were just as valid and the method was far superior in terms of ease and efficiency.

Measurement was a comparison between two things and Jonathan could refine it as much as he wanted without ever resorting to numbers, but this kind of measurement was not important in physics. It was the conversion of physical reality into numerical quantities that mattered because this was what opened the door to a mathematical universe. Physicists could start doing calculations and just forget about the original objects and all their inaccuracies and imperfections, their indeterminateness and endless indefinable aspects. Numerical measurements launched them into another realm, taking them to a place where they'd rather be, a fantasy world of pure, mathematical beauty where everything was as it should be, where everything was neat and precise and logical.

When Jonathan had tried to measure the azimuth of the sun, instead of a protractor, he could have just fastened two sticks together by installing a pivot pin at one end of each stick and used the crude device as a carpenter's miter, copying the angle by transferring it to the sticks. He would now have the precise angle in terms of sticks, but not in terms of numbers. The problem with a measurement of this sort was that it lead nowhere. Nothing more could be done with the sticks. The physicist wasn't satisfied with a single measurement, but wanted to tie all measurements together into a system of mathematical relationships. The introduction of numbers involved a change in the nature of things, not apparent at first, where the ties and connections started being supplied by the ideas of algebra and geometry and not by the concrete realities of the situations, but here Jonathan had already begun with geometry and the idea of a triangle, never for a moment considering the physical realities of the world.

When Jonathan compared two rows of plants with each other, he was able to take everything into account, not only the general coloration of the leaves, but subtle abnormalities, slight patterns of variations, tinges of brown or yellow along the margins or at the tips, the separate colorations of the veins. He was able to note minor deformities in the shapes of the leaves indicating a corruption of the plant's metabolism. He observed the leaves pointing either slightly up or down signifying perhaps an imbalance in hormone levels or perhaps a turgidity related to water uptake. He was able to check for insect predation, a reflection of the plants defense mechanisms, along with the plant's susceptibility to viral or bacterial attacks. Feeling the texture and thickness of a leaf, gauging its pliability or brittleness, told Johnathan something about the overall health of the plant. A mathematical analysis, on the other hand, emphasized the relationships between numbers, relationships he had originally learned not through a close and long-term association with plants, but through a close and long-term association with books and blackboards. Once the initial enthusiasm had worn off, quantification could be seen as a tradeoff, the exchange of one thing for another.

Attempting to express the numerous attributes of plants in terms of quantities and equations was a grueling and exhausting task requiring a great deal of mathematical invention. Quantification was the long way around, a detour through the endless expanse of mathematical imaginations, making things much more complicated than they needed to be. Physicists took relatively straightforward observations and they corrupted them, subverted them, convoluted them, initiating a lengthy digression into a labyrinth of cryptic abstractions—for the sake of what? Mathematics at any cost, no matter what it took, insisting that quantities become the relevant parameters of the phenomena. Mathematics was not on the road to anywhere. It was a destination rather than a waypoint—but where else was there to go?

The circuitousness of quantification and numerical analysis was clearly illustrated in robotics. Robots attempted to mimic human behaviors in strange ways, by performing complex computer calculations, even though that was not at all how humans performed the same tasks. Mathematical calculations were the most inefficient way of doing anything —outside of mathematics—but since this was the only way computers could function, scientists were stuck with trying to make this process work. The reason why robots could never rise above the behaviors of even the most basic organisms was because mathematics was so inefficient at capturing biological and physical realities. Robots were forced to do everything by manipulating large volumes of data, but biological organisms never made calculations of any sort because adding numbers together did not come naturally to living things.

A robot had no geometric understanding of anything. Walking to the fridge by going around the table did not involve geometry because the task did not require envisioning straight lines or calculating the angles between them. The robot's computer processors might very well be making calculations, but that was not what a person did in order to walk to the fridge. The computer scientist could say that the person heading for the fridge was making vague estimations of distances and directions and thus employing geometry in a crude way, but this could never be more than a metaphorical use of the word 'geometry' whose essential characteristic was precision and exactitude and the implementation of specific types of geometrical constructions. The robot wasn't really envisioning the points and lines of geometry in its mind as its creator had been when that person developed the software which animated the robot. The creator was envisioning geometry for the robot, utilizing those ideas to create circuitry which then embodied algorithms and the logical flows of data. The creator wasn't teaching the robot geometry, but instead using his knowledge of geometry to tailor the mechanical operation of the robot, telling the motors specifically what to do in response to the electrical outputs of the sensors. Geometry was necessary to design and operate a robot since the robot was a mathematical object and thus it demanded mathematics.

While robots could perform some basic functions, no one seemed to realize how far there was to go before anything resembling intelligent behavior could be realized. The robot was always doing everything the hard way through quantification and numerical computations. The enormous number of binary logical operations that were required for the simplest actions showed just how inefficient this process was. Yeah, it could be done, but it was a stupendous effort for the most meager returns. The deception was revealed when a relatively minor anomaly in the data caused the robot to suddenly veer off course and begin displaying the stupidest behaviors, forcing the operator to hurriedly turn off the power, demonstrating that the robot didn't really grasp the situation at all and its supposed intelligence was just a sham. The robot wasn't mimicking human intelligence, but only following a computer program. Likewise the physicist wasn't mimicking the actions of natural phenomena in his equations, but only following a line of mathematical reasoning.

Emma came running toward Jonathan with something in her mouth. She stopped several feet away and lowered her head, gently placing the object on the ground. Jonathan saw that it was some sort of small burrowing creature, probably a member of the large family of moles that ransacked the garden each year. When he and Sarah harvested the carrots in the fall, many of the tops came free with the slightest tug because the bulk of the carrot had been eaten away by these underground scavengers.

Emma did not lift her head, but maintained the close proximity of her mouth to her victim. At the same time she rolled her eyes upward to verify that Jonathan saw her prize. Her expression said "Look at what I've got!" She was proud of her accomplishment and knew that Jonathan was impressed by her prowess. The small animal was quite attractive, a darling little creature with a cute face. Jonathan stepped forward in order to get a better look at it but Emma immediately snatched her trophy and galloped away with it. "Uh, uh, it's mine!" She took the dead animal to bury it at some secret location to ensure that no one would steal what was rightfully hers.

Jonathan had become accustomed to the daily swings in temperature in his garden in the midwest and had developed a feel for the best planting times according to the flow of the seasons, but when he came out to the desert, he was duped by the abnormally warm afternoons. Not realizing the true extremes of temperatures in this novel climate, he planted his garden way too early. After observing the recurrent frost damage, he went over to the local nursery and purchased a max/min thermometer to investigate further. He set up the thermometer in a central location and constructed a wooden canopy over it to shield it from the sun. He monitored the temperatures as spring gradually turned into `summer. He was surprised to learn that in midsummer, on several occasions, the high temperature had been 93°F and the low temperature 38°F, an overall temperature swing of 55°F. Curious about his findings, he started measuring temperatures at other locations around the farm and discovered numerous microclimates: in the orchard under an overhang of branches, in the pocket between the house and the garage, along the fence line near the driveway, and down by the river amid the cottonwoods. Not only were the temperatures all different, but they changed differently over the course of the day and during the night. If someone had asked him what the temperature had been in the valley that day, he couldn't tell them. There were lots of temperatures and large discrepancies between them. To get a single value, he'd have to randomly select one location out of the limitless number of possibilities and arbitrarily make this the standard, the representative for the entire group.

He could similarly go around weighing all the rocks and calculating their volumes. He could measure the heights and spacings of the bushes, the angular slopes of the hillsides, the velocities of the winds. The scientific maxim that in order to know something you had to measure it stood in contradistinction to the fact that measuring things out in the world was pointless because the values had no significance. The world was a perpetually shifting jumble of quantities, a mathematical mess. So how could the physicist convert it into its opposite, a precise system of rigid mathematical laws? By creating a second world where quantities had meanings and measurements were important, a world where numbers added up to something. The physicist would invent a geometry which existed only on paper. He would fabricate a world which satisfied his demands for order. He would construct a mathematical analogue to the physical world, an imaginary realm of ideas, a network of symbols to replace the real objects, and leave the untidy mess behind.

Jonathan often made it a point to take Emma for a walk at the end of the day after the work on the farm had been completed. He noticed that the sun was setting so it was time to quit the irrigation. He blocked off the water, abandoned the field and began marching up the well-worn path out of the valley. The first part of the trail was rather easy, a steady but gentle incline over the relatively flat ground of an old road, but the last part required some heavy climbing. The greatest challenges were a couple of deep and treacherous rifts that hadn't existed when Jonathan first started walking the trail, but had greatly enlarged over the years due to the repeated runoff from the hills. Each time he followed this route, he invariably thought about cutting down some nearby standing deadwood and laying the trunks across the openings to serve as foot bridges, but he never seemed to have time to do this. All the while the openings got larger and larger and more and more difficult to cross.

Once beyond these annoyances, the trail was steady and unbroken, curving sharply as it rose, avoiding the precipitous walls of a towering sand cirque. Emma raced up the trail and disappeared around the bend. Unbeknown to Jonathan, a coyote was at the same time coming down the trail. The animal suddenly sensed their approach and erupted into an earnest song, belting out a steady stream of yelps and squeals, extremely upset that Jonathan and Emma had thwarted its efforts to access the valley and scavenge fallen fruit in the orchard, apparently a traditional evening ritual during the summer and fall months. Jonathan listened to the familiar melody echoing off the opposing hills. He would know by the coyote's voice when Emma had made contact with it.

Jonathan bore down on the increasing grade laboring his way upward, making slow progress toward the rim of the valley. He stopped again and waited, listening for a sign. There it was. The coyote's yapping abruptly went up in pitch and became a desperate shriek. Emma had found it. A long stretch of silence followed the brief initial outburst, and then the shrill notes began again, only this time they were very far away, coming from the canyon on the other side of the ridge. The sounds kept receding until the faint cries were barely audible. Emma must have really gone after this one and not let up on the chase. He slowed down and lingered on the trail, waiting for Emma to return before proceeding further up the slope. She quickly came looking for him, backtracking down the trail, wondering why he had halted, thinking that perhaps he had struck off in another direction. Emma hadn't chased the coyote at all. The animal had bolted in horror at the first sight of Emma and just kept running until it had gained what it deemed to be a safe distance, warning Emma the whole time not to follow it, while Emma paid absolutely no attention to it. She had met many coyotes over the past few years and they no longer held the same fascination for her as they once did when she was younger.

Since the hour was late, Jonathan settled for simply reaching the top of the ridge where he stopped to enjoy a few moments of relaxation before heading back down to the valley. For many years a patch of soft sand had provided him with a comfortable place to recline and mediate on the meaning of life, the course that his continued existence in the valley would take, the merits of the goals he'd set for himself, the progress of the work he'd undertaken. Jonathan abandoned the practice after Emma was born because she had absolutely no interest in pausing to reflect upon anything. She always chose to keep going, forcing him to continually follow her into the forests, crossing arroyos and mounting hillsides, tracing the ridges and descending into the lowlands. Emma was still too young to appreciate this ritual, a practice that was characteristic of those who were old and worn out, but someday she would sit with him and they would both rest their weary bones together on top of the ridge and stare vacantly off into the distance, looking for some meaning in the grand scene spread out before them. Emma immediately took off as usual and disappeared into the trees, but this time Jonathan unbuckled his pack and set it down in the sand, then sat next to it.

On the other hand, he and Emma might never get the chance to sit on the ridge together. Jonathan's many years in the valley were slowly drawing to a close. The world he'd known for so long was falling apart, the foundations of his previous existence crumbling as he helplessly watched them disintegrate. He and Sarah had been thinking about moving on to another life somewhere else. As he pensively looked down on the valley below trying to make sense of the lost years, the failed dreams, the blunders and mistakes, he wondered if the Curse had been real. His perception of this demonic spirit had come to him early on in his first years in the valley. The problems he had faced back then had not been limited to himself. Others had also struggled and failed, bad things had happened to them, situations rapidly became hopeless, lives were broken. He saw a general pattern here that was beyond any single individual, more than just an isolated poor decision, a wrong turn, a stubborn unwillingness to see things clearly. Something more was going on here than simple bad judgement. Forces were at work that consistently brought misery and despair, that compounded minor errors and magnified the effects of seemingly innocuous goofs. Everyone spiraled into the darkness and became inexorably tangled in a web of futility. He had managed to claw his way out with little or no resources to assist him, other than Sarah's wholehearted support, but the superhuman effort had also been accompanied by good fortune. Others had not been so lucky.

The midnight visitation of the Curse had been a classic example of a phenomenon known as sleep paralysis, a fairly common experience that had been recorded throughout history and traditionally attributed to all sorts things from witchcraft to alien abductions. Jonathan still couldn't believe that the howl had only been an hallucination, because by the time he sat up in bed he felt that he was completely awake and clearly cognizant of his surroundings. The voodoo style trance that he seemed to enter periodically where he did stupid things that endangered his safety and wellbeing could easily be explained by the sustained sensory deprivation of an isolated and monotonous existence leading to bouts of confusion and disorientation. The recurring feelings of dread elicited by the creepy house with its dark corners and labyrinthine, dungeon-like rooms might only mean that he was a person who was easily spooked, although he and Sarah had always referred to the house as "The Castle," at least in part due to a long series of gloomy and often disturbing sensations that they had both experienced. So perhaps in fighting this unseen entity he had merely been jousting windmills. He was the one who had caused the Curse to appear and he blamed this perception on the world around him, tying it to an imaginary history that he had invented in his head. Was he cloaking a collection of rather mundane experiences in the guise of an evil spirit that roamed the valley searching for retribution, or was he camouflaging a dangerous and powerful force in the benign terms of a psychology textbook, invoking the concepts of sleep paralysis, sensory deprivation and childhood phobia to brush aside the real situation, ultimately anchoring his explanations on the mysteries of the human psyche, things that were not really well understood at all? The fact that everything he did in the valley had been cursed, along with the deeds and actions of his neighbors, could not be denied, but the reason why this happened was still open to debate, and the issue would probably never be resolved.

A layer of cold air washed over the rim of the valley now that the sun had set and flowed into the depths, but the warm air above it remained stationary. Buoyancy was a fascinating phenomenon. The cold air hugged the ground so Jonathan didn't notice it while he was standing up, but lying down put him directly in its path and made him acutely aware of its presence. This air slid along the surface of the ground and eventually collected at the lowest points in the valley where it remained for the night, causing his neighbors who lived along the river to have their gardens freeze earlier in the fall than his. He had always wondered though, why the cold air stayed segregated from the warm air during the entire length of its journey, all the way from the rim of the valley to the river below. Cold air could not possibly weigh more than warm air since they both had the same mass. And shouldn't the motion of the cold air molecules agitate the warm air molecules alongside them causing the two masses to mingle and the whole acquire a median temperature? Yet the cold air remained an independent entity and shunned the warmer air which it pushed out of the way to make room for itself. In that case, shouldn't the friction between the two air masses cause a buildup of static electricity? Since he saw no signs of such a buildup, he could simply explain it all away by saying that the electricity dissipated as fast as it formed. He could always explain anomalies away, no matter what they were, but sometimes he wondered what was really going on with things that he couldn't see. Perhaps these things were not as he pictured them in his mind.

—8—

The light from the setting sun often cast an orange glow on the mountains in the evening, but on rare occasions, this rather ordinary phenomenon became very special. A unique combination of dust and haze in the atmosphere along with a peculiar type of clouds along the horizon turned the mountains into a deep red, which miraculously continued to darken until it was almost black. Everyone knew, of course, that when blood was exposed to air, it similarly turned black. Despite this shift in spectrum toward the infrared, instead of losing its power to illuminate, the light on the mountains took on a divine glow, reminiscent of an angel's nimbus. The light became otherworldly, a stunning radiance that caused everyone to stop and turn their heads. The peasant farmers of Northern New Mexico, being deeply religious people, naturally interpreted this phenomenon as an apparition of the Blood of Christ, a heavenly symbol projected across the glorious, often snow-capped mountains for all to see, an indisputable assurance of eternal redemption.

During the night, Jonathan dreamed of an old university acquaintance, a person he hadn't seen in many years. In his dream, this former classmate was now a talented mathematician working at a prestigious institution. He called Jonathan from his office in the Midwest and told him that he had been working on some equations regarding the mechanics of light and clouds. He found that when certain algebraic conditions were satisfied the Blood of Christ would appear on the mountainsides. Combined with a solid knowledge of the principles of cloud formation and a deep understanding of the local weather patterns, he could now predict the dates and times of these occurrences with the same accuracy that an astronomer predicted the dates and times of eclipses and conjunctions. He wanted to come out to New Mexico to verify his calculations.

Jonathan was greatly impressed. He told his friend that this discovery surpassed all the works of Kepler and Newton. The motions of the planets had never been a big deal as far as Jonathan was concerned, but his friend's theory unraveled one of the great mysteries of the universe and ranked among the most spectacular achievements of modern physics,

189

an amazing feat of pure rational thought. The mathematics had to be infinitely more complex than the simplistic formulations of Planck and Einstein, who could only account for rather mundane laboratory experiments. As his excitement mounted, Jonathan gradually reclaimed his consciousness and finally roused himself to put a pot of water on the stove.

Emma was still sound asleep thoroughly embedded in the ruffled covers. Jonathan donned his slippers and clumsily unlocked the aluminum door of the camper, then sleepily emerged from their cozy quarters, rattling the loose steps on the way down. Emma peeked out at him with one eye half opened, then immediately went back to sleep. As Jonathan surveyed the numerous weeds and grasses surrounding their camp, nestled in the midst of an extensive mountain park, the first light of morning steadily gained momentum, highlighting the black clouds floating above the eastern horizon. He turned to face the clouds, intrigued by their distinctive shapes and multitudinous details, when suddenly a few tinges of color magically appeared in the strangest places, the most beautiful magentas highlighting various pockets and corners of the assembled mass. He watched with anticipation as these themes developed, then morphed into new motifs as various yellows and reds were introduced. The hues built in depth and intensity as they spread across the black and white canvass, continually reshaping their respective parts, trading places, rising and falling in importance.

The sunrise appeared to be reaching a spectacular climax when, to Jonathan's dismay, it abruptly washed out and the sky became bland and diffuse. The patterns of each sunrise were always different and the sequences they followed were always unique. The other morning, for example, the eastern sky had been divided into two parts by a horizontal line elevated somewhat from the horizon. Below this line was a black base of horizontal furrows similar to the rows of a freshly plowed field. Pure white clouds had sprouted and grown above this fertile substrate, many of them with long stems and cauliflower tops, resembling exotic mushrooms. The contrast between black and white was so striking that it held Jonathan's attention for quite a while. Another time, the sky was full of triangular clouds resembling the bows of ships, a formidable armada sailing out to sea in tight formation to face an enemy, perhaps a second armada of differently shaped clouds. Yet on another morning, the sky was a very civilized pattern of the happiest clouds he'd ever seen, all cast in the most exquisite details, when suddenly they were overrun by a group of translucent clouds with vague forms and menacing demeanors and these marauders obliterated the fine structure of the original clouds, erasing their pleasing outlines and taking away their happiness, much in the manner of mongrel hordes pillaging and razing peaceful settlements. Jonathan trembled at the sight of such wanton destruction. The invaders eventually moved on leaving some of the original clouds unscathed.

If physicists could predict the unfolding of a sunrise, that would certainly be something to boast about. If a team of physicists could come out to Jonathan's camp in the dark of night, take some measurements and draw some maps, then show Jonathan a picture of what the sunrise will look like at 6:03 AM, not just naming the various cloud species and giving an altitude range for each one, but drawing the exact shapes and configurations of each cloud and showing the precise color patterns, and further provide him with pictures of the sunrise again at 6:10 AM and 6:17 AM, Jonathan would have to admit that physicists understood the mechanics of light and clouds quite well. Instead, all he got were excuses, They told him that the world was chaotic and unpredictable, but in Jonathan's mind, either the phenomena obeyed the laws of physics or they didn't. A physicist clearly understood the mathematics which he had imparted to his laboratory experiments, but a sunrise on the other hand wasn't built out of mathematical ideas. Nature didn't put mathematics into the sunrise in the same way that the physicist had put mathematics into his scientific apparatuses, the arrangements all based on geometrical configurations and the precise ratios of preconceived schemes. Nature didn't first come up with some notions about a sunrise and then fabricate the sunrise out of those ideas. So how then was the sunrise put together? Wasn't that the question that Jonathan needed to answer?

The physicist retorted that the structure of today's sunrise was meaningless and therefore not worthy of consideration, but what he meant was that the various aspects of the sunrise did not have the same purposes and meanings that the various aspects of his experiment had, significations that he had artificially given to them. A sunrise wasn't necessarily meaningless just because it didn't exhibit precise geometrical forms or lend itself to clearcut logical propositions. If nature was able to function without ideas, then all the phenomena were assembled by some other means, and if that were the case, then ideas were ultimately useless for trying to understand nature. Physicists couldn't get inside the mind of nature because it didn't have one. The minds of physicists then became burdens that they had to cast aside. Physicists had to stop thinking about the sunrise because the essence of the sunrise could not be thought.

No one could foresee the existence of things in the world by looking at equations. Quantum mechanics could not anticipate the existence of a piñon tree, a black-tailed jackrabbit, a cumulus cloud in the shape of a human face, or a companion in the form of a yellow dog. But why not? If quantum mechanics was the way the world worked and the basis for all physical realities, if quantum mechanics was a theory of everything, then shouldn't physicists be able to deduce something like the Blood of Christ from it?

The problem was that mathematics was not what caused things to happen. Nothing existed because of mathematics, so therefore mathematics could not explain why things existed. If Jonathan limited himself to mathematics and focused solely on the equations written in books, he could never envision the richness of this world. He could not arrive at any of the phenomena that he routinely took for granted. When physicists said that prediction was the key to a good theory, this was not what they were talking about. What they meant was that if they could design an experiment that hadn't been tried before and they could predict the outcome of that experiment beforehand, the results would support the theory. But what about all those people who lived in the real world? What could they possibly do with quantum mechanics?

Jonathan ascended the stairs and resumed his morning chores, turning his back on the unappealing remnants of the failed sunrise. The show was over and the sun was getting intense. He began taking his clothes out from the wardrobe beside the bed and preparing Emma's breakfast on the small, collapsable table in front of the couch. He poured another cup of coffee and got dressed. It was time for Emma to get up. The sleeping berth was located directly over the cab of the truck, about eye level from the floor of the camper. The distance was too great for Emma to jump down without risking injury, so he never allowed her the pleasure of this grand leap, although he was sure that she would relish the opportunity to attempt such a daredevil act. Emma playfully reared back on the bed as he beckoned her, doing her best to entice him in a game of keep away. He lowered his arms and disdainfully told her: "You know I'm not going to play that game with you." She realized that her ploy was futile and relinquished the puppyish urge to amuse herself. She marched forward in resignation, stepping past the edge of the bed out into the open space beyond it. Jonathan slipped one arm under her chest and the other arm under her stomach as she left the bed behind, seamlessly continuing her motion. The sequence ended with him standing in front of the bed cradling her in his arms. He then gently lowered her to the floor. Emma completely trusted Jonathan. She did not balk at the idea of walking out into mid air because she knew that he would catch her. The role Jonathan played in her life was a big responsibility for him. What if he lost his grip and she fell to the floor? She would never forgive him.

Emma traversed the small space of the camper and stopped in front of the door, her forehead nearly touching it. She stared impassively at the solid metal barrier, waiting for Jonathan to once again fulfill his obligations. He disengaged the deadbolt and swung the door open. Emma did not exit, but stood at the opening surveying the area from her privileged position. As a hunter, being low to the ground was a prime disadvantage, perpetually frustrating her attempts to locate quarry. For once she was freed from this debilitation and she did not want to give it up, instead intensely searching the area for some signs of activity, so Jonathan let her have it and went about his business straightening the bed and cleaning up the countertop.

Jonathan got his gear ready to go. He locked the camper door behind him. He noticed as he set out that the wind was stirring. On some days, the wind was sleepy and seemingly unable to rouse itself, managing only wimpy and short-lived rustlings. On some days, the wind was remarkably uniform, for hours nothing more than a steady and constant breeze. On some days, the wind rose and fell in slow rhythms, reminding him of swells on the ocean. On some days, the wind was fitful, an endless series of invisible walls slamming into him and knocking him off balance. On some days, the wind seemed to come from every direction at once, so that no matter which way he turned, he was always walking directly into it. Jonathan felt that the way the wind was building momentum this morning, by afternoon it would be howling.

He crossed the shallow concavity of the mountain park, surrounded by a ring of steep hills. Once a beautiful, grassy meadow, it was now a comparatively beaten down and degraded piece of land bearing many scars. A long history of grazing abuse had caused the park to become overrun with noxious annuals and it was now home to a wide assortment of rather nasty burrs, stickers, and sharp angular seeds, all of which got caught in his socks and pricked his ankles. He found welcome relief when he reached the margin of the traditional grazing area and the damage to the native grasses abruptly subsided.

Jonathan stumbled upon an old road that led him up into the surrounding hills. The tracks of this once active thoroughfare had faded after many years of disuse and the remnants of these shallow ruts were narrowly wedged between thriving stands of vegetation, barely allowing the passage of even an ATV. Up ahead, a massive tree limb had fallen and now completely blocked the road, forcing the ATVer's to detour around it. As Jonathan approached, he saw that the branch had ripped away from the central trunk of a tall ponderosa and crashed down onto the ground. Being fragile from decay and largely divested of its needles, the branch had shattered into a million pieces, scattering the fragments across the road in an intricate pattern. The physicist declared that the mathematical laws of motion gave him a precise description of the interplay of forces, and using this conceptual system, he could have predicted the exact outcome of this event.

Jonathan was dubious about this claim given that the event had been unimaginably complex. No structural analysis of the original limb, no matter how detailed, could have predicted exactly where the fissures would appear. The physicist could not use mathematics to pinpoint those fracture lines beforehand and determine exactly where the cell walls would pull apart prior to the actual pulling apart, since they gave no outward signs of their potential to do this. The limb was clearly not uniform in structure, but rather a composite of diverse materials. Portions of the wood had deteriorated, essentially turning to cork, and these areas fell apart with little or no effort. Other parts of the limb remained healthy and strong and these areas snapped only under great stress. The parts of the limb thus behaved differently according to their different constitutions, and this varied widely.

The limb had struck some of the other branches on the way down, including a couple of branches that were attached to adjacent trees, causing it to veer, twist and break apart. The pinball action of these multiple deflections was difficult to foresee, but the complete trajectory had be known in advance in order to specify the conclusion of the sequence. Since a fierce wind probably precipitated the breakage, this wind also played havoc with the trajectories of the falling pieces, meaning that the simple law of gravitational attraction was not applicable.

When the oddments finally hit the ground, some shattered further, some bounced, some spiraled, some tumbled, some slid along the ground and some remained still on impact. How each reacted depended on its orientation and speed at the moment of impact. The final resting places of the fragments were determined in part by the roller coaster outline of the earth below, including the exact shape and position of each individual rock and hillock. If a stick hit on one side of a protuberance, it caromed in one direction, but if the point of impact was moved only slightly, the piece caromed off in an entirely different direction.

The action took place simultaneously on every level of organization, from the atomic level up to the level of the intact limb. All kinds of things had to be taken into account: the overall configuration of the tree, the distribution of trees in the immediate vicinity of the tree, the design of the limb as a whole with its unique branching pattern, the collisions between the broken pieces thrown off from the original structure, the impact of the wind on the trajectories of the pieces, the adhesion between the individual cells in the tissues of the limb, the variation in density and weight of the molecular structure of the wood. Influences had been transmitted vertically between different levels as well as horizontally between the elements on a given level. All of these connections were synchronized to constitute one motion, the dispersal of the limb as it disintegrated into pieces. There was no way to put all of this together into a single equation, or even into a set of complementary equations. Marshaling the different kinds of data in ways that would allow for meaningful calculations was utterly impossible.

Jonathan circled the fallen branch and surveyed the debris from various angles, examining the extreme range of litter from the tiniest splinters to the heftiest logs, marveling at the bewildering complexity of the final result. No one could have predicted any of this using mathematics. Calculating the lengths and weights of every one of these jagged chunks of wood and determining their positions and orientations relative to each other simply by starting with the initial conditions of the intact branch was not even conceivable. The problem was not that there were mistakes and omissions in the equations of the motion. The mathematics was fine when viewed from a mathematical perspective. The problem was that the fall of the limb was not mathematical in nature and therefore it could not be understood in that manner. The event was not obeying mathematical laws and thus it could not be described by mathematical formulas.

The behavior of the falling limb was not at all unusual. The unmanageable characteristics of this phenomenon were typical of every phenomenon in the desert:

thunderheads, sunrises, lightning bolts, raindrops, and swirling winds. Whenever physicists told Jonathan that we lived in a mathematical universe, he told them they were crazy. None of this was related to mathematics in any way. To say that this was all done by Newton's laws was merely to invoke a supernatural agent, to rely on the invisible hand of an unseen entity. How did the physicist imagine that he was going to apply Newton's laws to something as complicated as a falling branch? By making a diagram and drawing arrows to represent the forces on a piece of paper? But those symbols were just convenient fictions. He might as well say that the fall of the branch was entirely the work of God. If Newton's Laws could not actually be used to determine the exact outcome, then belief in these laws was an unverifiable act of faith. The laws were the results of laboratory experiments, staged performances set as far away from the realities of the desert as one could possibly get, products of highly artificial environments with unnatural arrangements of fabricated objects. They did not carry over to this world—other than as mental deductions.

The behavior of falling branches in different situations would not be uniform in their sequences and actions. If the physicist failed to measure a falling limb correctly, he could not put the limb back on the tree and try it again. Every event in nature was absolutely unique, a once-in-a-lifetime experience. The event could not be replicated and the pattern of effects could not be reproduced. This was due to the infinite particularity of the physical world and it was an essential feature of that world, but an experiment, on the other hand, could be duplicated as many times as necessary.

In the laboratory, constraints and simplifications boxed in the wide range of variability that was found in nature. The endless combinations and limitless possibilities were held firmly in check, abnormally restrained, until all the wiggle room was gone. Every facet of the experiment was precisely controlled, leaving only one course for events to follow. Each trial was an exact copy of the the previous one, just as the workings of a machine. Every time Jonathan pushed the button or pulled the lever on a machine, the same thing happened because the mechanism was strict, well-defined, and unimaginative. This was how scientists turned nature into a machine. They forcibly limited the possibilities until nothing was left. They made the outcome predictable by eliminating all the choices, leaving only one. Physicists overcame the inherent incompatibility of mathematics and nature by turning nature into a machine, essentially by turning it into its opposite. Once that was accomplished, the mathematization of nature was rather straightforward.

But physicists couldn't get back to nature once they'd stripped it of its essential character. As soon as the freedom was restored to nature, the mathematics disappeared. As soon as the constraints were removed, nature reverted back to its former self: undisciplined and unruly, incalculable and unpredictable. Physicists could not carry over the mathematics they'd created because the mathematics wasn't real. It had been artificially created and was non-transferable. All the discoveries they had made in the laboratory were useless once they were back out in the desert. The real world had not been conquered by physicists after all—it had just been temporarily held in check, and thus the clever ruse had failed.

So what did physicists do when confronted with the failure of their ideas? They said: "Forget about the mechanics of this particular branch. Let's instead set up a game board and play a game. I'll draw a panel of squares on the ground beneath the tree. Pieces of the limb will fall onto the squares. The farther away from the tree, the fewer number of pieces will land in a given square. We'll assign numbers to each square to represent the average number of pieces over many trials. Doesn't this sound like a fun game to play?"

The pieces of the branch didn't interact with the panel of squares because the panel didn't exist. They didn't operate within this structure and thus they didn't put themselves into the squares—the physicist put them into the squares. The pieces distributed themselves on the ground and the grid was never a part of this process. The physicist added the grid to what he saw, so from this point onward he was only studying the grid and not the facts that had been originally handed to him, the realities of the fallen branch. His knowledge was now all about the grid, the one he had imagined and then fabricated. The probabilities he generated came from the definitions he had used to set up the game, that is, they referred to the game rather than anything in the world. They didn't help him understand the way the branch had fallen to the ground, nor did they provide him with a better description of the result. In looking at these probabilities, he was really describing the grid he had created and not the branch. He couldn't predict the probabilities he would get solely from the structure of the grid because he'd given the branch a role to play in his game, but the game wasn't really about the branch. The game was all about the mathematics of a non-existent entity.

Jonathan wasn't sure what textbooks everyone else had read in school, but his textbooks had always specified in very clear terms that probability only applied to identical trials. So what happened to this stipulation? Each branch was utterly unique, not only in terms of its size and shape, but also in terms of its internal structure as well. The height above the ground, the wind speed, and the contour of the ground—all varied considerably. Falling limbs were far from identical trials. The physicist replied: "If we squint, then they will all look the same." The purpose of this squinting was to take away the details so physicists could make room for a mathematics that wasn't there. Squinting represented a turning away from the world and this allowed mathematical ideas to take the center stage, a not-so-subtle replacement of reality with fabrications. The chances of a piece falling in a given square would vary greatly with each situation, making the probabilities determined by a large group of branches a meaningless number. Jonathan saw this clearly whenever he was presented with the meteorological predictions of storm tracks. The meteorologist said: "Today's hurricane had an 85% chance of moving up the coast by Saturday." When the storm didn't behave as predicted, the meteorologist had an out. The probability didn't apply to this storm, but only to a hypothetical storm he'd formulated in his imagination, a mathematical construction that he had arrived at by manipulating numbers in a computer. This particular storm happened to be atypical. Jonathan quipped: "Well, my dear friend, you know they all are."

Why wasn't the physicist—as well as the meteorologist—unhappy with playing childish games and creating a fantasyland out of numerical probabilities? Because in their minds they had started out with the physical realities of the world and that was all their method required, even though they then built all kinds of dubious and far-fetched structures on top of it, adding layers of fanciful constructs over all the others, and ended up in a no-man's land of sketches and abstractions. They told themselves that if they took enough samples, eventually the substantial variations would all cancel each other out and they'd have the underlying mathematical reality of the world. But if that reality had been obtained by blotting out the reality of the world, by erasing the infinite details of that world, then the procedure was contradictory.

Physicists could salvage the mathematization of the world by dispensing with the world's particularity, even though that was essential to its nature. They replaced the world with a mathematical form that divided the event into a series of "outcomes" and assigned a number to each one of them. A set of theoretical states now formed the universe of discourse. Although nothing specific was said about the fate of this particular branch, certain statements could be made concerning the mathematical set of all tree limbs even though real limbs exhibited a wide range of behaviors that didn't have much in common with each other.

The phenomenon now existed on an abstract plane and physicists were no longer talking about anything in particular. This was a clever trick. Mathematics rose up from utter defeat to make another assault on nature—only this time it wasn't nature anymore. Physicists now had an abstract parody of nature, a straw man that was easily toppled. The important point was that the reformulation allowed the calculations to continue. The game could still be played, although the rules had been changed and nature was present in name only. Physicists should have admitted defeat and gone home, but instead they threw the rules into the bushes and reinvented the game.

What role could the concept of probability possibly play in a world where every event was unique and nothing was reproducible? The fact that physicists made probability a cornerstone of their theories demonstrated beyond all doubt that they had turned everything into a game. Probability applied not just to cards and dice, but to every sort of game where the plays were made within a common, consistent framework. In the context of a rigid and well-defined system of rules and definitions, plays became the independent yet identical trials satisfying the necessary conditions behind the mathematical concept of probability, the strict requirements that must be placed on the phenomena in order for probability to have any meaning.

A blackjack player could count cards and observe the hands of the other players seated at his table, then make calculations concerning the probability that the next card would be the card he needed to complete his hand, but gamblers could also bet on more complicated events such as horse races. Bookmakers assigned odds according to the number of bets that had been placed on each horse. If few people bet on a particular horse and that horse went on to win the race, they could afford to be generous and make a larger payoff since most people had bet on one of the other horses and lost their money. Since the track was simply dividing up the pool of money they had taken in, they couldn't lose. A portion of the proceeds went to the winners, and the rest they kept for themselves.

Bettors, on the other hand, measured a horse's chances of winning a particular race based on its past performances in other races against different fields. Bettors poured over the forms and compared times and distances, track conditions, competitiveness of the field, then made decisions about what was important and what was not important, then tried to guess the outcome of a particular race. Marshaling the information was no longer straightforward and there were about as many strategies as there were wagerers in the stands, but people had fun comparing notes and pushing different theories and that was what to a large extent constituted the entertainment value of the races, along with the chance, of course, to win some money. But a given race was not a replica of any previous race and many things had to be taken into account. Sarah had quickly developed a strategy that worked pretty well for her. Instead of studying the forms, she watched the ponies parade past the fence just prior to the race and observed their demeanors. She knew horses well and she could tell which horses were ready to run that day simply by watching them prance around. She would pick the horse that looked most likely to win and this approach worked about as well—if not better—than any other approach Jonathan had observed during his years of attending the track with his father, and during that time he believed that he had pretty much seen it all.

A horse race had many more variables than a card game, yet Jonathan could move even further up the scale of complexity by betting on a sport such as baseball. Here the outcome depended on a host of unforeseeable events: fluke plays and lucky bounces, mental errors and awkward attempts involving a serious lack of judgement. The ball took a funny hop and skipped over the shortstop's glove or cleared the outfield fence by an inch. Jonathan remembered reading a book where a mathematician commented that according to his calculations, two baseball teams would have to play over a million games against each in order for all the variables to cancel each other out, finally enabling everyone to identify the true winner. But each game changed the knowledge and the abilities of each player. Every one of them learned a little something every time they got out onto the field. Every player honed his skills, gauged the players of the opposing team a little better, figured out how to execute particular plays more efficiently, learned from his past mistakes. With each passing day, every player was a day older, and for younger players that might translate into greater coordination and dexterity, but for older players it translated into a decline of stamina and physical strength, as well as motivation and desire, so with each game, the players on each team became different players.

Sportscasters often posed the question before the start of a game: "I wonder which team is going to show up today." Looking back over the course of the season, both pundits and fans alike understood that it was not always the same team who arrived at the stadium each day. The team that had opened the season would never play again. Circumstances had changed over the course of the season, but the rules of the game had not changed, and the mathematician could latch onto this fact and maintain that his calculations were still meaningful. But a card player never wondered which queen of diamonds would show up in the next deal. He didn't question whether a one-eyed jack would be ready to play that day. But what about the card player himself? Could he change his luck by altering his strategy? If a blackjack player always hit on 16 and stayed on 17, then his chances of winning were fixed, but what if he started playing hunches? What were his chances of winning then?

Beyond the athletes themselves, many other variables came into play and changed the nature of a baseball game. The weather was different from day to day, the temperature was hotter or colder, the wind was stronger or weaker and perhaps it came from a different direction, the crowd was larger or smaller, more boisterous and inebriated or calmer and more distracted, the infield grass was a bit longer or shorter, the angle of the sun was higher or lower thus changing the shadows and shinning into the eyes of the outfielders differently. Sometimes the players were tired after a grueling travel day or complacent after spending time at home with their families, hyped up from the previous day's stunning victory or despondent over their recent losing streak. One team might have an advantage under certain circumstances, making each matchup a whole new ballgame with an individual probability of its own. But each game could not have a probability of its own because probability referred to the law of large numbers. The conditions required for probability to have meaning were not even remotely met in the real world, and thus the world had no probabilities.

The situation was akin to weighting the dice differently on each throw or altering the configuration of the die faces so that they tended to land on different numbers. The probability of a given number turning up on a particular throw now depended on how the dice had been weighted and shaped and thus the probability was not always the same. Probabilities could still be calculated by tabulating the results of many repeated trials, but these numbers did not reflect the actual chances of throwing a given number on a given throw. They were not always the same dice and without knowing which dice they were in advance, an accurate prediction could not be made. Probability was the province of controlled experiments where such variations could be eliminated and everyone could rest assured that the dice were always exactly the same.

The variables within baseball games were not random and they did not cancel each other out, therefore the law of large numbers did not apply. Probabilities in the real world could not be determined because each event was unique and this uniqueness could not be eradicated. The mathematician could still proceed with his calculations, but the numbers were not the real probabilities of the next game. The probabilities were determined by the mathematics rather than what actually happened in the world. This should be unacceptable to everyone, including physicists, however the statement did not raise an eyebrow because physicists had been letting mathematics determine realities all along and they had no reason to start complaining now. The world of particulars withdrew into the background where it belonged, out of the way, and even though events were as inscrutable as ever, they still had these numbers to guide them, at least as long as they continued to believe in them, but why should Jonathan trust calculations that were based on false assumptions and abstract ideas?

Jonathan wondered if thunderstorms were really less complicated than baseball games. Perhaps they were, but thunderstorms introduced a new problem. The rigid structure of a baseball game allowed all sorts of outcomes to be defined in unequivocal fashions: "What was the probability of a given team scoring 5 runs in the 7th inning?" But thunderstorms offered no such definite outcomes. "What was the probability of lightning striking at this very moment?" The number of moments was a very big number, so let's instead say during the next minute. Even if physicists found ways to calculate numbers for the current storm, these numbers would not apply to the next storm because the next storm would be completely different from this storm. Every deal in blackjack had exactly the same form and only the dealt hands were different, but not so for thunderstorms. Mathematically averaging the probabilities from various thunderstorms accomplished nothing because the storms had to add up to something, an idealization, a typical storm, and not be a hodgepodge of disparate types with little to no relationship to one another.

Applying the ideas of probability to thunderstorms in ways other than the occurrence of lightning was hard to imagine. Well-defined outcomes were difficult if not impossible to come by and the physicist was forced to make rather arbitrary definitions regarding highly amorphous situations. As a result, physicists didn't waste much time calculating probabilities for thunderstorms. They would rather play games in the laboratory with imaginary spinning tops because the answers they got were precise and clearcut. Once they got probability out into the real world, everything wasn't so easy and straightforward. The mathematics evaporated and they were left with a drivel of numbers and symbols. In the laboratory, physicists provided a structure which they then analyzed. Horse races and baseball games were also structured events, but once outside the realm of human fabrications the whole program fell apart.

No one wondered which jack of diamonds would show up in the next deal because the jack of diamonds was a passive element whose behavior was fixed and determined. The role it played in the game was unequivocal and strictly defined, much as the cart in Jonathan's grade school physics class, the door and roof on his house, the electron in the double-slit experiment. Passive elements were not merely desirable but absolutely necessary in order for physicists to play games with them and then calculate the probabilities of the outcomes. The probabilities were a direct consequence of the structure of the game, determined by the mathematics of the setup. The cards in blackjack were nonentities, markers and placeholders, mere indicators of value and quantity, pawns manipulated not by the dealer who was also a pawn but by the inventor of the game, the one who had set the rules which then created the plays. The dealer blindly set the order of the cards by shuffling the stack prior to the deal and thus dictated the outcomes of the plays. He couldn't tell anyone what would happen since the exact details of the shuffling were too complicated for him to sort out mentally, yet these precise manipulations were what determined the winners and losers of the next game. The overall result at the table was completely predetermined in every way prior to the deal, yet everyone seated at the table first discovered the outcomes of the plays as the cards were distributed to them. They witnessed the dealer performing a game of blackjack much in the same way they might observe a physicist performing an experiment in a laboratory, where again the results were established in advance but only known afterwards. The physicist could never possibly keep track of all the details regarding his precise manipulations of physical objects, and thus he had to discover the outcomes during the course of the experiment.

The passive character of the cards produced an entirely mechanical result. Nature, on the other hand, was an active force rather than a passive element, an unbounded creative spirit that shattered the constraints of clearcut definitions and prescribed roles and consistently broke the rules. The world was alive with possibilities, continually conjuring up unique situations. The question of whether nature was actually alive could be put aside for the moment because as long as nature behaved in the manner of a living being, creating phenomena with wide variations that introduced novel characteristics at every step, natural events had to be treated as if they were living beings.

The accuracy of probabilistic calculations depended on repetitiveness, a consistency in the way the game was played, a structure that defined outcomes in clear and unambiguous terms. As physicists moved into the real world, the calculated probabilities became less real. Meteorologists threw probabilities around freely knowing that no matter what happened, they could never be proven wrong. No one could answer the question: "What was the real probability that this storm would hit a particular city on the mainland on Thursday?" The probabilities referred only to an imaginary mathematical structure that was totally irrelevant.

The turn toward probability in physics had been a desperate attempt to salvage the idea of a mathematical universe and a huge mistake. Physicists had gotten away with it so far only because they were not looking at the real world and they were oblivious to the failures of the method. They used instrumentation to define games and this was what allowed them to calculate probabilities. They created structures with unequivocal outcomes, but this was not the reality of the atomic world. Atoms did a lot more than simply perform stunts in laboratories. Everyone needed to remember that it was atoms who played baseball, having first taken on a breathtaking number and diversity of forms. How were physicists ever going to explain what the atoms were doing on the diamond today in terms of the comparatively simplistic games that they played in the laboratory? Physicists created the situations where probability could be applied, but that was not the real world. Baseball was the real world. Baseball was the complicated actions of real people with individual agendas in a stadium of intricate design before a large crowd of cheering spectators who could influence the results through their shouts of encouragement or their demeaning insults.

If Jonathan started with the game of baseball as it was actually played on a baseball field, he saw that the mathematics of probability was completely invalid, but the great inspiration of physics had been that if physicists started with the mathematics, they could invalidate the world instead by turning it into a ridiculous caricature, a mathematical entity with blank symbols and formal relationships. They could strip the world of its infinite subtitles and unfathomable interrelationships and just say that each baseball game was an independent, identical trial and that the numbers they calculated were the actual probabilities of the individual games. All they were doing here was assigning probabilities to the members of a mathematical set by calculating arithmetic averages—a standard procedure.

When Jonathan criticized the method by pointing out that the world was gone and the physicist was no longer talking about baseball, the physicist turned toward Jonathan and exclaimed: "What are you saying? I didn't throw the world out. The world is still here, see it, it's this "x" in this equation. There it is, right in front of your eyes, it's this symbol here, and all the numbers I generate by manipulating this symbol according to strict mathematical rules are the true realities of the world. You're wrong in what you say. The realities of baseball are not the details of the plays, but the letters and symbols written on a piece of paper." All Jonathan could say at this point was: "I don't believe anybody buys this nonsense. The reasoning is absolutely ridiculous."

Starting with mathematics was cheating. Physicists couldn't go around looking for mathematics in the world and then finding it if that's what they had started out with in the first place. If the world were actually mathematical, then they wouldn't need to start with their own mathematics. They could start with the world and the mathematics they sought would be right there in front of them.

Jonathan saw that the essence of existence was freedom and not constraint. The molecules had to be allowed to run wild and choose their own paths. Everyone had to let them decide what they wanted to do, and what they wanted to do was play baseball on grassy fields, build thunderstorms on summer afternoons, fashion colorful sunrises out of nothing other than sunlight and water vapor. The works of art that the molecules created were stupendous. Jonathan was so underwhelmed by the experiments performed by physicists, so unimpressed with their expertise. Their creations were really just material substances cloaked in mathematical forms. The builders of these laboratories truly lacked artistic inspiration. The real creators of the universe had vision, something to express. As with all great works of art, everyone could debate the hidden messages and embrace the conflicting interpretations of the symbolism because the works were infinite, timeless, eternally changing. No matter how long Jonathan stared at them, there was always the possibility he would see something he'd never seen before, a different angle, a fresh insight, a hidden truth. The mathematics of the laboratory was lifeless, but nature on the other hand was always alive and full of promise.

Physicists believed that everything around them was the consequence of something else. Supposedly the whole process started billions of years ago with a great act of creation, commonly referred to as 'The Big Bang', followed by an endless series of passive responses, all the objects and phenomena reacting in predictable ways to the initial impetus. Physicists lived in a world of effects because that was how laboratories worked. What Jonathan saw all around him on the other hand was the creativity of matter. Whichever way he turned, he found himself exclaiming "My God, look at what the molecules have done here!" He turned his eyes toward the fantastic geology of the earth and then up toward the amazing clouds in the sky and said to himself: "I can't believe any of this! Look at what the molecules have created today!"

This morning's sunrise had not been caused by yesterday's sunrise, and this afternoon's thunderstorms would not be caused by last week's thunderstorms. The molecules were making things up as they went along, improvising, creating the world afresh at every moment, No one knew what they would do next because even they didn't know. Everyone would have to wait to see what mood struck them and see how they felt about things. Physicists could not deal with the real world because physicists had no concept of the creative force in nature, the ability of nature to spontaneously invent the moment on the spot, turn events in a new direction on a whim, launch into an unexpected series of inspired works of art. All the phenomena produced in laboratories were simply the consequences of the physicists' interventions, but that was not how the world outside worked. Physicists had come to their conclusions not by looking at the world outside, but by looking at their own works. They saw only what they had made, and felt that everything should be based on that, that everything in the universe should mirror their own techniques.

Jonathan hadn't fully realized it at the time, but the first year he went to the mushroom conference in Telluride was an exceptional year for mushrooms. Rain had been persistent and plentiful throughout the summer and come the last two weeks of August, the forests exploded with magnificent specimens. He remembered crouching down and surveying the forest floor. Mushrooms were everywhere, hundreds of them in every direction, but they weren't ordinary mushrooms. They were spectacular. Jonathan had studied and collected mushrooms for many years going all the way back to his college days in Wisconsin, so he had observed them before, but that year they were different. The various species seemed more elaborate, more detailed. He noticed aspects of familiar species he hadn't noticed before, additional ornamentation, more frills and ruffles, extra bundles of fibers woven into incredible patterns, novel bands of mycelial segments adorning other elaborate fibrous creations, embellishments that could not possibly serve any purpose. The mushrooms were just showing off, finally achieving their full potential, displaying their artistic talents to a higher degree. He picked up individuals and held them in his hands. They were so fragile, the structures so delicate, and as he brought them closer to his face, more details appeared before his eyes. The mushrooms seemed infinite in design, as if none of the minutiae were unimportant and none of the things the mushrooms had done were meaningless. But why, Jonathan asked himself, did they go through the additional trouble? Why did they expend priceless metabolic energy on such trivialities? Because that year was a bountiful year and they had energy to spare. What were they going to do with all of that surplus energy? They were going to make more, more of everything, putting additional structures wherever they could find a place for them. They created simply out of the urge to create. They didn't have to do any of this in order to survive, but life wasn't about survival. If survival had been their only concern, then they would have taken the additional resources and devoted them entirely to making a greater abundance of spores. Instead it was all about making the most of themselves, providing the best possible examples for others. They were dedicating their efforts to the Great Creator, the one who motivated everyone and who was responsible for the existence of everything. The Great Creator would be proud of their unflappable dedication to beauty and hard work and look kindly on them for their sacrifices. How did they know this was the right thing to do? Somehow they must've known. Jonathan could find no other explanation. Except in the rarest of cases, their efforts would not be seen by anyone. They had no devoted audience of admirers, no public showings, no means of getting their artwork out to the world. Their tedious and minute labors went entirely unnoticed by everyone—except for the occasional mycophile such as himself who then saw only the tiniest fraction of all there was to see. Only the Great Creator knew what they had done. The mushrooms had no one else they could impress with their skills, nothing other than Creation itself. The atoms and molecules of the universe were also like this, doing what they did not simply for perpetuating their own kind, not merely for base survival, but solely for the sake of creating something higher than themselves, taking their places alongside the untold numbers of uncredited actors in a production of biblical proportions.

Just as Jonathan had suspected, the wind steadily increased in magnitude as the afternoon wore on and now it roared with terrible ferocity. He stood atop the western rim of the valley surveying the park below where his truck was hidden from view beneath the branches of a piñon tree, compelling him to firmly hold onto the brim of his hat. He inadvertently relaxed his grip and the hat took off, rolling on edge, bouncing up and down, doing flips in the air, just as he had imagined it had done during its long journey over the winter. He now saw that the hat was fleeing in the hopes that it might escape. The hat nearly succeeded, but Jonathan ran after it and caught it. Perhaps the wind was not responsible for the behavior of the hat after all, but it was the hat that longed for freedom.

Two hundred years ago, a mathematician named Pierre Simon Laplace made, in Jonathan's estimation, one of most most inane, incomprehensible statements ever uttered by any human being. Jonathan had often tried to understand it, but it baffled him. Laplace said: "From its largest to its smallest motions the entire material creation moves in a way that can be predicted with absolute accuracy by the laws of Newton." This was the philosophical doctrine known as determinism, and it has plagued physics throughout its history.

Jonathan dropped down to the basin floor to avoid the onslaught of the wind and kept to the trees along the arroyo for protection. Emma ran off to investigate something. He spotted her up on a hillside a good distance away, so he paused to wait for her. He noticed that the branches of a large piñon tree standing directly in front of him were describing complicated patterns as they were buffeted by the harsh winds. The branches were all swaying and gyrating: some were moving up and down, some were moving side to side, and some were moving in circles. Individual branches switched modes according to the momentary variations in the currents of air. The basic idea of determinism was that Jonathan could calculate the motion of each one of these branches given only the equations of motion of the wind. He could also precisely describe the individual geometrical tracks of each flake of bark on each branch relative to the ground and draw a graph of their positions. He could also predict the angle of the each individual pine needle as it was pushed in different directions by the variations in air pressure.

But that was not all. According to determinism, Jonathan could further predict exactly when each needle would fall off the tree, and the time required for a new one to replace it. By using Newton's Laws, he could then predict the paths of proteins transported within the cells of the tree, calculate the trajectories of water molecules in the xylem, determine the moment when the tree would die, and know exactly to the second when it would fall over. Then, in order to calculate the future of this one little valley, he would have to work out the unique equations for every tree. Jonathan looked around. There were lots of trees and at this very moment they were all swaying and gyrating wildly. To calculate the future state of just this one little biosphere, he would have to make calculations for each wavering blade of grass, each swirling particle of dust, each minute seed scattered hither across the land.

But then he would have to move on to the next valley, and onward across the entire surface of the earth. Jonathan thought about all the faraway lands that he had never seen and all the trees that must exist there. He pictured to himself the erratic winds buffeting the flailing branches. For determinism to work, he'd have to somehow specify all of these motions simultaneously and calculate all of the forces on every one of these trees, creating a giant mathematical snapshot of the whole earth. He'd then have to not merely estimate but to know in complete detail the consequences of every one of these actions, and then couple these influences together to arrive at a complete picture of the next moment. This laborious effort would then have to continue through the seemingly endless moments all the way to the end of time, a moment that would be entirely calculable from the data. The tiniest errors in the initial measurements would be magnified as time progressed, resulting in great divergences in the mathematically derived motions, so the calculations would not only need to be exact, but to be absolutely perfect.

The prospect of making such calculations was not merely a practical impossibility. The whole program was completely non-sensical. No sane person who had ever reflected upon this proposition even momentarily would continue to seriously entertain it, let alone write it down and publish it in a book. Yet strangely enough, the idea had not been abandoned in physics because it still lurked behind much of the current thinking. Determinism was still the narcotic pipe dream of the analytical mind, the illusion that the physicist could somehow construct the entire universe from a simple set of building blocks or list of axiomatic premises.

Laplace had been quick to admit that our minds were way too feeble to comprehend all of the mathematical equations that would be required for such predictions, even on a limited scale, but a supreme intellect such as God, on the other hand, would find it a cinch--assuming of course that God was in fact a mathematician, something which mathematicians generally took for granted. When Napoleon asked Laplace where God was in his system, Laplace supposedly replied "I do not need that hypothesis." Well, he lied. Laplace did need that hypothesis, because the inscrutable mind of God was the only thing that could give credence to this whole way of thinking. There was no other possible interpretation of determinism outside of this imaginary intellect, and that was what made the hypothesis pure fantasy.

The real problem with determinism was not the technical difficulty of collecting all the data, but the way in which determinism conceptualized the world. The world simply did not work according to an interplay of ideas. Only mathematics worked that way. The jet stream did not determine the chaotic movements of the individual air molecules any more than the erratic movements of the individual air molecules determined the jet stream. All kinds of things were happening at once, at every scale and every level of organization, and the influences were being transmitted in every direction, creating a great morass of causation. Single causes could not be surgically removed from this tightly knit body and held apart as independent entities. The heating and cooling effects of solar radiation, the pressure differentials between the swells and troughs of the ocean of air, the conflicting air currents pushing each other out of the way, the deflections by terrestrial obstructions, and countless other untold atmospheric processes, were all simultaneously working together, combining in infinite ways to give everyone the final result. There was no way to calculate any of this.

With the development of Newtonian physics, the idea had emerged that this immense complexity of events could all be reduced to a few simple rules and formulas. From the perspective of the sheer numbers of atoms and molecules and their complex interactions, one might suppose the opposite, that nothing other than confusion and pandemonium would arise out of such a wide range of free and nonaligned behaviors. The truth was that simplicity appeared nowhere in this vast network of events and what we had was complexity producing even more complexity. The strange appearance of simplicity out of such diversity would desperately need an explanation. Any theory purporting such a conclusion would require some form of justification. Many centuries earlier, Plato and Aristotle had begun to dream of a different world, a world of order and precise mathematical relationships. Medieval scholars took heart and set forth absurd maxims such as Occam's Razor when instead they should have said: "If we take the most complex explanations that we have and then complicate them even more, elaborating them as much as humanly possible, the truth will always be much more complex than that." The program of simplification faltered until Newton came along and gave rationalists renewed hope that perhaps the situation was not as hopeless as it seemed.

If Laplace's statement had been made late in the evening in some drunken beer hall to the uproarious cheers and laughter of its patrons, and spoken with the intent of upstaging the previous outlandish proposition that had been put before the bar, it would have made sense. Laplace would have certainly won the prize for the most ridiculous declaration of the evening to the delight of his comrades and drinking buddies. Unfortunately, the statement occurred in the context of a sober discussion of physics in order to introduce the notion of probability.

Determinism described how ideas fit together. The doctrine was basically a belief in the causality of ideas, the mainstay of all physical theories, but the world was so very different, so completely, unutterably, and unimaginably different, that determinism was nowhere to be found. The world was an infinite collection of immense sandcastles, each one perpetually in motion, each one recreating itself at every moment via minute interactions that were enigmatic and unobservable. The world turned the neat, precise interplay of ideas into utter nonsense because the world wasn't based on ideas. So how then could Jonathan possibly understand the world? The magnitude of this problem could not be overstated. The world might, after all is said and done, simply be incomprehensible.

Why did all the theories of physicists flop so miserably once they were out in the real world, when they tried to walk without their crutches, without the simplistic laboratory setups and man-made artifacts with all their built-in mathematics? Physicists complained that the world was simply too complex for them, and while that was certainly true, it was only part of the problem. The world was too complex *and* the world wasn't based on ideas —not the physicists' ideas, not Jonathan's ideas, not anyone's ideas, and certainly not the mathematician's ideas. The second part of the problem was a much greater stumbling block than the first.

The real absurdity of determinism was the delusion that the world followed mathematical laws. Laplace had gotten swept along by the general exuberance and optimism following the publication of Newton's ***Principia*** and he hadn't been thinking clearly. The mathematics sounded so good and he had gotten transfixed by it because it was extremely clear and straightforward. He got so caught up in theory that he lost his head, but a consideration of the real world—even for a moment—would have quashed his unbridled enthusiasm and brought him back to his senses.

Jonathan loved the high country in summer. Pine forests and aspen groves diligently stood guard over the ground, providing a haven for exotic mushrooms, lush grasses and delicate flowers, a place where these organisms could live out their entire lives sheltered from the harsh rays of the desert sun. He established a nice camp overlooking rolling hills of grasses and shrubs, situated along the edge of a large tract of forest. He left camp before sunrise, as he often did, and hiked three miles in the fresh air, entering into the forest then breaking out of the trees to find the Rio San Lorenzo flowing across another broad patch of open country. The river would have been considered a creek back in Wisconsin, but here it was given the status of a full-fledged river. Jonathan swaggered up to the lively water and marveled at its enthusiasm, while Emma plunged into the center of its stream and immersed herself in its reality, letting the currents caress her body as they slipped past her. The idyllic scene caused Jonathan to abandon his plans for the afternoon and instead devote his attention solely to this lovely brook.

He began to follow the river downstream where the land became more rugged. After a quarter mile or so the gushing water knifed into a wall of solid rock and disappeared into the dark recesses of the narrow gap, but the constriction was only temporary and the way opened up again on the other side. He peeked through the doorway and imagined an invigorating hike on the floor of a splendid canyon. The entrance to the canyon was wallpapered with crystal inclusions, a myriad of unblinking eyes similar to the eyes of cats, each one noting the unexpected appearance of intruders. He slipped into the opening and entered a cavernous amphitheater. This was just what he had expected, walking among tall grasses with plenty of opportunities to admire the engaging faces of the rock walls. But after crossing the open space, he exited the palatial enclosure and this was where the river took a downward turn and Jonathan's progress became more tedious and strenuous. He found little room along the banks for footing as the forest and the canyon closed in on both sides.

Jonathan scrambled down bouldery inclines, lowering himself by sitting down and sliding over the bumpy surfaces on his backside, frequently catching his pack on snags and branches along the way, then having to extricate himself by twisting around and reaching backward to unlock the barbs and spikes. The fading possibility of making a return trip via this route caused him some concern, but he shrugged off these qualms by reassuring himself that it wouldn't be necessary for him to come back this way. He could simply forge ahead and traverse the entire length of the canyon to triumphantly emerge at the opposite end where he would renew his delightful stroll through the placid forests.

Further progress over the thin margin of land alongside the rushing water became impossible as the shelf abruptly ended, dropping out of sight. The only way to continue was on the other side of the river. Jonathan surveyed the slender strip of boulders and deadwood for a place to land, took a few quick steps to build momentum, and nimbly hurdled the cascading water, falling awkwardly on the irregular contour of the alternate bank. Emma followed suit and they continued onward until his progress was once again stymied at a point where the rocks rose slightly then disappeared into the depths below. He crawled up onto a narrow perch at the termination of the stone shelf and peered over the edge. The water was plunging ten feet straight down into a circular pool. Jonathan crouched on his hands and knees directly above the vertical axis of the falls, watching the water rapidly retreating and violently crashing into the surface of a heaving, spattering pond, then pouring over the lip and rushing downward into even more treacherous sprays and cataracts. Emma jumped up beside him and pressed her shoulder against his. The tiny ledge held barely enough room for the two of them. They both stared in amazement at the falls, their heads nearly touching, similarly transfixed by the remarkable sight of the thundering water.

Jonathan gradually rose to his feet, careful not to disturb Emma's precarious foothold on the rocks beside him. He gauged his objective across the empty chasm and bent his knees, lowering his body, then with all the effort he could muster, he catapulted into the air. Emma anticipated his action and leapt in unison with him, not wanting to be left behind. They arced together over the void, the river having departed into the dark shadows of an indeterminate emptiness, and simultaneously landed on the opposite shore, once again standing side by side. If Emma had stumbled or misjudged her leap, she would have fallen into the pool and been immediately swept over the side, continuing downward, tumbling and struggling, helplessly caught in the raging torrent of water as it poured through crevices and plunged over precipitous brinks.

His distress steadily increased as Jonathan continued his trek down the canyon. The few places where he could now find footing were separated by distances that he could not traverse. Emma did not have quite the same problem and managed to work her way along by leaping from one niche to another. Several times Jonathan found himself self lying on his stomach and lowering himself feet first down the curvature of a gradually increasing slope, pushing himself backward until the friction of his clothes released him and he broke free, sliding uncontrollably the rest of the way down, trusting that he would land on some horizontal surface at some point. He no longer harbored any hope of making his way back up to the place where he had started his journey.

Scrambling along the banks was tedious and he didn't get very far before he had to once again cross the river and return back to the other side by executing another dangerous vault. He finally reached a place which afforded him a good view of the canyon up ahead and noticed that the downward gradient of the river steadily increased and the banks were impassable. His situation was hopeless.

He abandoned his original plan of reaching the end of the canyon and began to search for another way out. The canyon walls were steep but not impossible to climb. Gravel and dirt had accumulated against the sides and filled in the hollows between the rocks forming a ramp that was sharply slanted, a matrix of obstacles and pathways with thorny bushes and grassy shelves and rocky outcrops, yet it had enough of an angle to allow him to claw his way upward. He repeatedly faltered on the loose rubble, but with great effort he steadily ascended toward the rim only to discover that the ledge was unattainable due to a final vertical stratum of rock. He clumsily backed down the slope and crawled horizontally along the gravel apron for a while, often finding small protrusions that he could use as steps to facilitate his progress, then made another attempt to reach the top. He encountered the same insuperable barrier as before. Several more attempts revealed a similar situation. He was trapped. There was no way out. He had really done it this time. He should have known this would happen when he heedlessly climbed down into the canyon.

As he labored to catch his breath, crouching halfway up the steep grade, he could see that a side canyon joined the main course of the river only a short distance downstream, so he decided to look for a favorable chance to escape his predicament.

He could only catch an occasional glimpse of the river below through the brush and trees and outcrops of rocks because it had dropped a considerable distance, yet he could still hear the incessant rushing of the cold water. He maintained his elevation and stepped sideways, frequently upsetting the loose detritus and slipping on the eruptions of freed gravel, being carried along with the stones as they flowed down the radical pitch of the canyon wall. He then had to painstakingly recoup the lost ground. In this zigzag fashion he laboriously made forward progress and eventually arrived at the smaller secondary canyon where he rounded the spur, clinging to a narrow path along the side of the canyon —until he reached an impasse. The side canyon did offer a potential way out just as he had suspected, but the route was effectively blocked by a large ponderosa standing right in the middle of a narrow opening in the rocks. The tree had obviously been dead for a long time. After many decades of barring all commerce along this route, the tree had finally succumbed to old age. The bark slowly disintegrated, the limbs fell off, the wood silvered and blackened, yet the lifeless trunk remained in place to carry on the legacy of the living tree. Despondent, Jonathan made an awkward attempt to sit down, but he had precious little real estate to work with so he settled for squatting in an uncomfortable position, relishing the slight relief this offered from the strain of holding himself upright.

This secondary canyon had been his last hope. Not willing to give up on it, he considered the notion of trying to move the tree out of the way, simply from the lack of other options. The idea seemed completely crazy, to think for a minute that he could possibly budge this massive trunk, even just a little bit, but the tree had been dead for some time now and he only needed a small space to slip around it. A small hollow was conveniently located in the rocks directly opposite the tree, granting him just enough room to slip into the pocket with his back against the rough stone and his feet firmly planted against the blackened bole—provided of course that he kept his knees pressed tightly against his chest. Jonathan took a deep breath and heaved, pushing against the tree with all his might, his entire body tensing and straining against the anticipated effort of doing the impossible.

A resounding crack echoed off the canopy of branches above his head, comparable to the discharge of a large-caliber gun. The enormous shaft of wood jolted, dropped down a foot, staggered, turned slightly to one side, then slowly fell away, the huge tree plummeting downward, crashing into the unknown depths with a fantastic boom. Jonathan clutched at the rocks nearby struggling not to follow the tree into the abyss, then leaned backward against the canyon wall as he adapted to his uncertain and newly found footing. He remained in the same position for a long time, stunned by this remarkable event, peacefully listening to the eloquent soliloquy of the rambling water wafting up from the stream below, now following its rhythmic cadences with great interest. Finally he turned his head and examined the area up ahead. The withdrawal of the tree had uncovered what appeared to be a trail embedded in the side of the canyon, leading to an idyllic scene —the lush, green grass of an aspen grove.

Jonathan blinked his eyes in disbelief at the apparition of this strange and magical land. The clear path was easy walking and he quickly found himself standing in the midst of the most magnificent stands of aspen trees. Emma loved his new discovery and gamboled with delight, chasing butterflies and romping through the extensive fields of wildflowers. In a single, unbelievable act of insanity, Jonathan had set them both free and all was right with the world once again. The secret path led to an odd gateway, a pair of ancient junipers arching over a series of stone steps that culminated in a circular altar circumscribed by two large tree roots. A third root bisected the circle forming a large yingyang symbol inlaid on the ground. Jonathan stood on the altar with one foot in each sector and closed his eyes, stretching his arms outward and clenching his fists, trying to reconcile opposites by pulling them toward one another, but the effort was useless and nothing happened.

When Jonathan opened his eyes, he saw for the first time that the altar was actually a doorway to a hidden sanctuary. He hesitated for a moment, but then stepped into this otherworldly realm. A sky full of ghostly photons poured over his head, streaming down in geometrically perfect lines. He drifted through the knee-high grass beneath the quaking aspen leaves, while Emma loped and galloped off in the distance, running for no other reason than the pure joy of it.

The photons rained down in limitless numbers, thick bundles of needle-like trajectories. The incident radiation of the sun filtering through the leaves resulted in showers of photons exploding outward from the carpet of grass, much as fountains of sparks at a fireworks display, only the photons never got tired. They didn't quit and fall to earth like the metal particles ejected by pyrotechnic devices. They were more like a laser show at a rock concert—only infinitely more complex. They splashed off the trees in every conceivable direction. This crazy behavior allowed Jonathan to see the shaded backsides of the trees that did not receive photons coming directly from the sun.

Photons bounced off the stones and dirt granules and blades of grass in the manner of highly energetic yet extremely tiny rubber balls. They ricocheted off the corners of the wafers of bark, caromed off the sparkling crystals of mica embedded in the rocks, fanned out from the delicate hairs of stems and leaves, thrown in every direction. Some of these photons rebounded from these convoluted surfaces and completely reversed direction, traveling all the way back out into space to create earthshine. This light, visible from outer space, was actually the blurred images of trees and rocks and blades of grass.

Photons sprayed radially from every pinpoint and showered over everything. If they hadn't done this, Jonathan wouldn't have been able to see the sharp edges of these objects from every conceivable angle. He looked around and he could identify individual blades of grass for 20 or 30 yards in every direction, all wavering in the wind, broadcasting photons in shifting patterns. The blades didn't reflect the sunlight at particular angles as they did in laboratory experiments, but scattered the incident radiation in every possible direction, otherwise there would have been dead zones, places where the blades were invisible, his eyes receiving no photons from that location.

Other photons flooded across the course of every single photon, the huge tides attempting to wash the minuscule speck away, thoroughly swamping its fragile motion with other motions that were contrary in nature. Even though the path of every photon was crisscrossed by innumerable other photons at each point along its entire course, the unimaginable congestion caused by all this simultaneous traffic did not produce the slightest deflection in any of the trajectories of the photons themselves. If the photon was anything at all—even if it had only the frailest existence and the most tenuous substantiality—then this whole scenario would be utterly implausible. But if the photon was only a convenient fiction, a mathematical abstraction whose utility lay in its ability to make the equations come out right, then the scene was quite understandable.

When Jonathan stopped to think about it, he realized that photons couldn't possibly exist. They couldn't possibly be anything at all because if they were, then the staggering number of conflicts and crossings implied by ordinary daylight would be utterly impossible. The irrefutable reality of such a situation left Jonathan with only one conclusion: the photon was just a mathematical form, a design in an invisible and unknown reality that was too small to be observed. The photon was nothing in itself, a chimera of mathematics and materiality. As with the aspects and so-called properties of an object, it was a ghost, insubstantial and fleeting. The photon was only an illusion, an observable pattern, an empty form, a mere appearance—the essential characteristics of all mathematical constructions. It had been created out of the mathematical arrangements of the laboratory and took its identity from these ideas. The photon was an artificial phenomenon that did not exist in nature, the creation of the mathematical mind. Did anyone honestly believe that molding material objects into mathematical forms did not create a reality of its own? Did anyone think for a moment that these objects did not impart their own mathematics to the phenomena they produced?

The waveform of light was nothing but a mathematical form. Saying that light was a wave was comparable to saying that a baseball was a sphere. While the bare outlines were interesting and non-trivial, telling Jonathan that a baseball was a sphere didn't tell him what a baseball was, nor did telling Jonathan that light was a wave tell him what light was. He was just given the profile of something that remained anonymous. In the case of light, the entities that were waving were also mathematical constructions, vector fields, graphs of potential measurements of the concepts of magnetic and electric force. So why couldn't physicists say that light was just a system of mathematical ideas? Because in that case light became one of the masks of Dionysius where no one peered out through the mask. Nothing had been truly explained here, but only superficially described in rather empty terms.

Jonathan wondered if anyone else ever thought about what light really was? If Jonathan could only answer that question, the knowledge would explain everything to him and for the first time he would actually know what he was talking about. He would understand why light sometimes took on the appearance of a wave and at other times it took on the appearance of a particle. Grasping the infinite particularity of light would satisfy Jonathan's curiosity and at long last give him the answers he needed, but the mathematical characterization of light was as much an enigma as the physical experience of light itself. The mathematics only created paradoxes, raising more questions than it answered.

The constancy of the speed of light was a rather puzzling idea, forcing physicists to complicate their theories and develop all sorts of new explanations, but instead of an intriguing juggling act, tossing around the stock concepts of physics, deforming precise arrays of imaginary lines, stretching and compressing space and time into seemingly impossible configurations much as a circus contortionist assuming a series of unnatural positions, perhaps the idea of light as an object traveling through space was incorrect. Perhaps light was a completely different type of phenomenon that could not be understood in terms of mechanics and linear motion. Perhaps light was stranger and more complex than anyone had ever realized. It might simply be the unfolding of something, the realization of a potential that was already there but previously invisible, as throwing a switch caused a lamp to illuminate, the structure of the lamp having already been in place just waiting to be activated. The movement of light from one place to another might just be an illusion and light was the manifestation of a deeper reality that was perceived the same way by everyone, their relative motions being totally irrelevant. Perhaps light was already everywhere but simply surfaced into the realm of the observable and took on the appearance of mathematical forms, then receded into the darkness once again, the pure forms dissolving into nothingness and revealing their true natures, that they were never anything other than images. The physicist saw the manipulation of time and space—or rather his ideas about time and space, the nonexistent mathematical formulations he had made up in his imagination—as the only means for resolving the issue of the constancy of the speed of light, but perhaps the problem was in thinking that light was a mechanical phenomenon. Perhaps there was another explanation for the constancy of the speed of light. Physicists had no idea what light was so they made up all sorts of assumptions about light which weren't necessarily true and perhaps they needed to rethink the whole affair. If photons were merely convenient fictions, then there was nothing to be moved, no traveler to be transported.

Although physicists couldn't figure out what light was, they had figured out how to collimate a beam of particles and direct it at a crystal lattice thereby projecting a diffraction pattern onto a screen of sensors situated behind the object. Jonathan looked for similar events in nature but he couldn't find any examples. Nature didn't seem to use diffraction for any of its phenomena, the things that he saw everyday on his walks in the desert. The precision of the beam, the exact angles of incidence, the perfect structure of the target, all contributed to producing the mathematics of the result, but in nature particles didn't align themselves in this way and irregular objects didn't diffract anything, so the whole setup was meaningless. He couldn't construct phenomena out of diffractions or put the idea of diffraction to use in forming explanations of natural occurrences because diffraction was not a part of nature. As soon as the mathematical arrangements of the laboratory were removed, the mathematical results vanished along with them. The mathematics wasn't real because it had been fabricated. Since mathematics did not exist in the desert, the events that occurred there never produced any mathematics of their own.

The laws of polarizing lenses, diffraction gratings and other optical devices were not the laws of sunrises, or the laws governing how the twilight transformed the landscape in the evening, or the laws determining the way that light filtered down through the domes of pine needles in the cathedrals of ponderosas. To understand these phenomena, Jonathan needed to make up his own laws, create his own explanations and form his own observations. What did the behaviors of laboratory devices have to do with anything? Physicists tried very hard to make connections between the real world and the world of the laboratory, but the results of experiments performed in bleak rooms full of metal boxes bound together by bundles of spaghetti-like cables could never account for the perpetual beauty and splendor of the ever-changing desert landscape.

Jonathan realized that the experimenter in the laboratory played a role that was similar to a lion tamer in a circus. The animal trainer found ways to make the lion sit on a chair, jump through a hoop, and parade around in a circle, but this was not the way lions behaved out in the wild, and this was not what lions were all about. They hunted and mated and fought for dominance over the other lions. They laid in the grass and surveyed their domain and gorged themselves on raw flesh. Jonathan was sure that a seasoned naturalist who had observed lions over many years would be able to carry on and on about the endless quirks and idiosyncrasies of lions and dwell on the nuances of their personalities and daily routines, but sitting on a chair was definitely not one of them. Everyone marveled at the knowledge and skill of the trainer and his uncanny ability to find ways to make lions perform stunts, but if Jonathan looked at these tricks as a means for understanding what a lion was, then he was grossly misled.

Physicists had gotten atoms and molecules to perform all sorts of tricks on the stages of laboratory setups. No one could deny that these presentations were entertaining and often quite fascinating. The problem was, physicists claimed that they were revealing the true nature of the world in these performances. Its true identity was exhibited in these displays and not in the displays that nature regularly put on for everyone. Physicists insisted that the genuine character of the world was unveiled in these choreographed routines, even though the world took on strange and unfamiliar forms that Jonathan had never seen before. Perhaps the stage shows in the circus also revealed another side of lions that no one had previously known about, but this side of lions didn't exist in the wild, patterns of behaviors that didn't appear until the trainer took measures to create them. Obviously, the lion had the potential to become a trained animal, just as the world had the potential to become a laboratory experiment, but did the animal trainer know the lion better than the passionate naturalist or the veteran hunter, better than the educated biologist or the native tribesman? Had the lion's capacity to perform these stunts existed within the lion all along and the only thing the trainer did was unmask it, or did the trainer create something new, something that hadn't existed before? Had the lion possessed this identity buried deep within its soul, an identity based on a relationship with a trainer without a trainer ever having been in the picture, a set of behaviors that was dependent on a person without the lion having ever met a person before, or did the trainer add something to the lion by giving it a new personality and a different identity?

Back in the early days of scientific investigations, when modern physics was still a novelty, experiments were actually presented to the public as stage shows. Various notable personalities in the scientific community became renowned for their theatrical skills—among them Michael Faraday, the creator of the Christmas lectures at the Royal Institution in London. In this capacity physicists doubled as entertainers. They put together elaborate acts and then performed them in front of live audiences, showing the patrons, as the circus placard might say, "Things never before seen by anyone." All of that stuff was old hat now and Jonathan didn't think that the firing of a collider would particularly appeal to a general audience. He imagined that the gadgetry hummed and the lights flickered, but he doubted that there was much else to see beyond the impressive arrays of tubes and wires, the banks of multicolored pilot lamps flashing on and off giving life to the tall stacks of instrument panels, and the bewildering number of interconnections. A detailed explanation of the various roles played by all of the metal boxes would probably elicit only yawns. The crucial awareness that an experiment was essentially a stage show had somehow gotten lost.

Jonathan saw light, but he did not see photons. No one could actually see a photon, only the presumed effects of photons. If Jonathan threw a rock through a window, he could say that the rock had broken the window because he could hold the rock in his hand. The rock was perceivable, tangible, obtainable. But if Jonathan didn't have such privileged access to the rock and it were unknown to him, merely a deduction, when the window broke, how could he be certain that it was the rock that broke it? The window simply broke. There wasn't anything else he could do in that situation. He studied the glass shards in exquisite detail and then wrote a treatise on the nature of rocks. Although Jonathan applauded miraculous feats of reasoning, he did not champion the impossible.

With no direct experience of rocks, Jonathan had no choice but to make up the cause of the broken window in his mind. The cause was an empty reference and the broken window pointed to nothing specific. Ultimately it had been an idea that had broken the window. Jonathan could still verify that the idea had been the cause, but being verifiable was not the same as being knowable. Verifiability could not fill in the blank spaces and supply the details. It could not be a substitute for the experience of reality. Physicists were forced to settle for the lesser criterion because the stricter one was well beyond their reach.

The myth of Sherlock Holmes has always captivated scientists. His ability to use a few scant clues to reconstruct the entire crime scene was a model for scientific thinking. Unfortunately the world was convoluted and things were seldom what they seemed. Sherlock's conclusions were really just giant leaps of faith. Jonathan wondered how many innocent people were now sitting in prison because of this fallacy in reasoning. Still Sherlock was a superhero to the rational mind. Through him logical thinking was deified and granted unrealistic powers. Reason became absolutely compelling, flawless, omnipotent. Its beauty became perfect and unblemished—much as the visage of a supreme being.

Jonathan knew that in truth rationality was more like Pascal's rubber nose: it could be pushed in any direction one chose. Rationality led to errors because it was just made up of ideas and ideas didn't necessarily relate in any way to the world that everyone lived in. How did Sherlock verify his wild theories? The criminals invariably confessed that Sherlock was always right, much as the yes-men in the Platonic dialogues.

Sherlock had been a masterful observer of human nature and well versed in the foibles and frailties of human behavior, partly because his subject matter was close at hand and accessible to scrutiny, and partly because he was a human being himself and could easily understand human motivations and duplicities. At various times he had probably wanted to kill someone himself, embarrassing incidents where perhaps Watson had gotten under his skin and pushed his buttons too many times. He resisted the temptation to rid himself of this annoyance once and for all but knew all-too-well the urges to which others had succumbed in moments of weakness. He could predict how people would act in varying circumstances with uncanny accuracy because he knew both the people and the circumstances firsthand. The human world was not merely a deduction, but an infinite reality of particulars spread out directly before him.

No one knew what it was like to be a photon or an electron, or to coexist in intimate quarters with so many particles of like and contrary natures. Sherlock possessed a great advantage in having his subject matter directly available to him. He knew people inside and out. But physicists couldn't get inside photons and electrons, nor could they observe them at close range. In fact, all the properties of these particles were inferred from large-scale phenomena—mostly produced by devices, instruments, and highly complicated machines—and no one ever actually saw what they were made of. The world of photons and electrons could only be imagined. Physicists didn't realize that Sherlock was not their true hero. Their professional lives were modeled instead on the careers of P. T. Barnum and David Copperfield, showmen and illusionists who could make the world conform to their own visions.

The subatomic world could never be a part of anyone's experience. In fact, it was so completely unlike anything anyone had ever known that no one could understand it—not even metaphorically. This was the main reason why metaphorical reasoning had been largely abandoned in physics. But physicists could still write equations and they could write equations without the slightest idea of what they were talking about. They hadn't a clue at the moment as to what dark matter and dark energy might be, but they were already writing equations and making calculations regarding these mysterious, unknown, and perhaps nonexistent entities. The x's and y's in formulas were just symbols and the letters required only abstract assignments. They were quantities generated by mathematical objects, divorced from the corporeal bodies that had originally given birth to them. They now existed independently in their own right. Physicists only had to deal with the numbers.

Jonathan and Emma stumbled back to camp tired and weary. A black cloud snuck up behind them and drifted silently overhead, its features cast in exquisite relief and infinite detail. The delicate structures were not the results of well-defined parts put together in precise ways creating sets of fixed relationships. Precision alignments were not what the cloud was all about and Jonathan could not understand its true nature by thinking of the cloud in terms of the accurate, interlocking parts of a machine.

Scientists created something entirely different in the laboratory. They made precision their goal and they manufactured everything toward that end. Precision was what made experiments unique. Precision was what set experiments apart from nature. Using precision, scientists had found ways to fabricate events that were fundamentally different from the works of nature.

Under the right circumstances, a cloud could produce raindrops that caught the light of the sun and by refraction these globules separated the sunlight into colors, a continuous spectrum commonly known as a rainbow. However, the bands of color in the sky were fuzzy because the falling raindrops were distorted by the wind and pushed into various shapes. They had different sizes and trajectories making the phenomenon of refraction vague and imprecise. The raindrop was no substitute for the geometrical prism of ground glass manufactured according to mathematical criteria.

A spectrometer in a laboratory was a machine that also split light into a spectrum, only here, in contrast to the rainbow in the sky, the lines separating the wavelengths were sharp and this allowed for precise measurements. Carefully marked linear distances along an axis could be correlated to exact wavelengths of light, and everything could be conveniently expressed in terms of numbers. A rainbow in the sky could not be used in this way because the rainbow in the sky was fundamentally different and did include the mathematics of the spectrometer, even though refraction was a central feature of each one.

Everyone naturally assumed that the rainbow was a fuzzy, approximate version of the precise mathematical relationships that were exhibited by laboratory phenomena, but the mathematics of the laboratory was a direct result of the precision with which the mathematical objects had been fabricated and the rainbow in the sky could never be based on that. Mathematical exactitude was an artificial phenomenon that was peculiar to the laboratory and it was not found anywhere else in the universe. The rainbow was not an adulteration or corruption of the laboratory phenomenon because the rainbow was not built around the mathematics of the spectrometer. Physicists turned refraction into a mathematical object in the same way they turned a scribbled line into a mathematical object, by surrounding it with a geometrical construction that they then used to define the original phenomenon. Regrettably it was not the same phenomenon anymore but an entirely new phenomenon with its own unique properties.

In physics, the spectra produced by the innumerable objects of the universe, comprising a wide range of fiery stars and glowing gases, became the basis for grand cosmological theories. Spectra were geometrical constructions, the direct results of the mathematical designs of the instrumentation. Astrophysicists then related the geometries of these spectra to the geometries of distances and velocities, and then to the geometries underpinning the laws of physics. This was easy enough to do since all geometrical constructions were related to one another through the ideas of geometry. Geometry was not just the central focus of this enterprise but the sum total of all knowledge, and nothing else was important. No one needed to think beyond the ideas of geometry, and that was strange. The assumption that geometry was capable of providing a complete picture of the universe was not only unproven, but a task that seemed clearly outside the province of such simplistic and unrealistic mental fabrications. As demonstrated by many terrestrial examples, the universe could not be understood through geometry, yet the ideas fit together so beautifully and this harmony made the theories compelling. As long as the geometry made sense, physicists were perfectly happy with the results. Physicists failed to remember that geometrical constructions could not possibly copy the realities of the universe, but Jonathan found that coaxing them away from these tangles of symbols and funny looking drawings was virtually impossible.

Trying to apply the precision of the laboratory to the clouds in the sky was utterly futile because that was not what the clouds were all about. Mathematical exactness was not an essential part of their character, therefore a mathematical approach was grossly inappropriate and could not help Jonathan understand the clouds. The high standard of precision in the laboratory was abandoned wholesale whenever investigators moved out into the real world, and this was nowhere more apparent than when scientists attempted to predict the complicated behaviors of clouds commonly known as the weather.

A weather forecast should not only tell everyone exactly how much of the sky would be covered by clouds at each moment—down to the nearest percentage point—but it should also specify the genus and species of each cloud, and not only that, but the unique size and shape of each cloud. If a weather forecast had the same level of exactitude as a physics experiment, meteorologists would be able to predict all sorts of distances and velocities that could be used to specify the exact patterns and arrangements of the clouds, then accurately plot the courses of individual clouds on a map. Instead the precision of the laboratory was glaringly absent. Meteorologists and cloud physicists were in a different world, unlike the world of mathematical relationships and accurate measurements that had been fabricated under the highly usual conditions in a laboratory.

The daily newspapers typically presented the five day forecast in the form of a cartoon. Each cell was a fanciful drawing summarizing that particular day's weather and nothing at all like a photograph or a genuine likeness of the sky. Jonathan, to satisfy his curiosity, once made a daily ritual of cutting out the cartoons and pasting them on a sheet of paper, shifting the successive strips to the right so that the five forecasts for a given day formed a vertical column. He then put the actual weather that day below each column. Not only were the five forecasts almost always different from one another, they almost never matched the actual weather. Predictive power was largely non-existent and the failure was magnified by the fact that the forecasts themselves were only figurative drawings that were almost entirely devoid of content.

Instead of providing an accurate picture of the weather each day, meteorologists instead divided the days into broad categories with generic titles: partly cloudy, mostly cloudy, overcast, or clear, also windy, rainy, snowy, or sunny. A correct prediction amounted to selecting the proper category out of the handful of choices. Computers later allowed forecasters to generate animated maps that they then displayed on their websites and relayed to television stations for broadcast during news programs. The meteorologists cleverly had the machines draw the maps to resemble actual radar images in order to give them an air of reality, when in fact they were nothing more than pure fictions. These maps never had anything to do with the actual events that would follow, but it was easy for everyone to believe that they were watching the day's weather magically unfolding right before their eyes. The weathermen told their viewers to tune in later and they would know more about the approaching storm front when it was finally right on their doorstep and they no longer had any doubt about what it was going to do.

These kinds of atmospheric predictions were not limited to the current week's weather, but freely expanded to include seasonal, annual, and long term changes in the earth's climate. Extrapolating trends from the past was a common practice in many other fields as well, both inside and outside the sciences, analyzing everything from stocks markets to sports teams, despite the fact that the method didn't work and couldn't possibly work. These investigators really had no choice. What else could their predictions possibly be based upon? Those past trends were the only data anyone had, the only hard evidence available on which to make a judgement. But the future would not be based on the past because it was a whole new ballgame. History would not repeat itself and what had happened in the past no longer mattered. In order to make a reasonable prediction about the future, investigators would have to anticipate influences that didn't exist at the moment and hadn't been in play previously, unseen and perhaps unforeseeable factors that would determine the future course of events. But if these factors didn't exist now and had never existed before, how could anyone determine their structures and forms and use this information to gauge the impact of their presence in the world, an inevitable presence that would surely come to pass?

Following the usual procedures in the sciences, the investigators would turn to mathematics. They would concoct a mathematical model, plug in a bunch of numbers, then crank out a bunch of other numbers. They would delude themselves into thinking that the model mirrored the events of the real world and that the numbers it produced were meaningful. When it turned out that the truth was very different from what they had predicted, they would console themselves with the notion that all they had to do was tweak the equations or adopt a different set of assumptions. They never considered that the very method itself was at fault. If the world wasn't mathematical in nature, then every mathematical analysis of it was invalid and every model of it was an utter waste of time.

These investigators never came to this conclusion because mathematics was a sacred cow and not something to be questioned. They could point to physics where mathematics enjoyed the status of a god, where it stood on a pedestal as the supreme ruler of the universe. If there was one thing they could all agree upon, it was that mathematics always led to the truth. The simple step of introducing mathematics—any kind of mathematics— immediately conferred weight upon the results, but the mathematics didn't just lend a significance to the approach. The mathematics was a necessary first step because in fact there was no other way to proceed, no alternative method to fall back on, no other options. If mathematics couldn't do the job, then nothing could do the job. If the mathematics failed to predict the actual course of events, they would make light of the situation and slough off the inadequacies, but what were they supposed to do? Conduct a non-mathematical investigation of events? Such a program was unthinkable, primarily because thinking outside the box was not one of their strong points. They had already invested a large amount of time and energy studying mathematics and they weren't about to throw it all out the window because it didn't work. They'd find a way to make it work, or at least pretend that it worked by turning their backs on the glaring discrepancies.

Placing a method above all else, regardless of its faults and shortcomings, and demanding that everyone follow this method anyway, was moralistic. The method was right even when the results were wrong. Jonathan, on the other hand, felt that the method should be judged entirely by how well it worked. But where and when did the scientific method work? By what criteria should the method be tested? In no case did the mathematics ever reveal the actual mechanisms or the detailed events of nature, so in that respect the method never worked and could never possibly work. Instead the method was usually tested by making calculations. If the calculations of abstract quantities could be successfully made, then physicists were satisfied that the method worked. The calculations really only worked on mathematical objects, since these objects were required in order to define quantities. Out in the real world, applications were severely limited. Most of the events of nature were utterly beyond mathematics.

In the world of dog training, positive reinforcement was championed by many people because it was morally correct, the right thing to do in all cases. These people claimed that all forms of punishment, even the simplest reprimand, were bad in themselves and not to be tolerated. But positive reinforcement was touted as also being the most effective training method, something that would naturally achieve the best results. The fact that the most morally correct method was also the most effective method was the strangest coincidence, given that these factors were entirely unrelated. What a remarkable state of affairs! Morality and pragmatism coincided perfectly and the two ways of thinking were essentially one and the same. Advocates chimed: "Of course it would work! It was the right thing to do. It was the proper way to approach every type of dog in every type of situation." A similar state of affairs existed in physics. Mathematizing events was morally correct and the right thing to do in all cases. Not only was it the proper way to approach every problem, it was the most effective way to attain an accurate and comprehensive knowledge of the world.

Well, positive reinforcement did not live up to its hype. Certain dogs with peculiar traits and dispositions performed much better than others, but dogs like Emma flopped miserably. Labs were generally written off by these people as being stupid and the failure of the approach was thus explained away. Positive reinforcement proponents could make the method work to a certain degree, and by using ingenuity and determination they invented new schemes that remained true to the moral code yet achieved limited success in the field. They wrote off the remaining problems by telling the owners to just keep performing repetitions because someday these exercises would start to have an effect. Of course, that wasn't true, but at least the trainers saved face and got rid of the dogs and their owners. The owners went home and kept banging their heads against the same walls while the trainers paraded the dogs who had responded well as testaments to the superiority of the method.

Jonathan had made the mistake of playing tug-o-war with Emma from the time he first brought her home. The game seemed harmless enough and he was eager to interact with his new pup, so consequently they played it quite often. Emma loved playing the game and growled fiercely as she yanked her head from side to side. Little did Jonathan realize that he was creating serious problems that would come back to haunt him later.

When he went to leash walk Emma down the street, she immediately thought that this was another game of tug-o-war. He couldn't even get her to the curb as she tried to jerk the leash out of his hands. The positive reinforcement people told him that he could never say "no" to his dog—a rather bizarre statement given that they were clearly saying "no" to him. He tried luring her with treats, but these offerings were only temporary interruptions in one big overarching game of tug-o-war. Feeding her cookies didn't seem to tell her anything important.

When Jonathan complained to the trainer, he was told to purchase a leash with a two foot chain at the end, apparently the brainstorm of someone in the positive reinforcement movement. When the dog turned to grab the leash, its teeth would clash against the metal links creating a highly unpleasant sensation, thus discouraging further behavior. Emma never thought twice about it. She lunged forward and grabbed the leather leash at the point where it attached to the end of the chain. The answer was so simple and obvious. All she had to do was ignore the chain and focus on the leash. She had no problem figuring that one out.

Undaunted, Jonathan purchased a six foot chain leash which had only a small leather loop at the end. Again, Emma didn't hesitate for a moment. She lunged forward and tried to pull the leather loop out of his hand, nearly knocking him to the ground, a response that once created quickly became firmly entrenched in her mind. After a few more failed attempts, Jonathan unhooked the chain leash and threw it in the garbage. The use of this leash had created a behavior that was far worse than the one he was trying to correct.

Fed up with the whole program, he fastened his regular leash to Emma's halter, firmly grabbed it a few inches above where it attached to the metal ring and started to lead her up the driveway. Emma immediately turned around and tried to grab the leash in order to initiate a game of tug-o-war, but Jonathan gave her a firm "no!" The abrupt reprimand caused Emma to turn her head away and face forward, and with Jonathan gently but firmly leading her ahead, she took a step. In a very soothing tone Jonathan said "good girl!" She turned around and again tried to grab the leash, but Jonathan again interrupted her with a sharp "no!" Emma took another step forward and Jonathan softly praised her. By using this technique, in a matter of about ten minutes, Jonathan had Emma parading up and down the driveway like a champ. Alternating and contrasting the commands of "Yes" and "No" had made the exercise crystal clear in Emma's mind and she immediately got the idea, whereas previously she had not been able to understand what was expected of her. The essence of dog training was successful communication. In order for Emma to grasp the meaning of the words, they had to be placed alongside each other so she could see the difference for herself.

For positive reinforcement advocates, whether or not the method worked wasn't the real issue. The important thing was that everything be done according to a prescribed way of thinking. Adherence to principle took precedence over practice, over everything else in fact. The vast complexity of particular trainer-subject interactions and the wide variety of dispositions and personalities in dogs were happily ignored, but in truth, every dog and every situation was unique and no one method was appropriate for all. The training techniques had to be tailored to fit the particular dog at hand.

The positive reinforcement people were laboring under the misconception of a universal truth, a concept they'd picked up in elementary school at a very young age while studying mathematics in class. Universal truth meant not only that the method worked in all cases, but that all other methods were false, since they were contradictions to the one correct method, or in other words, the worship of false idols. When counterexamples arose within the sanctioned method, these frustrations were curiously dismissed as anomalies, quirks of canine nature, or misunderstandings of some sort. If an opposing method worked perfectly well in a specific instance, the trainers would still maintain that there was something wrong with it, a hidden problem or shortcoming that was not immediately apparent. When looking at nature, physicists similarly focused on the phenomena that most closely matched their method: the motions of projectiles rather than frisbees, the trajectories of falling cannonballs rather than old tree branches, the orbits of planets rather than spinning dust devils. These other phenomena were just plain stupid—a lot like Labrador Retrievers.

Jonathan remembered standing in his garden late last summer working the ground, hoeing and weeding, and for no reason in particular he raised his head and looked for the first time into the face of the advancing storm. He recognized it immediately. He knew exactly what those clouds meant and what they were going to do. Although he had never seen anything quite like it before, he had no doubt in his mind. Jonathan couldn't say that it was simply the blackness of the sky. He was only given a glimpse of the clouds through a narrow opening at the mouth of the canyon, a thin slice of sky between opposing cliffs, but that was enough for him. There was something about those clouds, the way they were structured, that told him the truth. The storm was coming fast and it would be devastating. He had no time to pick produce to salvage it, no time to cover the plants to protect them. The whole summer long endeavor, the product of many days of hard work, was all going to end shortly. He knew at once that his beautiful garden was doomed.

This was not a prediction. If Jonathan had stood in the garden earlier in the morning when the skies were still sunny and the storm was nowhere to be found, at a point when the storm hadn't even been born yet, and said that the garden would be buried in precisely six and a half hours under tons of hailstones piled knee-deep in places, that would have been a prediction. Going to a website and watching the storm approach on a doppler radar image might have given him a little more warning, but it was still just another version of his visual identification of the clouds as he stood in the garden. The images were merely the confirmation of something that was already happening, the observation of events that were already in progress, yet this was what passed for "prediction" among meteorologists and weather forecasters.

What Jonathan wanted was a precise, mathematical prediction of specific meteorological events before they developed, the kind of prediction talked about in physics and executed to some degree in the laboratory. If the weather had actually been calculable, then it would have been possible to pinpoint each individual thunderstorm on the map before it appeared in the sky, and draw exact lines tracing its path over the course of the day. It would have been possible to chart the variations in precipitation along its entire track and graph the pattern of wind speeds around it prior to the onset of the storm, but still these charts and graphs and quantities would not be the actual storm. They would not allow Jonathan to picture the storm in his mind. What Jonathan saw approaching him that day was the actual storm and no one could have provided him with the series of images he perceived as he watched it coming toward him, then spreading overhead and engulfing the valley, then unleashing its fury on both him and his garden.

The problem was that thunderstorms were not mathematical objects and thus they could not be calculated in principle. They could not be generated out of a set of preconditions by a chain of deductions. This was not a matter of insufficient data. This was because thunderstorms did not follow precise rules, and if they were not based on mathematical precision, then every mathematical analysis of them would be a bad fit.

The world was very different from the laboratory. What puzzled Jonathan was that physicists saw the laboratory as a means of approaching the world and not something that caused them to retreat from it. To the physicist, the laboratory was more real than the real world. The laboratory was the way the world should be, a perfect place of precise relationships, an ideal environment for mathematics to flourish. Physicists felt that the laboratory represented the essence of everyone's world in spite of the fact that it was so unlike their world in every way, a radical departure from the chaos and confusion of ordinary events. The physicist's proclamations concerning the outside world all stemmed from his bizarre experiences inside the laboratory. The oddities and abnormalities of laboratory phenomena were then mentally transferred to the larger domain of ordinary experience. The practical application of this quirky knowledge to the phenomena of nature was sketchy at best. Well, Jonathan admitted, let's be honest: it was an abject failure. As soon as the physicist stepped out of his theater of staged performances, he was lost. He floundered outside his element, gasping for breath. The precision was gone. Even the concepts he held most dear, such as uncertainty and entanglement, did not cross over to the things that everyone knew, the world around them, and that was baffling. How was it that the essential qualities of the quantum world with all of its weirdness did not carry over to the world at large, especially if quantum mechanics was the basis for that world?

Physicists clearly had misgivings about their work in the laboratory, vague inklings that the results of experiments were limited to the apparatuses and therefore the conclusions they had drawn from them didn't pertain to anything else. Given the gravity of this shortcoming, they made deliberate attempts to show that the laws of physics applied to everything, and took steps to expand their compass beyond the narrow realm of mathematical objects and laboratory phenomena. They managed to demonstrate Einstein's "spooky action at a distance" at a macroscopic level, by which they meant a tiny disc of silicon wafers about the width of a human hair, and this was a huge step in the right direction. So now, given this startling new discovery, physicists demanded that everyone start taking entanglement seriously because they had proven that entanglement was found everywhere in the world. Well, that wasn't exactly true. In order for entanglement to be found everywhere in the world, physicists would have to show that entanglement existed between non-mathematical objects, because the natural world was comprised entirely of non-mathematical objects. It was the mathematics that entangled the objects, so without the mathematical forms, entanglement had nothing it could latch onto, and there was no possibility for the objects to be connected together in this way. Our world could never exhibit entanglement because the phenomenon had been created by the mathematics of the laboratory and thus could never appear anywhere outside the walls of that domain. Entanglement did not arise from ordinary everyday mathematics, but required a very special and elaborate mathematics found nowhere else. The problem was not that quantum mechanics applied only to the submicroscopic world, but that it applied only to a bizarre, otherworldly mathematical realm of detached abstractions that had been mimicked by molding physical bodies into precise mathematical forms, and no one lived in such a place—that is, other than physicists.

The forest dwindled away as Jonathan emerged from its interior and he chose a path through open country. The sky was now completely obscured by thick clouds. A subtle change in their appearance told him that the sun was setting somewhere. The exact location was impossible to determine, yet a couple of red clouds indicated that the event was taking place at this very moment. The clouds reminded Jonathan of a black and white photograph he had seen once, the portrait of a woman with vibrant red lips photoshopped onto an otherwise colorless face. The changing light magnified the contours of the gently floating bodies above his head, augmenting the contrasts and amplifying the details. With his head facing upward, he twirled his body around trying to take it all in, sweeping the heavens with his eyes. The patterns were incredible. The outlines of countless shades of black and gray were painted with such a fine brush as to form a picture with apparently infinite resolution, every secret of the floating bodies revealed down to the tiniest wisps of vapor.

Clouds were unquestionably great works of art, but at this very moment they appeared to be sculptures molded out of metal rather than paintings brushed on a canvas, strange shapes cast in aluminum and iron, frequently inlaid with tufts of steel wool, multiple layers of hammered metal plates, each cut with ragged edges as if someone had gone crazy with an acetylene torch, handfuls of silver medallions welded onto a weathered chassis of twisted black beams. He peered into the intricate designs and became lost in their depths, falling into a vortex of particularity. The clouds seemed to be as complicated as the surface of the earth, an infinite world of precise detail. Somehow all of these structures had suddenly precipitated from the vague solution of air and water by a process that Jonathan did not understand. He wondered if the clouds were always like this, only he didn't normally notice their true natures.

The desert was no longer just silent. It seemed that the very possibility of sound had vanished and he had entered a world where sound didn't exist. He stood absolutely still and held his breath. He was no longer standing on the earth, but inside of a strange room with incongruous furnishings. The outlandish decor forced him to blink in utter disbelief. He knew this area well, but now the terrain was unreal, transformed into a shadowy vault with diffuse light sneaking in through the shuttered blinds. The huge chamber had been stripped of ordinary air and filled instead with some vaporous elixir that transmitted and reflected light by a completely different set of rules, making everything appear other than what it was. Emma was also enchanted by this place. She scampered up and down the gentle slopes in the general direction of the truck, which was still hidden from view, lurking somewhere in the bushes a quarter mile up ahead. Jonathan was glad that Emma had gotten the chance to see this, because it was something that was very much worth seeing.

Physicists told everyone that the world was not what it appeared to be. Inside of every stone was a fantastic circus show of motions and energies. So why was it, Jonathan wondered, that clouds were nothing more than what they appeared to be, diffuse puffs of water vapor with no secret activities inside? Clouds might have marvelous bodies that were ordinarily encased in fluffy fleece jackets, but on rare occasions those fuzzy cloaks conformed to their inner bodies and revealed the complex anatomies inside, clinging to the underlying structures much in the way that wet t-shirts clung to the bodies of swimmers, thus exposing the extraordinary physiques of clouds, and granting Jonathan a clear view of these beings in their full nakedness.

—10—

Jonathan parked the truck next to a line of bushes a few miles from the rift of the Rio Grande, an otherwise undistinguishable spot in the middle of a vast, open plain. Hiking away from the river took him into unbroken country, the land climbing imperceptibly until it reached a high plateau somewhere far off in the distance. He marched across this barren, seemingly endless landscape, dwarfed by the sheer magnitude of it. Earth and sky exploded into gigantic dimensions, mercilessly swallowing up ordinary creatures such as himself and causing even the largest of objects to compress into insignificance. The truck slowly disappeared from view and he found himself alone with Emma. She seemed to be enjoying the experience as much as he was. They were both free, untethered, severed from their ordinary existences, lost in a treeless wilderness.

The sole occupants of the limitless sky were a handful of innocent clouds, isolated by many miles of emptiness between them, each going about its own business with no regard for the others. As the morning heated up, the clouds all grew in stature turning into menacing adolescents, a smattering of black splotches hanging in an otherwise vacant sky. Jonathan steadily deflected his course to one side, tracing the circumference of a large circle. The truck reappeared as a fleck on the horizon directly in front of him. There were no trees as far as he could see, nothing higher than tall grasses and scattered bands of low shrubs. He had cranked the camper top up prior to his departure and now the tall structure stood defiantly on the flat surface of the grassland, directly beneath an angry black cloud, the only cloud for many miles in any direction. The cloud was highly agitated and as Jonathan watched from a distance, the cloud sent a bolt of lightning to the ground in front of his camper, apparently being careful not to hit it.

Jonathan kept walking. He figured that the cloud would move on by the time he reached his destination and not be a problem. Jonathan looked up and saw another electrical discharge streak down in the vicinity of his precious abode. He was getting closer now, but the cloud wasn't leaving. Convinced that the cloud would soon drift away, he put down his pack and waited. The cloud didn't seem to be going anywhere. Another bolt of lightning issued from the little devil and he was afraid to approach any closer. He became agitated himself. He didn't like being intimidated in this way.

Jonathan was also hungry and weary from his morning hike. He picked up his pack and decided to circle around in order to sneak up behind the cloud, hoping to slip into the safe quarters unnoticed, but the cloud appeared to come toward him. It abruptly hurled another frightening bolt to the ground as a warning not to advance any further. The crazy, jagged streak suggested a state of absolute hysteria. Maybe he shouldn't mess with this cloud. He backed off and tried a different approach, but the cloud once again intercepted him. Apparently the cloud had appropriated his camper and it was determined to defend its newly acquired prize. Jonathan felt as if he were confronting a mythical dragon belching fire as it lorded over a priceless treasure. He was struck by the absurdity of the situation. Nothing was around for miles—just the cloud, the camper, and him and his dog, caught up in this intense confrontation out in the middle of nowhere. The cloud stubbornly held its ground and did not retreat, continuing to issue more bolts of lightning, while Jonathan watched helplessly from what he deemed to be a safe distance.

The fact that none of the bolts hit the camper flew in the face of conventional wisdom. Being caught out in the open was always considered to be the most dangerous place in a lightning storm, yet the camper was the only structure of any significance for as far as Jonathan could see. The camper was the epitome of being out in the open. The cloud was directly above it, so if being out in the open mattered in any way, then here was the perfect opportunity to verify that claim, but the often repeated cliche was clearly refuted here. Lightning bolts were hitting the ground everywhere, while completely ignoring the most obvious target standing right in front of them.

The situation also refuted the commonly held belief that metal attracted lightning. Hikers had even been advised not to carry metal-framed backpacks. Everyone remembered the incident where the golfer raised his metal club and lightning struck it at that very moment, but if the golfer had not raised his club, would he still have been hit by lightning? Everyone took it for granted that it was the raising of the club that drew the lightning, but that was just an assumption. Maybe the lightning strike had nothing to do with the raising of the club and the two events were merely a coincidence. Maybe just as the lightning was about to hit the golfer, he happened to raise his club. Jonathan remembered the story about the horseback rider who was struck in the back of the head by lightning. The bolt traveled down his back to the ground, killing his horse. If the rider had dismounted, would the lighting still have struck him? Or would the lightning still have killed his horse?

Jonathan had always read that he should crouch down on the balls of his feet in a lightning storm, but he couldn't believe that a distance of two or three feet would matter given the vast dimensions of earth and sky and the great expanse of clouds overhead. He felt stupid crouching down, much as he had in grade school when his class performed a drill of what to do in the event of a nuclear bomb, cowering under his desk with his hands over his head. Such a posture would have had absolutely no value if a bomb were actually dropped, and he also felt that it would be of absolutely no value if he were actually struck by lightning. He was only making a fool of himself, listening to the inane advice of people who liked to make stuff up. He might as well enjoy the storm and just keep walking, which is what he did anyway.

Jonathan had been warned not to stand next to a solitary tree in a lightning storm, yet solitary trees rarely if ever showed signs of lightning strikes. The problem was that lightning didn't always produce visible damage to the tree. Trees that had obviously been hit, on the other hand, were often located close to other trees. People felt vulnerable and exposed out in the open, so Jonathan could understand how the idea had gotten started. Being hidden within a dense forest of trees seemed safer—but was it really? Obviously the trees would not provide any protection in the case of a direct strike and the process of determining a point on the earth to receive the bolt would strictly be a matter of cloud positions and the contours of the underlying terrain, something that did not take Jonathan's presence into account. Over the years Jonathan had experienced as many near misses while walking on the valley floors as he had while walking on the ridges and he didn't feel there was any significant difference between the two, but he was definitely more scared when he was up on the ridges because he felt vulnerable and defenseless, but what did that feeling have to do with anything? The fact was that lightning hit the lowlands as much as it hit the highlands and one did not appear to be any safer than the other. All the books and articles with maxims for outdoor safety were based on overly simplistic ideas about lightning and the concise rules they set down needed to be elaborated, the statements qualified and refined before they would even begin to have genuine applicability in real life situations. No one had the necessary knowledge and experience to do this. Six miles from dangerous lightning activity was generally considered to be a safe buffer, yet he was already within a quarter mile of the immediate danger. He knew that he should turn around and run for it, but six miles was a long way to run.

Instead Jonathan stood still and surveyed the scene. There was really nothing more he could do. He could simply ignore the danger and walk up to the camper, but he was aware of the difference between bravery and foolhardiness, and the problem was not just for his own safety. He had to think about Emma's safety as well, so he turned around and walked away. The cloud could have his camper for the time being. The cloud couldn't stay there forever. He'd just wait it out. He kept walking until he found a nice spot overlooking a shallow arroyo and laid down in the grass with his head on his pack, while Emma busied herself with some of the local scents.

He carefully studied one of the other clouds floating off in the distance, watching it peacefully drifting along, apparently searching for someone to harass. He was trying to figure it all out when suddenly the world flipped and he saw everything differently. He realized that clouds had rudimentary intelligences, perhaps similar to those of small children, or maybe more like simple beasts, yet they wielded this terrible power which they barely understood. The important thing was that he could talk to them. Whenever they approached him with their mischievous designs, he should look directly at them and scold them as if they were wayward children: "You go away! Don't you dare strike me! Do you hear me? You leave me alone!" If he treated them as a superior would treat an underling, and let them know that he was aware of their evil intents, they would mind him and not lash out at him.

The technique was similar to what he would do when confronted by a threatening bear or mountain lion. He would make himself as large as possible and not back down or run away. He would show the animal that he was not afraid and that he was prepared to defend himself. Perhaps he could use the same strategy as a deterrent against the clouds.

Jonathan realized that clouds were truly like sheep grazing in the fields of moisture-laden air, and this was not just a metaphor for their outward appearances. Clouds in fact were roaming herds of animalistic spirits in the sky. Although they were often placid and serene, like wild animals they were capable of extreme violence and could turn on him unless he faced them squarely. Thinking of clouds in this way made Jonathan very happy because now he could deal with them. He could relate to them. His newfound perspective was not just some abstract, cerebral association, a notion that he entertained in his consciousness. Now when he turned his head upward and looked at clouds, he saw strange creatures suspended in the air, large populations of living beings conducting their entire lives in the vast domain of the atmosphere, perpetually feeding on the limitless bounty of invisible water droplets, just as he might look out across an open landscape and see a flock of sheep grazing on the green grass of a lush pasture.

Years ago Jonathan had been a passenger in a car traveling westbound on Interstate 10 in Arizona. He'd been staring out the window at the stark terrain to the north when the world suddenly jumped back in time. He was not looking at the world as it was on that day, but at a world that once existed long ago, long before the dawn of civilization, when there was nothing. He became gripped by a terrible vision. He was no longer riding comfortably in a car anymore, but stranded in some forgotten past, standing all alone and having nowhere to turn, facing the sheer magnitude of creation with nothing else to fall back on. He had no cozy home waiting for him, no asphalt highway or metal vehicle to transport him, no cities or towns to find twinkling in the evening light, no people or friends to meet. He could walk forever, but he would never see anything other than the bleak desolation of an empty landscape. He was completely alone on the entire surface of the earth.

Jonathan had experienced similar hallucinations at other times when he looked up at the sky. A cloud was not just a thing in itself, but a part of something much greater. Ordinarily he saw the cloud in the sky as a piece of his world, sharing that world with airplanes and motor vehicles, televisions and computers, city lights and blinking radio towers, but on rare occasions, the cloud became the cloud of another world, the cloud of ancient Greece with its stone temples and coliseums, its burning fires and open air plazas, or the cloud belonging to the earth before civilization, a world without roads or cities or humans. The cloud over his head today was the cloud of his world and he saw it for what it was, but sometimes when he emerged from his camper in the moonlight, he saw the cloud of another time and place. He saw someone else's cloud, someone long forgotten and gone from the face of the earth.

Sometimes Jonathan thought that he could see the entire world by looking into the mirror of a cloud-filled sky. He witnessed the clouds performing the same age-old rituals, the ones they had done for so long, since the beginning of the earth, only now they were doing these things above horse drawn carriages and thatched roof villages. Such roads and villages had once been what people saw when they turned their gaze back to the earth after surveying the sky overhead, and sometimes he thought that this was what he too would see when he turned his eyes away from the clouds. The sky had always been the same, while the human world had gone through many radical changes, meaning that he could couple the sky to many past worlds and to the people who had previously existed on earth. He attached the same sky to different earths, so the clouds became a leaping point, a means for him to enter into those lost worlds.

When someone looked up at the clouds, what they saw was entirely up to them. The physicist told Jonathan that when he looked up at the sky, he saw forces. This idea made Jonathan smile, because when he looked up at the clouds, sometimes he saw the marble temples and sandal-clad citizenry of a place far away and a time long past. Perhaps vestiges of that world were in fact still up there, coded in the molecules, the vibrations of these tiny entities still carrying the faint suggestions of entire civilizations, the specks of dust holding onto the remnants of past events and echoing the clouded memories of once great cities and heroic undertakings. The smoke from the ancient funeral pyres, the burning of incense in the sacred temples, the torching and pillaging of entire towns—all of these materials went up into the atmosphere, molecules that had once been a part of something else, perhaps a living body or an inhabitable structure. Some of these particles had come down to earth only to be picked up again and carried aloft by the winds once more. But the molecules of those lost worlds all remained in one form or another—the dust, the carbon dioxide, the moisture that had once been contained within human breaths —perpetually circling the globe, carried by the jet stream to all parts of the world, eternally populating everyone's sky.

So what was Jonathan looking at when he looked up at the clouds? He was looking at the very same molecules that had been exhaled by the famous orators of ancient Greece, along with their eloquent words. He was looking at the air that had carried the voices of the great philosophers across the room as they sat around tables drinking wine and debating the issues of the day. He was looking at the vapors that had accompanied the cries of victims and the shouts of victorious armies. He was looking at the residues from the smelters of the iron age. He was looking at the campfires of nomadic bands of hunters as they wandered through desolate lands. He was looking at the evaporated sweat exuded by the slaves who built the pyramids. When Jonathan looked at the clouds, he was looking at the fragments of all those ancient worlds. Everything that had ever happened was still up there, in tiny bits and pieces, all mixed together. But there was one thing Jonathan wouldn't see drifting along with the debris of the ages: he wouldn't see forces. He wouldn't see forces because forces were concepts that had been wrought by the imagination. Jonathan had smiled to himself earlier because the physicist had chosen the one thing that wasn't up there. If the physicist had chosen almost anything else, he would have been right.

When Jonathan questioned the concept of forces in the sky, the physicist asked him: "Why then do the numbers agree? What other explanation could there possibly be to account for this fact? Forces must be physical realities and their behaviors dictated by the laws of nature. The clouds appeared the way they did today only because forces were at work, entities that acted in predictable ways even though the clouds did not." Jonathan looked up at the sky. Right now the clouds were a fantastic celestial landscape of intricate designs, but the patterns were everywhere unique, irregular and non-mathematical. Jonathan asked himself in what ways were the clouds obeying the laws of physics? The physicist projected his mathematical concepts onto the clouds, drawing isobars and lines of force, mapping currents of air, plotting temperatures and humidities, introducing the concepts of condensation, vapor pressure, thermals and buoyancy, putting everything on numerical scales and then tying all the numbers together. The physicist vainly attempted to sketch the same mathematical framework into the clouds that he had set up earlier in his laboratory, because that was where the mathematics came from, including the mathematics of the physical laws themselves.

The physicist refused to accept the fact that the clouds were not obeying the laws of physics. Jonathan told him that the clouds were not obeying the laws of physics because the clouds were spontaneous and erratic. Most people would recoil at such a statement. Everyone assumed that it was a given fact that the laws of physics applied to everyone and everything, that these laws existed everywhere in the universe, but the laws of physics were not the laws of nature. So when Jonathan said that the clouds were not obeying the laws of physics, he was saying that their behaviors were not mathematical in nature. If the courses that the clouds followed today were not mathematically determined —and if they had been, then they would have been calculable and predictable—then obviously the clouds were not following these laws at all.

The laws of physics were the laws of mathematical objects. Jonathan could construct a seesaw and then come up with a formula for its behavior, something physicists had already done, a formula commonly known as "The Law of the Lever," a mathematical relationship stating that the moments of torque must be equal. Jonathan might easily mistake this formula for a law of nature, but he wasn't studying nature when he was observing the action of the seesaw. The mathematics of motion was a direct result of the mathematics of the setup. Whether the beam was made of wood or steel, whether the fulcrum was a ball bearing or a sharp edge, whether the seats were contoured or flat, whether the parts were assembled with carriage bolt or hex bolts, whether the participants were people or dead weights—none of this mattered because the Law of the Lever wasn't about the actual seesaw, the device that sat out in the playground. The Law of the Lever referred only to the mathematics behind the seesaw. It was all about the configurations and dimensions that had been deliberately built into the seesaw, the bare forms described by skeletal relationships, the mathematics that had been artificially introduced when material objects had been molded into mathematical shapes and put into geometrical arrangements.

The physicist retorted that the functioning of the seesaw depended on gravity and no one had created gravity, therefore the seesaw was in fact about the world, but the seesaw employed gravity for its own ends. It was deliberately designed to use gravity to its advantage, to incorporate gravity into its operation, in the same way that it capitalized on the rigidity of the wooden beam. The world could not be understood through fabricated objects because in studying these objects, the designer was only studying what he had made. He was only looking at his own ideas. The world didn't include these objects for precisely that reason. Mathematical objects were solely the products of the human imagination. If Jonathan wanted to study the world, then he couldn't focus on things that were not a part of that world, yet in order to come up with the mathematical laws of physics, that was exactly what he had to do.

As a child Jonathan had quickly outgrown the rather mundane experience of the seesaw and never thought that it compared with the excitement of slides and swings, but physicists were able to improve upon the simple game provided by the crude apparatus. Physicists correctly surmised that it was the mathematics that gave the seesaw its unique character, so in order to make the device more interesting, they added more mathematics to it. They fastened a pointer to one end of the beam, which defined a precise spot. They then erected a vertical board with a grid work of evenly spaced horizontal lines, all drawn with the precision of a strict geometrical construction, and placed it directly behind the pointer. By putting different objects on the seats, they could get the pointer to point to different lines on the board. Now wasn't that a more fun game to play? What kind of person would ever think so?

The action of the seesaw could now become a property of the object that was sitting on the seat, but as with all properties, the property was a mere phantom. Jonathan could not study an object and foresee the behavior of the seesaw, still Jonathan could start comparing objects by placing them on the seats. He could go around weighing everything, but what was the relationship of what something weighed to what something was. Weight told Jonathan no more about the nature of existence than a number told him about a person's happiness or pain.

Jonathan imagined the physicist handing him a small picture frame and a pencil holder and asking him what they had in common. Of course, Jonathan was stumped. The objects had absolutely nothing in common. Then the physicist let him in on the joke. They balanced each other out on the seesaw. Jonathan immediately responded: "So what? What did that have to do with anything?" The seesaw was obviously not essential to his understanding of these objects and therefore it was not important, but physicists were going to turn this all around. They were going to put the seesaw first. They were going to make the seesaw the most important thing in the whole world. All objects were now going to depend on the seesaw. Each object, no matter what it was, would now be defined by the seesaw. Jonathan realized that physicists didn't care about the objects. They only wanted the mathematics.

Seesaws would tell Jonathan something about the world if seesaws were natural phenomena that existed of their own accord, that is, if they were created by nature and represented nature's design, or revealed a mechanism or structure in nature. But people had constructed seesaws already knowing that they would function according to a set of preconceived ideas. They knew about gravity and they had figured out a way to take advantage of it. The seesaw was a deliberately constructed device and therefore the seesaw only expressed the ideas that had been built into it. In a world without gravity, a seesaw wouldn't make any sense. A person could use a seesaw to see if gravity were present, but that already presumed a knowledge of gravity. Without the existence of gravity, the idea of a seesaw would never have occurred to anyone.

All laws of physics were produced in a manner similar to the Law of the Lever. The laws of physics were only the laws of mathematical objects. This wasn't a problem for physicists because they had fully immersed themselves in a world of mathematical objects, so therefore the laws of physics applied to everything in their world. Nature, on the other hand, had no mathematical objects, so therefore the laws of physics did not apply to it. The physicist imagined in his mind that mathematical objects existed everywhere in the universe, however in truth these objects didn't exist anywhere. By fudging the equations he could make the numbers agree somewhat, but ultimately the whole program broke down. Eventually he would have to admit that the laws of physics could in no way explain the phenomena of nature. The physicist was trapped in his laboratory and he could not escape. That was alright with him because he really didn't want to go anywhere. He could just stay in the laboratory forever and be perfectly happy with his own creations.

The physicist stubbornly tried to squeeze a few drops of mathematics out of the whimsical clouds, and labored to elicit some semblance of a mathematical design, to find some hint of lawful behavior or locate some signs of obedience, but the efforts didn't really amount to much and the clouds were always full of surprises, happily going their own way, putting the infinite details of existence together in highly original ways, creating shapes and arrangements that had never been seen before because they had never existed before. The clouds continually proved to everyone that there were no laws of nature.

For mathematics to exist within the clouds, a vector force at each mathematical point had to be aligned with an acceleration along an axis that coincided with the orientation of the vector. The laws of motion depended on this geometry and could not exist apart from it, but the clouds did not have geometrical shapes. They did not display mathematical tendencies and had no geometrical configurations of their own. Saying that the clouds obeyed the laws of physics meant that the clouds were actually infinite arrays of straight lines. The physicist believed that these arrays copied the realities of the clouds and students were told that this was the way everyone should think about clouds, but all sorts of ways were available to think about clouds and students had no guarantee that this was a correct representation or even a valid analysis. Only a blind faith in mathematics instilled it with this power, a faith based largely on its logical consistency and rational beauty and not on its appropriateness for capturing the essence of clouds. The constructions could be put absolutely anywhere, freely introduced into every situation without restrictions, so their appropriateness was determined by the nature of the concepts themselves and not by the world. The ideas made sense from an internal perspective, so nothing could possibly be wrong with them—unless the world wasn't mathematical, in which case students had to turn elsewhere for answers.

As with all mathematical constructions, from the structure of a crooked line to the plays on a Monopoly board, the system of forces within a cloud was a new reality, a reality of arbitrary definitions that had been fabricated entirely out of mathematical imaginations. These constructions merely used the clouds as instruments in a devious plot, a scheme concocted by physicists to perpetuate the illusion that clouds were really mathematical objects. But even if Jonathan had been willing to admit that the concept of force had some legitimacy here, this was not what accounted for either the behavior or the appearance of the clouds today. Even if the mathematical relationships surrounding forces were true within the mathematical structures of the laboratory, these were not the reasons why the clouds looked the way they did today or why these clouds were different from yesterday's clouds or why they were completely unlike the clouds this morning. There were more to clouds than a few simple mathematical formulas—much more in fact. Jonathan wasn't satisfied with some bland generalizations or a few simplistic outlines. He wanted to know the reasons why he saw exactly what he saw today, why the clouds had shaped themselves into these exact forms and how these forms would lead to the complicated and intricate—and at this point unknowable—forms that he would see later in the day. The physicist had to understand that the sky was way beyond the laws of physics. The concepts fell hopelessly short and there was still an infinite abyss to be crossed. The forces and energies that had been so simple in the laboratory had unfolded and multiplied to such a degree that the clouds completely transcended these idealized interconnections. Condensation, vapor pressure, buoyancy and crystallization were all fitted together in such intricate ways and it was this fitting together that had to be understood and here was where the gesturing and the hand-waving came into play. The physicist left Jonathan to put it all together in his head, but if Jonathan tried to do this as he studied the infinite details of these particular clouds, he couldn't. When he stopped and looked closely at the specifics of what was actually happening, when he focused intently on the complex processes unfolding in the atmosphere at this very moment, they were mysterious and incomprehensible. He could put the ideas of physics together easily enough, but he couldn't put these concepts together in such a way that they actually formed the sky that he saw above him right now. All of the processes melded in fantastic ways that he couldn't understand, somehow the infinite and undefinable aspects of the atmosphere all wrapped themselves up in ornate packages, the inseparable agencies coexisting together in the same place still found ways to conflict and harmonize with each other, taking every opportunity to mesh and blend together or fight and oppose each other, and so he was left to scratch his head and say "I still don't see how any of this happened, and you know, you haven't really explained anything." Jonathan could sit at his desk in his study and be satisfied with all of the physicist's grand ideas, how one idea caused another idea to exist, but once he donned his pack and hit the trail and he looked up at the sky, all of a sudden he said to himself "Wait a minute! I guess I don't see how all of those ideas could possibly have produced what I see over my head right now. The whole system of ideas sounded so good when the physicist was talking about them and for a moment the physicist had me totally convinced, but now when I get out here, I realize that talk is cheap, and that it was all just talk."

Physicists believed that the deliberate mathematics of the laboratory could be found in the clouds as the inadvertent mathematics of natural processes. The clouds were composed of strict mathematical relationships that were then arranged haphazardly in non-mathematical ways. The lack of mathematics in the clouds as a whole could be explained as the dissolution of the underlying mathematics, a scrambling of the precise mathematical components to achieve imprecision in the overall designs and patterns. The delicate relationships fell apart and became a morass of incidentals all behaving badly, yet the mathematics was still in there somewhere—at least in the minds of physicists. They envisioned the clouds as assemblages of laboratory experiments, but if Jonathan started with the clouds, he could not find these experiments anywhere within their chaotic behaviors. The experiments depended on precise mathematical arrangements that were no longer present once Jonathan transferred his attention to the clouds. Physicists had to start with these arrangements and then insert them into the clouds in bewildering and impossibly complex schemes, mentally constructing the clouds out of these artificial building blocks in a deliberate attempt to make the clouds appear as if they were mathematical in nature. They could fudge the equations and obtain approximate correlations—but what was the point?

No one had made the clouds so who could say what had gone into them. There might be other sides to clouds that no one knew about, alter egos with strange idiosyncrasies, a form of personhood caused by networks of hidden structures, invisible electric circuits and ulterior motives. Jonathan saw that clouds were infinite and he didn't pretend to know what they were capable of doing. At the very least they had artificial intelligence, meaning that they could respond in seemingly intelligent ways to the world around them, carving out a place for themselves by tapping into the flows of energies and the atmospheric resources of their world to invent themselves and perpetuate their kind. Without knowing what consciousness was, Jonathan couldn't say for sure whether they possessed it or not. They might have the awareness of an ant or a bacteria—or even the intelligence of a barnyard animal. They might have a different kind of consciousness altogether, a sentience not comparable to anything found in the realm of terrestrial organisms. Clouds were very strange and deceptively complicated. Jonathan didn't know what they were doing up there in the sky, but he felt that it was more than just making rain and providing shade.

Jonathan knew for a fact that a computer would never become conscious. He knew that because he knew what a computer was and he knew how it worked. The clouds he wasn't so sure about. Physicists disparaged clouds as much as they disparaged lightning: a cloud was nothing more than water vapor and lightning was just the discharge of static electricity. Jonathan couldn't turn to physicists for knowledge because they didn't understand anything about the world he lived in. Physicists idolized the players in their laboratories, consistently glorifying the performances of these actors in their puppet shows, and at the same time they demeaned the inhabitants of the desert, writing them off as trivial occurrences not worthy of their attention. Physicists were master cartoonists and they turned everything they saw into caricatures, sketching diagrams and illustrations that portrayed the lives of characters who weren't real in such a way that they appeared to be real. Jonathan didn't want a graphic artist style drawing depicting the flows of water vapor —he wanted to know what a cloud really was, but how could he possibly find that out?

His new insight regarding the nature of clouds awakened an open-mindedness in Jonathan that he previously hadn't possessed. He couldn't wait to test his theory on the pesky storm cloud guarding his camper. He rose and mustered Emma. Emma followed him as he made the journey back. As the truck came into view, he saw that the cloud had left and the camper was unattended. He searched the sky in every direction, trying to figure out which way the cloud had gone, hoping to spot it somewhere off in the distance, but the cloud had mysteriously vanished and it was nowhere to be found.

Jonathan could not call the cloud back for an encore performance. No one could entice the cloud to do once again what it had done earlier, yet this fact did not mean that what he had seen wasn't real. Reproducibility meant nothing because almost nothing was reproducible. He could pick up a rock and drop it on the ground. Each time he did this the result was the same—but so what? The event was unimportant, and paled in comparison to the visit by the angry cloud this morning. Dropping rocks was boring and the consequences meant nothing, yet physicists prized such mundane occurrences and placed them above everything else because such trivial affairs captured the essence of mathematics. Jonathan was not able to prove his theory about clouds, since every test was utterly unique and unlike the previous examples. The next cloud would not be a replica of the last cloud, so he could not count on the same response he'd gotten before. Physicists went ahead and dismissed the events on those grounds. Jonathan's theory could not be verified because the experiment wasn't reproducible, even though reproducibility obviously didn't mean anything and the criterion could not possibly be used to establish the validity of a theory.

Jonathan was disappointed that he had missed out on his chance this time, but if he relied on his theory in the future whenever he was harassed by these potentially dangerous beings and he didn't get hit by lightning, his success would serve to bolster his confidence in dealing with them, however it wouldn't prove that his ideas were correct. This was a universal problem with theories and ideas. Jonathan needed to know what something was before he could say for sure that his notions about it were correct. Were clouds really sentient beings—or did they just act that way sometimes?

If the day ever came when Jonathan was nearly struck by lightning issued from a cloud that he had previously rebuked, then he must complicate his theory. He could say that clouds had different personalities and therefore he must identify the temperament of this particular cloud, then mold his actions according to the category it fell under. Some clouds might respond to pleading, while others might be amenable to a persuasive line of reasoning—both being impervious to harsh scolding. Since experience was always open ended, the question could never be settled once and for all. Scientific theories were no different. All he could say was that his ideas had been consistent with his experiences so far, yet that fact was always subject to change in an instant.

The physicist ridiculed Jonathan for his theory of clouds, yet he could offer no theory of his own. He hadn't the slightest clue as to when the next bolt would appear or from which point in the clouds it would issue or at which point on the earth it would strike, yet this absolute ignorance was sufficient to prove Jonathan wrong. His faith in the ideas of mathematics was unshakable, just as the faith of the preacher was impervious to cogent lines of argumentation. Laboratory experiments could not possibly explain the behavior of clouds out here in the desert, so why should Jonathan waste his time with them? Yet the behaviors of clouds were the kinds of things that Jonathan desperately wanted to explain.

Jonathan realized that in order to move forward, physicists must stop asking the questions they could answer, the relatively simple questions about distances and velocities, and start asking the questions they couldn't answer—the tough, meaningful questions that would at least lead them in the right direction and get them to stop thinking in routine ways. The methods that had been adopted by physicists over the past few centuries had not led them to an understanding of the world because their methods did not even address the world. The desert was as much a mystery today as it was before the invention of physics and all the theorems of mathematics, before numbers and counting dominated everyone's thoughts, before the insane idea that reality was made up of quantities captivated the physicists' imagination. Jonathan wanted to go back and start over. How could he begin to answer questions such as "What was this object?" without resorting to an exposition of the mechanics of that object in response to external influences?

Just to humor the physicist, Jonathan tried to analyze clouds using the standard scientific approach. He saw that clouds must have a force of cohesion holding them together, otherwise they would dissipate into a uniform haze throughout the sky. They must have another force keeping them aloft, otherwise gravity would bring them down to earth. Then they must have the force of the wind driving them along, causing them to sail across the sky. Jonathan was unhappy with the results. Clouds obviously could not be understood in terms of forces, but Jonathan saw a greater problem. Wasn't the physicist merely investing clouds with nonexistent entities, just as Jonathan had done in his theory? Could the physicist make these entities real by figuring them into equations? The physicist told Jonathan that the physics of clouds was well understood, only he couldn't make accurate predictions regarding this particular cloud. In other words, the physicist had an assortment of ideas that didn't work in practice, but he believed in them anyway, because they made sense to him. Jonathan wished that he could share that sentiment, but the physicist's ideas fell short of his expectations. They could not account for what Jonathan had witnessed this morning.

Indigenous peoples around the world and throughout history have typically settled on animistic explanations for natural phenomena. Over a lifetime of outdoor experiences, Jonathan found himself gravitating toward these types of stories himself. In some strange way they seemed to fit his observations. He could use these accounts to make sense of what was happening around him. He wasn't sure whether he believed in supernatural powers or not, but after spending a lot of time in the wilderness, he had become sympathetic to that approach. What was it about the desert that made him think that it was full of living, conscious beings? Was it simply the unfathomable complexity of the world that caused it to act as if it were alive—or was it more than that? Were the behaviors genuine—or just mimicries?

Jonathan could understand that the physicist's experiences in the laboratory had led him to the conclusion that the world was a machine. The physicist made the world act as if it were a machine, but it was all just a puppet show. Why weren't all of these choreographed performances also just mimicries?

Given all the facades, Jonathan asked himself: "What was the world really like?" He understood that the world was pliable and he could make it into anything he wanted—up to a point. But if he pressed his inquiries hard enough, the explanations always broke down and the world dissolved into nothingness. No matter how he approached it, the world invariably disintegrated into mere appearances. Whatever it was, the world was weird. It was more than anyone could ever imagine. One thing he knew for sure: its facades would never be penetrated by mere reason. This approach had been tried repeatedly over the centuries and during that time the world had led everyone on a grand chase. They ran after it until they were all tired and panting for breath, but they did not catch it. Jonathan felt that no one would ever figure it out, but they would always kid themselves into thinking that they had.

Jonathan opened the camper door and stood his pack in a corner. He was very grateful for the opportunity to lie on a bed and rest after his morning ordeal. Emma curled up beside him and immediately started snoring, thoroughly exhausted after many miles of strenuous exercise, hunting and tracking the bounty of small critters living under rocks or in the back ends of dens and burrows. They both whiled away the afternoon, then Jonathan gathered himself to relocate camp further down the road. He quickly found a beautiful spot nestled within a large stand of trees and bushes. He replenished his pack with fresh water, stuffed a long underwear top next to the stainless steel bottles, activated and reset his GPS, then set out for an evening foray in the direction opposite from the one he had taken this morning.

The terrain along this second route was more varied, rolling hills of sagebrush broken by patches of forest, similar to the cultivated fields in the midwest bordered by bands of trees where the farmer's axe had never intruded. He and Emma crossed these narrow bands of woodlands always to emerge upon another open expanse of low vegetation. The pattern seemed endless—until he encountered a line along the river where the land fell away abruptly, dropping a thousand feet straight down.

There was no discernible warning, no gradual shift in the nature of things, no hint of the edge approaching. The demarcation was completely invisible until he was standing right on the brink looking straight down into nothingness. He could not linger at this point for very long without feeling vertigo because the ground was suddenly very far away and he was way up in the sky with no parachute and nothing to break his fall. The objects below were the most minuscule trivialities, the tall and mighty trees nothing more than tiny specks, the massive outcrops of rock only small pimples on the face of the earth. The drop was one fifth of a mile down with no slope, no intermediary. He could not see the cliff from the top and could only imagine what it looked like. The world around him just ended, suddenly vanishing into thin air.

Emma showed absolutely no fear. She walked right up to the termination of this towering stone wall and stood with her toes hanging over the edge, surveying the scene below with a devil-may-care attitude. Jonathan wondered if Emma fully understood the extent of this monstrous precipice. Did she not see what this was? Watching her, he was convinced that if she saw something down there—an elk or a deer— she would leap off the edge to chase after it, not realizing just how far down it really was. Her casual approach to this extreme danger unnerved him and he felt that he could not trust her. Her judgement was clearly impaired. He felt obligated to step in and bring sanity to the situation, so he backed away from the abyss and made a hasty retreat, beckoning her to follow him.

A rough lane paralleled the shear drop-off, set back a short distance from the veiled boundary lurking in the vegetation. The lonely thoroughfare obviously entertained no more than an occasional visitor, but it provided a convenient pathway for Jonathan to walk through the often dense stands of sage. He quickly noticed that this was rabbit country. Emma had already discovered this fact and had dramatically ramped up her hunting routine, running at a faster clip and hurdling minor obstacles hoping to surprise a victim hiding quietly in the bushes. Normally submerged in the sea of greenery, Emma broke the wavy surface in long, arcing leaps as if she were a salmon swimming upstream and fighting the rapids, only to plunge back down into the unseen depths beneath the leaves. What if one of her acrobatic vaults took her over the rim? Or what if a rabbit bolted from its hiding place and in its frantic attempt to escape Emma's clutches, it inadvertently catapulted into the canyon? Emma, not thinking, would surely follow it to its death, but then Jonathan considered that no rabbit was dumb enough to do this. They were capable of executing hairpin turns while running at all-out sprints. The rabbit would dart the other way at the last possible moment and Emma, not being so quick, would be carried forward by her momentum and thrown over the edge, the rabbit thus cleverly saving itself and disposing of its pursuer at the same time. The fact that Jonathan trusted a rabbit's judgement more than he trusted his dog's judgement certainly disparaged Emma's intelligence, but perhaps it was his own judgement that was impaired and Emma would never allow herself to be killed in such a ludicrous fashion.

Jonathan realized that he hadn't seen Emma for some time now so he stopped and waited for her, turning his head to catch the faint sound of panting or the cracking and rustling of branches off in the distance. To his dismay, the world around him was absolutely silent. He called Emma's name and waited for a response. He didn't hear anything. Reluctantly he continued to walk the path alone as it dipped into a shallow depression then rose back up, finally arriving at an excellent lookout. He climbed to the top of the broad mound of flat, compacted rocks, neatly fitted together and stacked on top of one another, then slowly rotated his body in a complete circle, carefully sweeping the terrain with his eyes, looking for some sign of Emma. She was nowhere to be found. She might have silently gone over the edge after all. Her demise would have been an accident through no fault of her own. He couldn't blame her for forgetting about the drop off because no one would ever expect the ground to suddenly disappear like that. He stood on the mound for quite a long time, feeling abandoned. Despondent, he began the return journey back to camp fearing that she was gone, yet holding out on the possibility that he might run into her along the way.

He cast a mournful glance toward the gorge and realized his mistake. He should never have put Emma in danger like that. He should have gone back to the truck when he had the chance, when he first came upon the cliff, and looked for another place to walk. Jonathan was angry. He was angry with himself for not doing the right thing, he was angry with nature for creating absurd hazards that served no purpose, and he was angry with Emma for being so obsessed with hunting that she didn't pay attention to anything else. The correct solution always made itself known after a time, but often it was too late to go back and remedy the situation. He should have been thinking more clearly.

He looked around at the peaceful desert landscape. He was all alone and he had the whole place to himself. He took a deep breath. The fresh air added greatly to his enjoyment of the evening as he absorbed the wondrous views in all directions. Suddenly he caught a glimpse of something over by the gorge, a brief flash of white, maybe a bird fluttering in the bushes getting ready to nest for the night. Perhaps he only imagined it. He looked more closely. The fields of sage were absolutely motionless. No, there it was again, the tip of Emma's tail pointing straight up, quickly passing between two bushes. She had been hunting rabbits along the precipitous ledge just as he had feared. She stopped running for a moment and her tail began jerking and twitching, standing stiffly upright. She was undoubtedly consumed with the task of excavating a burrow. Jonathan whistled. Emma looked up and noticed Jonathan's about-face, then came running toward him. She was a whirlwind of energy as she flew past him and continued on down the road. She didn't get very far before she plunged back into the bushes and disappeared, heading once more toward the dreaded brink of death, where apparently all the rabbits were waiting for her.

Jonathan was too worried about Emma to allow her to return to the ledge and put herself in jeopardy again, so he struck off into the deep sage opposite the danger of the canyon, abandoning his pleasant stroll down the smooth, open path, beckoning Emma to follow him as he went. Trail blazing was much rougher than he had expected. He zigzagged through the maze of tall obstacles and bulldozed between them, his arms swinging and grabbing as if he were fighting gangs of stout ruffians, stiff branches tearing at his clothes and jabbing at his legs. He felt better though, no longer having to worry about Emma's safety, yet his progress was agonizingly slow. Emma negotiated the terrain with comparative ease, slipping through the narrow openings between the bases of the tapered bushes. If it weren't for his backpack, he might also find it easier to get down on all fours and take advantage of these spaces.

He'd managed only a hundred yards or so when the light began to fade and he found himself hopelessly imprisoned in an impenetrable jungle of wild shrubs. He breathed deeply and redoubled his efforts to escape, crashing through the resilient branches and clawing his way over the dense tangles of sturdy stalks. He finally reached a patch of higher ground and the sage gratefully thinned out, permitting him to walk more normally, albeit in a circuitous route around the now waist-high plants. Frequently he was forced to backtrack after finding himself trapped in a cul-de-sac and this took up a lot of valuable time. Finally he came upon an unused and mostly overgrown path cutting through the otherwise unbroken fields of harsh vegetation and happily followed this narrow track in the general direction of the truck, basking in the last remnants of the diminishing twilight.

He accelerated his pace as the light failed completely, striving to get back as quickly as possible. The truck was parked on the opposite side of one last patch of forest. Jonathan charged into the enclosure of trees and entered what seemed to be the darkness of a cave, but he recklessly maintained his zeal to reach camp, now running mostly blind. He flew through the night with long strides, sailing along in the cool air at breakneck speed, the black shapes and dark hollows vaguely catching his attention as he quickly passed them by—until the toe of his shoe caught on the barbed wire of a downed fence, located in a place where he never would have expected to find a downed fence. The abrupt snag propelled him forward and slammed him down on the ground. He sensed objects nearby in the darkness. His head was enclosed on either side by two large rocks. Luckily, he had landed in the narrow gap between them. If he had fallen just a few inches to either side, he would have broken his jaw or cracked his skull, but as it turned out, he was unscathed, comfortably laying face down on a soft bed of conifer needles. He remained on the ground for quite a while, catching his breath and meditating on the nature of pure chance. Unable to reach any conclusions, he collected himself and proceeded more cautiously this time, not willing to tempt fate twice. He finally broke out of the forest and found his dark and gloomy camp waiting for him, the truck half submerged in the thick vegetation, encircled by domineering bushes and dwarfed by overbearing stands of young trees. Without hesitation he mounted the stairs and unlocked the door, then entered the interior, immediately bringing his little abode to life by turning on the cabin lights.

—11—

Symbolic logic could be based on different sets of axioms, as long as these statements didn't contradict one another, but from Aristotle all the way up to Russel and Whitehead, treatises on logic had typically included the fundamental axiom known as "The Law of the Excluded Middle." The proposition was this: for any statement A, either A must be true or its negation A⁻ must be true, but not both. Not everyone in ancient Greece had subscribed to this notion. Although we have only fragments of pre-Socratic writings, many of these philosophies appear to be incompatible with such an idea.

Apparently many of the pre-Socratics weren't willing to sacrifice the world at the alter of logic in order to appease the gods of mathematics. Jonathan admired them for that. Plato held the world down and Aristotle took up the axe, obsessed with the idea of splitting everything into two mutually exclusive alternatives, but some adherents to these earlier philosophies must have undoubtedly rushed onto the scene and exclaimed: "Wait! You can't do that!" Someone must surely have protested, and made the point that this act was a crime against nature.

The statement "Tom entered the room" seemed simple enough and Jonathan felt that he should have no difficulty determining its truth value. Either Tom entered the room or Tom didn't enter the room. But what if Tom stopped in the doorway and surveyed the room from there? Was the doorway part of the room or not part of the room? Tom had been looking for his coat and he saw that the coat wasn't there. Even if Tom had remained outside the doorway, he entered the room in a manner of speaking because he visually interacted with the objects inside the room, so in a way he was inside the room even though his body did not cross the boundary of the room.

The statement "The sun rose this morning" was equally straightforward. But what if the skies had been overcast and there had been no sunrise, only a gradual lightening of the sky? Jonathan needed to know the meaning of the word 'sunrise' in the statement and the word could have multiple meanings. By sunrise Jonathan had meant the change in the position of the sun due to the rotation of the earth and since the earth had kept rotating and a new day had dawned, there was in fact what Jonathan would call a sunrise. No one could know that this was what he meant by the sun rising, and that he wasn't referring to an actual sunrise in the sky.

The problem was more than just the endless ambiguities in the meanings of words. Jonathan picked up a polished stone in the shape of a coin and said to himself: "This stone is circular." The stone was circular only to the extent that it partook of the concept of circularity, and because it was not a perfect circle, the stone also partook of non-circularity as well. Some stones were rounder then others and although each one had a certain

234

element of circularity, none of them had the exactitude of a mathematical circle. So each of these stones was both a circle and not a circle at the same time.

Aristotle's Law of the Excluded Middle was a statement about symbols and the formal relationships between abstractions and not a statement about the real world. The mathematical logic was clear and precise, in contrast to the indistinct world of complex events and corporeal bodies with all their hazy boundaries and multiple characters. The law was only true in a purely mathematical world and it made little or no sense in Jonathan's world, which was the standard of truth for him. Although the statement was clearly false, it remained a cornerstone of symbolic logic and therefore of set theory and the remainder of mathematics.

So why wasn't mathematics pure nonsense? Because the statements were logically consistent. This was the reason why mathematics made sense—well, not to Jonathan, but to the mathematician. The logic of mathematics made sense from a logical point of view. But the logician demanded that this logic must make sense to everyone. Everyone must agree with the rules of logic. Yet given the amorphous character of the world, nothing fell into neat, clearcut categories and logic was nonsense. Logic only made sense to someone who already believed in logic, namely someone who was not thinking about the world as it really was, but someone who was thinking only in symbolic terms, someone who had turned the world into a caricature, simply using the world as a pawn in their interplays of logical propositions. Logic said nothing about the world and did not conform to the realities of the world. Logic held the world hostage and forced it to participate in devilish schemes, tortured by constraints that kept it in unnatural and uncomfortable positions for very long times—forever in fact—a punishment consistent with the concept of eternal damnation.

Logic was still touted in academic circles and commonly understood by the general public as being the correct way of thinking, but mathematicians failed to finish their sentence and complete the thought. Logic was the correct way of thinking about symbols, but if the world could not be reduced to a set of symbols and understood in those terms, then logic was the incorrect way of thinking about anything other than a system of formal relationships. Logic was categorically false when addressing the infinite particularity of the world, which by its very nature could not be covered by these types of ideas. In this broader context, logic was clearly the wrong way of thinking.

If people wanted to learn how to think correctly, the last thing they wanted to do was study logic. Logic would only drag them down into a clutter of meaningless squiggles and scratch marks, the muck of mathematical thinking. They needed to intuit the truth and that required a clarity of vision, a keen insight into the nature of things. Playing games with logical propositions would never lead them to any kind of understanding of the universe. They would just be wasting their time, caught up in the endless distractions of simplistic statements that in no way described the intricacies of the real world.

But for the logician holed up in his self-made castle of mathematics, the application of logical principles to the barbarians and the unstructured societies that flourished outside the castle walls could be downplayed. The mathematician could only make this statement because he was immersed in a mathematical world. He could keep the outside world at arm's length and push it away, then the discrepancies didn't seem so serious. But as a barbarian in the wilderness, Jonathan saw the castle differently. The problems with logic were insurmountable. They were deal-breakers. Their presence was intolerable.

None of this was new to the logician, it was just that the logician demoted these problems to a secondary level. For him, logic was beautiful and all-encompassing. The confrontations with nature occurred around the periphery of his domain. Here in the desert, Jonathan was outside the boundaries of logic. He was lost in an irrational world and trapped in a vast wasteland of incongruities and contradictions. Clearly he was over the edge and if anyone ever asked him about logic, he would say: "Off with the king's head! I will not be ruled by such a foreign power!"

As mathematicians developed mathematical ideas, these ideas acquired a force of conviction to them and they accumulated a momentum of their own. The concepts were so enticing, the reasoning so convincing, the conclusions so inevitable, and everything proceeded according to plan. All was fine, but then when physicists examined the details of their equations and contrasted them to the realities of the world, they encountered for the first time the hint of a problem. But since the discrepancies came at such a late stage in the proceedings and the mathematics had already become so firmly entrenched in the minds of mathematicians and physicists, the mathematics had the power to bulldoze over what seemed in comparison to be minor obstacles. Once the mathematics had gained a head of steam, slight mismatches and insignificant deviations from experimental data appeared much smaller than they actually were. Perhaps physicists could simply leave them out of the discussion altogether and refer the reader to an appendix at the back of the book. They could place an asterisk after the equations directing the reader's attention to a footnote cast in much smaller print stating that this wasn't all as neat and tidy as the equations had indicated. Such trivial matters could not undo the great accomplishments of pure rational thought. After all, the logical arguments were still valid, and the tightly knit structure was not so easily toppled. That was why the scientific method proceeded in this fashion, by establishing the mathematics first, without controversy or reservation, before the theorist had to confront the world full in the face. By this time the mathematics was very powerful. What was the world at this point? A trifling, an afterthought. Mathematicians could bask in the glory of their accomplishments regardless of the petty objections put up by a messy, disheveled world. Such an unkempt vagabond was not going to crash their party or wipe the smiles off their faces.

When Jonathan began instead with the realities of the world, mathematics had no way of ever getting off the ground. Mathematics was dead in the water before it even had an opportunity to learn how to swim. Mathematics? What kind of nonsense was that? The equations were awkward and inept. The world scoffed at mathematical formulations. They were jokes. If the physicist debated Jonathan's philosophical positions, the two of them would be like stand-up comedians at a roast. The laughs would come quickly and easily because the whole thing was funny when you thought about it. When the physicist drew a line from point A to point B, everyone should start howling and fall off their chairs in spasms of cackling and guffawing. Comedy was based on absurdity, and the comedian would find no better material than the scientific method.

Summer in the northern mountains typically ended on or about Halloween. The increasingly cold nights of late October turned the grasses a dull and dirty brown and caused the once colorful autumn leaves to fall to the ground. The numerous stands of thriving weeds and lush foliage laid down on the ground where they quickly rotted and decayed, the myriads of tiny flowers and their attendant swarms of bees were silent and gone. The skies became pale and were often obscured by layers of ominous gray clouds. The vibrant summer landscape was replaced by a colorless world with black trees framing dormant fields and somber homesteads surrounded by dilapidated outbuildings, piles of old junk, and toppled fences that had previously been obscured by thickets of vegetation, plumes of woodsmoke issuing from the motley assortment of rusted metal stovepipes.

Suddenly the scene was very grim and would remain that way for the next six months. Jonathan celebrated the passing of summer with a bonfire out in the field next to his Airstream. He always tried to stay up until midnight to officially bid farewell to both the month and the season, but the toxic smoke usually prevented him from doing this, causing him to become rather dizzy and lightheaded. This year he passed out under the stars while relaxing in his reclining chair, then woke abruptly, thinking that he was inside the Airstream and the burning coals were the charred remains of his wooden floor. He became alarmed since he was supposed to have been carefully tending the fire in the middle of his living room, but he had fallen asleep and the fire had gotten out of control, and now the floor was ruined. He realized that he was hallucinating. He wasn't inside the Airstream at all, but sitting out in the field in a lawn chair.

As he had correctly predicted, the first days of November were wintery and the temperature dropped to ten degrees on successive nights, the coldest nights he'd ever spent camping in the Airstream. The snows would be here soon and the time had come for him to leave. Emma, on the other hand, just wanted to stay, as winter was her favorite time of the year.

Finally being back in the low desert was a homecoming for Jonathan, a return to the place where he truly belonged, the land he loved the most. Jonathan pulled the truck off the road onto a generous gravel turnout and parked at the periphery of this rough patch of ground in order to let Emma run around for a few minutes. A large green sign nearby announced that there were no services ahead for the next 121 miles. He noticed a metal box propped up against one of the posts. Two rubber hoses issued from the device and stretched across the asphalt, spilling onto the opposite shoulder. The whole area was deserted in every direction as far as Jonathan could see, save for strings of puffy clouds floating in the sky and casting shadows on the ground, enticing him to linger in order to follow them with his eyes.

The highway department was obviously trying to determine how many people used this road, probably as a requirement for funding. Jonathan walked over to the box and alternately stepped on each of the hoses. There was no response from the box, making him wonder if his actions were being recorded. He looked up at the sky. The clouds had all moved away. They had also changed shapes and he could no longer identify the ones he had intended to track.

The connection between the blips in the box and the traffic on the highway was not an illusion. The relationship was perfectly clear. Every time something compressed the hoses, an electric pulse was generated, so there was a one-to-one correspondence between the two events and thus in the minds of physicists, the output of the box was an accurate depiction of the events that were taking place on the road. There was no sign of any traffic at the moment, but let's say five vehicles had passed over the hoses in the last hour. This was a statement of fact and no one could deny it. The physicist considered it to be an account of the reality of the world.

Jonathan pointed out, however, that the problem with the statement was that there were no vehicles to be found anywhere in the world. A vehicle was simply a definition and it could be defined in many different ways. A vehicle could be a M1 Abrams tank or a tricycle, a roller skate or a skate board, a paw or a foot, a rolling rock or a mechanical robot, a sled or a snowmobile. "Vehicle" was just a marker, a symbol, an empty reference. The term "vehicle" told Jonathan nothing about the reality of the world and thus it was not a statement about the world. The introduction of the term "vehicle" turned everything into a game where pieces moved about on a game board. The markers could be counted, the numbers added and subtracted from one another, and mathematics developed. The physicist was convinced that since the numbers had originally been obtained from the realities of the road, they therefore faithfully represented the realities of the road, but basing definitions on the world only resulted in a reality of definitions, a new reality that was based entirely on the imagination. The physicist effortlessly leapt into a conceptual universe and immediately started playing games with logical propositions and never thought twice about it. The infinite particularity of the world was quickly abandoned, but that was quite understandable given that the infinite particularity could not be thought.

The sequences of blips were nothing at all like the drama unfolding on the highway. The real-life travelers on the highway weren't a series of markers. They were constellations of concrete objects, each one unique and immensely complex. Casting them as markers misrepresented these objects, changed the very nature of their existence, and reinvented what was actually happening on the road, transforming the whole situation into a parody. This travesty was allowed and even welcomed because mathematics was so much better than reality, making the exchange more than just a fair trade, but rather a great deal. Once the mathematics was safely in hand, reality could be cast aside and forgotten, setting the physicist free to parade around with his formulas.

In broad outline at least, the metal box detector sitting alongside the highway was not unlike the detectors used by physicists. An object triggered a response in the electronic circuitry which then recorded the event. The detector could not tell Jonathan what the object was, but only that "something" was there. The pressure tubes could not tell the difference between a Mack truck and a Peterbuilt. They could not distinguish a Lexus from a Tacoma, or a full-size pickup pulling a double axle trailer from one car closely following another. The tubes could not tell him how fast a vehicle was going because the distance between axles varied greatly with different vehicles. If two vehicles crossed the tubes simultaneously going in opposite directions, the event would generate a pattern of impulses, but the detector could not decide which pulses belonged to which vehicles. Jonathan imagined a troupe of bicyclists crossing the tubes in tight formation and wondered what that would that look like in terms of the activities in the box. How could the operator of the device use the recorded data to differentiate between a bevy of cyclists and a parade of pedestrians? Even Sherlock Holmes could not deduce the actions taking place on the road using nothing more than the actions taking place inside the box. Sherlock could cheat and walk the right-of-way looking for clues, drips of black water from a leaking RV septic tank or plops of fresh manure bounced from the wooden platform of a cattle trailer, but these clues could never even begin to tell the whole story. Therefore the series of impulses produced by the box was not a portrait of reality, just as a diffraction pattern was not an image of an atom, a ribbon diagram was not a depiction of a protein, and the flow of neutrinos streaming out from the sun was not a picture of the sun's core. The reality of the road was a rancher hauling a trailer full of cows to his range allotment up in the mountains, or a retired couple pulling a fifth wheel Montana looking for an RV park. The blips did not capture these events or even point to them. The blips pointed instead to the operation of the mechanism itself.

But the reality of the road was more than just a couple pulling an RV. The reality was that the driver had a tumor in his brain and he was unaware of it, a scar on his face from when he was hit by a baseball bat as a child, red blood cells that had acquired thick coatings of sugar molecules, bustling colonies of Helicobacter pylori thriving in his intestines. The reality was that seven kernels of popcorn were lodged in the metal slides under the driver's seat, twenty three tiny flakes of potato chips were embedded in the pile

238

of the carpet, and 196 white dog hairs were affixed to the upholstery, mostly in the inaccessible places between the cushions. The doors and fenders had intricate patterns of scratches and dents embossed on the surfaces, the upholstery had other patterns of scuff marks and a host of practically invisible tears, imperfections in the threads of the fabric. In order to account for this particular vehicle and to list all of its distinguishing features, Jonathan would have to meticulously examine every bolt and bracket, every hose and wire, down to its molecular structure. Or he could simply replace the vehicle with a set of concepts, things that he could more easily deal with, and just forget about the actual truck.

Jonathan could not even begin to specify the cellular structures within the bodies of the driver and his passenger, or the chemical reactions occurring within these structures. The occupants not only sat in an ornate palace of finely crafted works of art and exquisite assemblies of exotic structures, they were themselves ornate palaces, and to top it all off, each item was simultaneously in motion, exhibiting the most intriguing overarching display. The blockbuster show traveling down the road was certainly a head-turner by normal standards. By the time Jonathan began enumerating the qualities and characteristics of each part, the whole constellation had changed and he must start over.

Jonathan could complicate his ideas all he wanted, but he could never approach the truth because physical reality utterly transcended every form of conceptualization and escaped every attempt to categorize it and pin it down. He could never know the details of what just went by, never comprehend the sheer magnitude of it. The more he examined the component parts, the more he focused on the particulars and fine details of these objects, the more the objects escaped him, the more he realized just how much there was to know. The box by the roadside produced an electrical impulse that was not itself a part of this grand circus rolling down the highway, and the empty mathematical forms told him nothing about the reality before him. How could he draw a picture of the truth or grasp the reality of existence by starting with the mere outlines of empty concepts, patterns of events that hadn't even existed until someone had made a box to create them? He couldn't. All he could do was forget about the truth and make the output of the box his reality. In this way he could hold onto the mathematics, but the possession came with a terrible price. He would have to sacrifice the entire world in exchange for it. He would have to turn his back on everything that existed and withdraw into an imaginary world of mathematical constructions. The physicist told him not to worry because he could always leave a trail of conceptual crumbs so that he could find his way back to the real world anytime he wanted to go there.

The scratches on the fenders of the truck had no connection to the scuffs on the seats, the crack in the windshield bore no relationship to the stripped bolt head on the oil pan, the burned out taillight had nothing to do with the dirt beneath the floor mat. The various attributes of the truck were not connected together. The truck was a constellation of independent objects all traveling together in unison. The parts did not share a common origin, a similar outline, a shared behavior, or resemble each other in any other way. The problem was not limited to the truck. Everyone lived in a thoroughly disconnected universe.

On the other hand, the pistons fit precisely into the cylinders, the patterns of holes on the rims matched the lugs on the hubs, the gears in the transmission meshed cleanly without binding. These relationships existed only because the truck had been manufactured using mathematical concepts as a guide. The parts fit together because people had made them that way, after thinking about them for a while. Did someone think about the metabolic processes in a living cell and then construct the organelles so that their functions meshed together like the gears of a transmission?

Photon detectors were just another example of blind sensing, a different version of unknown objects crossing a fixed point. The physicist could not see the photons and could only surmise their existence by the triggering of the device. The device was telling him "something" about reality, but the details had been omitted and there was no clear picture of it. The physicist filled in the blanks with abstractions and mathematical concepts, but what good was that if the world wasn't mathematical? The picture the physicist developed in his mind was not a true picture of reality because an infinite number of essential facts had never been brought into account by the detectors. The solution to this problem was to just say that the abstractions were the only things anyone needed to know.

The physicist hid the fact that he was trying to see the world entirely in terms of blips in boxes. The absurdity of such a perspective caused him to mask the true state of affairs. The physicist decided not to see the blip in the box for what it was. Instead the blip in the box was actually a photon flying through space, or perhaps a weird bundle of mass and energy called an electron, or a complex constellation of particles called an atom, or some other product of his imagination. The blip assumed the role of a symbol in an equation. It referred to something else—essentially whatever the physicist wanted it to be.

Jonathan imagined that someone had inadvertently wandered in off the street and entered the physicist's laboratory by mistake. All this person really wanted was a croissant and a cup of coffee at the commissary in the faculty commons, but once this person was inside the room, he became captivated by all the clandestine activities. This impartial bystander recognized the event of a detector going off for what it was, a noise emitted by a metal box, a light on a control panel coming on, a change in the numbers displayed on a computer screen. The unwitting observer was ignorant of the special meanings of these events and took them at their face value. The physicist reassured him that the light wasn't important in and of itself, but only for what it represented. He explained that somewhere in the mass of metal boxes a photon had been absorbed by an atom in a crystal lattice. The event could not be seen by anyone. There was only the light turning on. The physicist could open an access door to a section of the apparatus and point to where the event supposedly took place, but there was no point in looking because the photon and the atom were not there. The physicist explained to the lost individual that he had to picture these entities in his mind, or draw a diagram of them on a piece of paper. The photon and the atom were ideas, symbols within arrays of other symbols, streaks of graphite on a notepad. Did that mean that nothing happened when the light came on? Surely something happened, but perhaps the event hadn't unfolded exactly the way the physicist had pictured it in his mind. Perhaps the meaning of the light was more complicated than the physicist had realized, and the reasons for its behavior were more obscure. Perhaps the little diagram he drew for his guest wasn't actually a picture of the truth—but none of that really mattered. The light was the photon. The physicist turned toward the stranger and assured him that the world wasn't made up of blips in boxes. The world was made up of a menagerie of imaginary objects that corresponded through purely formal relationships to what in truth were nothing more than blips in boxes.

The physicist paused to reflect on this statement. Maybe the unexpected visitor had a point. The physicist's conscience began to bother him. He was just making all of this stuff up, picturing objects and events in his mind, and he knew that he shouldn't be fantasizing about reality in this way, but he could solve the problem easily enough. He could simply dispense with all the vain imaginations and just write the equations. The mathematical equations were the only things that mattered anyway. Why bother to picture anything at all? He no longer felt the need to justify the mathematics with other kinds of visions. The mathematics was its own justification. Ironically, instead of dispensing with the imagination, the physicist had taken it to a new level, because seeing mathematics as the ultimate reality of the universe required more imagination than anyone had ever mustered before. The physicist's fantastic new insight was an unprecedented leap of fancy that took him to new heights of detachment and abstractionism—in other words, even further away from the world he had originally sought to understand.

As Jonathan stood beside the road wondering about the box, an old school bus came barreling around the corner, careening wildly over the pavement, heading toward some unimaginable destination with an air of determination and conviction. The outside was hand painted in bright colors, the artist having no talent or expertise for drawing, the childish designs showing not even the slightest attempt at true craftsmanship. The windows had curtains on them so Jonathan could only guess at the interior decor and the scene inside, yet he could replace the multiple levels of organization superimposed over one another, the intricate constellations of details, layers upon layers of particulars descending into an abyss of even smaller details, with an "x" in an equation or an electronic signal in a metal box, and then treat the diverse collection of objects and attributes as a simple unity. He could then measure the velocity of this symbolic "object" and plot its trajectory, but these concepts could never help him understand what he was looking at.

The monstrosity zoomed past, then pitched to one side as it rounded a particularly sharp curve in the road up ahead, still traveling flat out, teetering precariously on fatigued and worn out leaf springs, severely testing the integrity of the bald tires. Jonathan shook his head. He muttered to himself: "What in the world was that thing?" Starting with nothing more than the existence of atoms and molecules, how could these tiny entities have assembled themselves into such a complex device complete with driver and a load of passengers? What elementary forces could have possibly molded such a contraption? The physicist was silent. The existence of the world was absolutely astounding. The physicist had never run into anything like this in his laboratory, but if the physicist was going to explain everything, then he was going to have to explain that bus, and do so in terms of the forces between atoms, the shapes of the orbits of electrons, the spins of subatomic particles, the probability curves of gravity waves, and all the other concepts of physics. The driver was probably telling everyone right now to hold on tight. How could the physicist explain that using the stock concepts of physics? How could he account for the mechanical vibrations of the air molecules inside the bus right now, the material correlates of the driver's voice reaching the ears of his passengers and pounding against their eardrums in rhythmic patterns conveying a specific message? Given the curve in the road and the bus's rate of speed, was the driver's admonition to his passengers a law of nature? Was the bus ever going to arrive safely at its destination? Who could tell? The molecules would decide all of that later.

Jonathan's observations of the bus did not allow him to deduce either the inhabitants or the activities inside, yet he had attended more than a few Grateful Dead concerts in his time and he had danced in the indirect glow of the stage lights with people who had driven vehicles similar to this, so he had grounds for speculation. But the metal box alongside the highway told him nothing. Even if the operator had better instrumentation that gave him a value for the momentum of the bus, he would still know nothing. The physicist chimed in at this point and said that momentum was certainly not nothing. Momentum told him something about the bus.

Compared with all that there was to know about the bus, compared with the totality of its existence, momentum was infinitesimally important at best—as close to nothing as one could possibly get without actually being nothing. Physicists always replaced one expression for another in a mathematical equation when the two terms had approximately the same value, so why couldn't Jonathan simply replace the importance of momentum with zero? Why couldn't he just drop the term entirely and say that momentum didn't figure into any of this? The quantity was completely unimportant and of no consequence whatsoever. When the physicist used such a ploy to get what he wanted, he loved the technique, but when Jonathan used it for his own purposes, the physicist objected. Momentum was not nothing because momentum provided the physicist with a number that opened the door to the mathematical analysis of the mechanics of the bus, and that was the whole point of physics. But mathematics was what gave the momentum its importance—not the bus.

Long ago physicists had set sail on the seas of the imagination where they discovered a whole new continent, a vast realm of unexplored ideas, a place where they felt they would undoubtedly find fantastic treasures, priceless gems of knowledge and wisdom, and reap a rich bounty of profound truths about the world around them. But in fact they were still separated from their goal by an entire ocean. Mathematics was not the reality of the world. The knowledge they were seeking was mysteriously locked up inside infinite bundles of particulars and they had no way to reach it. A theory of everything? What were they talking about? They were talking about their own little world, the confining and claustrophobic studio apartment of the laboratory, the closet-sized cubbyhole hemmed in by walls and floors and ceilings with no views of the outside world whatsoever—and they were having trouble mastering even such a minuscule environment, really nothing in comparison to the totality of existence. Outside galaxies were spinning, dust devils were dancing across the open range, decaying limbs were falling from trees and shattering on the ground, and busloads of aging hippies were driving recklessly down deserted highways. No one could explain it—least of all the physicists.

Emma had wandered into the adjacent valley, totally engrossed in her own adventure. She looked back and saw Jonathan mulling about the parking lot. She didn't understand why he was so fascinated with that spot. She had already checked it out and found nothing of interest there. Jonathan often dallied for no reason and frequently took an inordinate amount of time to get going, but she had to keep an eye on him because sometimes he took off all of a sudden.

Jonathan watched the bus disappear and pondered its fate. He realized that he hadn't been paying attention to Emma. He abruptly scanned the nearby tangles of branches looking for her, then suddenly caught sight of her far off in the distance threading a course between bands of trees, holding her nose to the ground and sucking up scents like a vacuum. He whistled for her. She abandoned her pursuit the moment she heard his call and came galloping toward him, ready to move on to something else. He unlatched the door of the cab as she approached and exclaimed: "Ok, let's go!" He climbed onto the seat after her and swung the truck back onto the highway, the tires slapping the rubber hoses as the truck accelerated down the broken asphalt. He sincerely hoped that the highway department got the money they needed to fix the road.

Jonathan met no one the next 20 miles and arrived at a turnoff where he left the quiet and empty highway behind. The gravel surface of this new road was wide and well maintained. Being the only point of access for the entire area, the road was used by everyone out of necessity. After five miles, Jonathan steered the truck onto a smaller, secondary dirt road and entered a shallow valley where he crossed a large arroyo. The original road had washed out long ago, but a makeshift detour had been established nearby. The hills were all a bright yellow, thickly crowded with the desiccated stalks of an earlier crop of lush summer grasses. After negotiating the rough and treacherous detour, he continued to follow the dirt track across the rolling landscape, the road lazily tracing the sweeping contours of the hills, until he reached a very nasty arroyo crossing, a large metal culvert jutting out ten feet into empty space, indicating where the road had once been. He noticed that ATV riders had created a trail off to one side that appeared to be well used, but it was too narrow and bumpy for a full-sized pickup. The considerable effort that would be required to get his truck across this arroyo wasn't worth it, so he parked short of the obstruction. The remainder of his journey would be executed on foot.

Emma liked this place and she was all excited, tearing off to explore the new territory while Jonathan gathered his belongings and locked up the truck. He stepped into the sea of swaying yellow grasses without paying much attention to them as they surrounded and engulfed his legs, instead scanning the area up ahead looking for an interesting destination, a landmark or something. After a short distance he heard a strange sound that he did not recognize. His thoughts came to a halt and he stood perfectly still in order to listen to the unusual sound more carefully. The noise was faint and far away. Suddenly recognition took hold of him. The noise was the sound of Emma screaming. He immediately started running toward the noise. His pack was unwieldy and lurched madly from one side to the other, bouncing up and down and gyrating in circles. He tried to steady it by hooking his thumbs into the shoulder straps and putting additional tension on them, pressing the pack more tightly against his back, but this had little effect. In desperation he unclasped the breast strap and let the whole pack slide off his shoulders. The hefty bundle fell noiselessly into the dense grass as they parted company.

Running was easier now but not particularly faster. What if a bear or mountain lion had Emma in its clutches? He might want to have the gun in that case, but he'd just left it behind. He couldn't go back for it now. He needed to hurry and support Emma, particularly if she was caught up in a confrontation with something that was attacking her.

He finally spotted Emma half hidden in the tall grass. He didn't see anything nearby. He meticulously scanned the area as he continued running toward her, but he could not locate any sign of danger. She seemed to be fighting something on the ground. He couldn't imagine what that was—perhaps a snake or a lizard. When he got closer, he realized the problem. Emma's front leg had gotten caught in a leg hold trap and she was frantically trying to free herself by bucking and twisting and yanking on the chain. He knelt down bedside her and gently put his hands on her shoulders, speaking softly at the same time, telling her that everything was going to be alright. Emma understood this and stood still, waiting for him to take care of the problem. He tried to compress the release, but her leg didn't come loose. He looked at the trap more closely and noticed that there were two releases. Using both hands this time he pushed firmly down on each side and the trap sprang open. Emma bolted before he could grab her. She took off running down the arroyo and quickly vanished into the trees. Blood had been running down her leg but judging from her all-out sprint, the leg wasn't broken. Jonathan called her name loudly and whistled wildly as he ran after her. He was worried that there might be more traps in the area.

He couldn't find her anywhere and feared that she had gotten caught in another trap, this time far enough away so that he didn't hear her cries. There was a lot of territory to cover and locating her might be a problem. He finally came to a halt and shrugged his shoulders in resignation. There was nothing more he could do. He didn't know where she was or in which direction she had gone, so he didn't know where to even begin looking for her. Jonathan zigzagged between the trees, scouting the terrain with his eyes, waiting, staring at the ground, looking for evidence of her trail. This was such a beautiful place. The time passed slowly and the day wore on with no further sign of his missing companion. He ambled about aimlessly, lost in a daze, picking a meaningless path through the scattered congregations of piñons and junipers, soaking up the serenity of the silent hills. This was not the kind of hike he had imagined when he first set out today.

Jonathan whistled again, but the pitiful sound was muffled by the enormous emptiness. Emma had panicked and probably just kept running with no particular destination in mind. He swept the hills searching for a tiny white dot advancing up the craggy slopes. She would be difficult to spot at this distance and he might have already missed her escape. In despair he called out her name, then bellowed it repeatedly. He followed each outburst with an extended period of silence, straining to hear a noise off in the distance, something, anything.

Jonathan was relieved to see Emma coming toward him gaily scampering along the far bank of the arroyo, bounding through the tall grass in her usual merry fashion. She was still highly agitated and wouldn't let Jonathan capture her, but he directed her to follow him by saying "C'mon, Emma, this way!" and they headed back to the truck together. Emma traced out crazy patterns in the grass but kept in sight. Finding his pack was not an easy task as it was thoroughly camouflaged by the tall grass. After wandering around for a while, he accidentally stumbled upon it in a spot where he wasn't expecting to find it. Jonathan was glad when he finally got Emma loaded onto the front seat and firmly closed the door behind her.

The unfortunate incident was all his fault. He'd noticed signs of an unusual amount of recent activity on the road, highly suspicious for such a remote location, a place that ordinarily received few visitors. The gates were all down and there were numerous tire tracks in the dirt, but Jonathan had dismissed it all as nothing of importance, telling himself that the heavy traffic was probably due to a hunting camp, or maybe a field trip for a group of geologists, or maybe an outing sponsored by an off-roading club. Earlier he'd spotted a large fifth wheel hidden in the trees along Murphy Canyon with a couple of flatbed trailers and all terrain vehicles parked in the camp circle. They must be the trappers.

Undaunted, Jonathan drove the main road further into the wilderness, gaining elevation the whole way. All of the side roads exhibited the same degree of utilization and all the gates were down. He and Emma had nowhere to go. The road made a sharp turn as it climbed the wall of Spencer Canyon and Jonathan happened upon a place to park, a rarity in such rugged country. He pulled right up to the edge of a sharp drop-off and pushed the parking brake all the way to the floor, added insurance that the truck wouldn't roll forward into the canyon while he was gone. He dragged his gear out from the back of the truck and struck out cross country to intercept a hiking trail that would take him into the mountains. He felt safe hiking on foot. Without roads, the trappers had no way to access the area. They'd have to get off their ATV's and walk long distances over rough terrain to set their traps, and that wasn't likely to happen.

After traversing a broad, gently rising expanse of sparse forest, Jonathan reached the principal arroyo of the canyon and accessed a vague and seldom used footpath, a trail he'd hiked before, although he hadn't had the opportunity to do so in a few years. The sand was littered in places with piles of bear droppings. He nudged one with the toe of his shoe to estimate its age and found the mass of digested seeds to be fairly dry and crumbly. He similarly tested each new heap he encountered along the way. The samples were of varying ages, but none of them were fresh, indicating that the bear had left the area—at least for the time being.

Emma flanked Jonathan to his right, about halfway up the side of the canyon, frequently disappearing into the dense brush and slipping behind rows of large boulders. From this distance, Emma appeared to be nothing more than a white dot moving between the rocks and shrubs. Jonathan couldn't distinguish her legs or head, and certainly none of her finer details. If he didn't know who she was, he wouldn't have been able to tell exactly what he was looking at. Similarly, when viewed from a great distance, galaxies were nothing more than points of light. From such a vantage point, an observer might conclude that they were simple entities, particles in fact, with specific sets of properties. But on closer inspection, these galaxies were really immense collections of smaller objects. As the observer zoomed in further and got inside one of these galaxies, he found from this new perspective what appeared to be more particles: stars, black holes, quasars, supernovae, planets, moons, comets, asteroids. But as he got closer still, reducing the scale even further, all of these objects were themselves complex worlds in their own right.

The earth was a bewildering assemblage of diverse items, all interacting in highly complicated ways. If the physicist took one of these terrestrial objects and looked at it independently, he was at the same place where he had started. The object was an intricate world in itself, an immense collection of interacting particles. According to physicists, as they went sailing along, free falling down the scale of magnitude, they suddenly hit a brick wall, screeching to a halt at the point where they met the elementary particles. This was an abrupt and unexpected end of the line for what had seemed during the entire voyage to be an infinitely repeating pattern. Examining the overall design of the universe, the notion of a set of fundamental building blocks did not appear to be a rational supposition. After all, the idea came from mathematics and not from the actual universe.

Physicists could argue against an infinitely receding universe by using the mathematics of physics in a couple of ways. The equations for energy and momentum came out exactly, which suggested that there was nothing more to consider, no hidden factors requiring an additional account. But the flaw in this reasoning was that physicists were not bearing in mind the further change in scale. From an extraterrestrial point of view, the energy contribution of a single rock on the surface of the earth was in principle undetectable, and the contribution of a single molecule in that one rock was so minute as to be unimaginable at the level of the entire earth.

The argument seemed to suppose that unobservable submicroscopic entities would exist independently of what was observed and thus they would contribute additional mass and energy which needed to be added to the equations, but in truth they were what constituted the observed phenomena and thus they had already been figured into the equations, although not explicitly. They were the basis for what was observed, not something else. They were beyond observation in the sense of being themselves unobservable, but not in the sense of being outside of what was observed. But the fact was the equations weren't coming out right and there was all this mass and energy that could not be explained, labeled 'dark' in the sense of being invisible. In the same vein, perhaps the cosmic background radiation was not a remnant of the Big Bang after all, but rather an indication of some unseen underworld that had only a ghostly presence in the macroscopic world.

A second argument was that the known particles were interwoven with electromagnetic radiations and these could not become arbitrarily small. But particles and waves were only mathematical forms. Physicists then took these forms to be the ultimate realities of the universe, however, these forms could not stand alone and must point to some other reality, something that gave rise to them. The universe could not be composed of pure shapes and the ultimate realities of the universe could not be mathematical forms. The waves on the ocean could not exist without the seawater, just as the particle formed by a baseball could not exist without the windings and jacket. This relationship held true for every object in the world. Jonathan could see that mathematics could not exist by itself, but must be generated by something that had genuine existence. Physicists wanted to change all that and reverse this essential relationship so that mathematics could generate physical reality instead of the other way around. Everyone lived in a world of force fields and potential energies even though these concepts were derivative—highly mathematical constructions that depended on actual events in complicated ways.

If Jonathan kept the earth in mind as a prime example of a particle, the fact that it was made up of an infinity of smaller particles didn't alter the mechanics of the earth as a whole. The earth had fixed dimension and mass so physicists could go ahead and calculate its trajectory, but internally and on the surface, the planet was in a state of flux and only its overall mathematical form was constant. If physicists weren't satisfied with the simple mechanics of the earth and they wanted to analyze it in terms of its constituent parts, how could they possibly take the earth apart? They couldn't remove a mountain range or an ocean, a desert or a thunderstorm, yet these were the things that made up the earth. The infinite particularity of the earth meant that it didn't exactly have parts. The same problem existed with ordinary objects, all of which were infinite particularities. "Parts" could be aggregates that lacked cohesiveness, fragile structures without permanent form, in some cases nothing more than dynamic equilibriums in the flows of other minuter particles, yet these things all existed and they were important aspects of the object as a whole.

Elementary particles were generally analyzed through collisions with other particles. If the earth collided with another planet, Jonathan wondered, what would the result be? A big explosion with fragments flying away in various directions. Were these fragments also particles, the constituent parts of the earth, all held together by equally well-defined entities called forces? Not really. Not at all, in fact. The impact would generate all sorts of debris, clouds of dust and emissions of hot gases, and probably sprays of rocks, all thrown off in certain directions. These phenomena would be detectable because all the remnants had mass, but what if mass vanished as the observer descended the scale of size? Entities still existed, but mass was no longer a primary aspect of their being. In fact, they couldn't possibly have significant mass because if they did there would be nowhere to put all of that mass. Mass was something that emerged out of the chaos of the underworld as an overarching form, a property of larger sectors of the sub-universe when considered as wholes, an attribute of our world. Saying that the mass was too small to be detected was essentially equivalent to saying that there was really no mass to speak of and mass was not an important part of these events, not a way to understand what was happening at that level. But if mass didn't figure into anything, then what did? Mass had always been a mainstay of the calculations, a central component of all mechanical interactions, but if anything actually existed at this scale of size, mass would necessarily disappear and no longer be relevant. The same would be true for distance.

Jonathan said: "Let's presume the electron did have definite parts and that it resembled a machine rather than an organic entity such as the earth. A strict procedure still had to be followed in order to take it apart." For Jonathan to take his truck apart, for example, he had to first remove the hood and then unbolt the engine mounts, then unbolt the block from the transmission. He had to pry off the hubcaps, undo the lug nuts and pull off the wheels, then unbolt the long u-bolts connecting the axles to the leaf springs. Or there was another way. He could put an explosive charge inside the cab and blow the truck apart. Rather than trays of bolts and metal brackets and neat piles of parts segregated according to the original assemblies that had made up the truck, all carefully ordered on the shop floor and on the work table and labelled with marker pen notations written on masking tape tags, he instead had a pile of rubble, broken pieces, scraps of metal, and tangles of wire. Examining this debris, it would not be clear to him how the truck had originally been constructed and definitely there was no way to put it all back together. Taking something apart was not the same thing as blasting it apart.

The explosive charge still broke the truck up in predictable ways and Jonathan could piece the truck back together in a crude way, just as NTSB investigators could lay out the twisted and broken wreckage of a plane in an airport hanger after a crash. In the case of the truck, the doors flew off, the tires remained attached to the rims and together they were catapulted into the air and landed a certain distance away, the axles and engine block survived somewhat intact, but the glass windows shattered and essentially disappeared. The object divided into chunks, complexes of parts, severed at various weak points in the structure, but these were not the functional parts of the truck, just broken pieces, incomplete fragments from different assemblies stuck together as if they had been one.

Physicists had no idea how to take an electron apart, so instead they accelerated them in large numbers until they achieved very high velocities and plowed them head long into a stream of equally energetic positrons. Physicists ended up with pieces of these particles rather than parts of them, a distinction that assumed that they actually had parts, that is, in the sense of machines. The idea that particles were comparable to machines had never been proven. If the pieces could only be identified by their mechanical properties, then no one could be certain that two pieces possessing the same quantum numbers were actually the same pieces. The resulting collisions threw off all kinds of fragments and debris, the trajectories were recorded and the pieces categorized according to the kinds of tracks they made. Many of these projectiles disintegrated almost immediately and had no independent existence. In fact, the more fragile components of the original particle might have been instantly vaporized by the high energies and never showed up in the detector at all, but the stouter residuals assumed their rightful places in the zoo of elementary particles that had been created by this technique. The mathematics got very messy, perhaps mirroring the situation itself.

Rather than reevaluating the overall approach and perhaps thinking of a better way to go about understanding the constituents of matter, physicists simply turned up the power in order to pulverize the matter into ever finer ash. No one questioned whether this would work. This was the apparatus they had so they were sticking with it. Again, knowledge was subservient to the existing technology and the results were dictated by the available equipment.

The wonderful thing about accelerators and drift chambers was that they provided fodder for a proliferation of mathematical concepts and the possibilities for seemingly endless calculations, filling books and blackboards at universities all over the world with long strings of symbols. The mathematics was the whole point of the endeavor anyway, physicists doing what they loved most in the world—scribbling. Jonathan supposed that if physicists knew of a a better alternative they would surely take it, but physicists could ignore the fundamental problem inherent in this strategy and still go ahead and base their knowledge of the atomic sub-universe on a technique that could not possibly yield meaningful results.

In multiple collisions involving identical or at least very similar earths, the patterns would probably be the same in each case so the results would be reproducible, at least in the general sense of simple mechanics. Colliding earths was an experiment that could not be performed so no one knew exactly what the results would be, but if the electron was also an infinity of particulars, it still wouldn't behave erratically as a whole, jumping around, popping up and then reappearing at another location. Physicists would still be able to determine its position precisely. But what if the electron itself was a dynamic equilibrium, a stable form under certain circumstances, but flowing and dissolving away in other cases, only to have similar equilibriums reform and reemerge when the conditions were ripe? Physicists always imagined electrons as hard yet elastic little balls bouncing around, partly because such an image mirrored the mathematical way of thinking, the unity of ones in counting and the monolithic symbols in equations, but what if electrons were just patterns in something else, forms composed out of a smaller reality that was unknown simply because it was too small to be detected? In that case, electrons would behave differently from ordinary macroscopic particles and be capable of all sorts of transformations, appearing and disappearing like storms on the ocean. The metaphor fell apart because storms grew and diminished gradually, broke apart and coalesced in a continuous series of gradations, while electrons seemed to have a form that was fixed and constant, at least under most circumstances.

If the world were actually bottomless, could the physicist say with certainty that the inhabitants of these sub-universes were also the inhabitants of our world? These unknown entities would necessarily have rules of their own. They would be fundamentally different from the larger objects physicists studied and detected with their instruments. This was not an impossibility, but did anything suggest that this might actually be the case? The answer was: if the explanations physicists had created fell short and failed them, then they knew there must be something more. But how much more? Perhaps infinitely more. And where would they find these unaccounted infinities? In the intricate universes residing inside the particles that were inside the molecules that were inside the floor that everyone so casually and thoughtlessly walked upon each day.

Protons, neutrons and electrons were once thought to be the basic building blocks of matter, but further investigations revealed that they too had structures and component parts, and these parts were also made up of yet smaller parts. The assumption of ultimate particles out of which everything else was built had never been demonstrated because as the scale shrank, the particles disappeared from view. No one knew how far down it went. Everyone assumed that somewhere, somehow, they reached some kind of bedrock and arrived at the ultimate foundation of matter, even though as technology advanced, the horizon kept receding. What if when physicists looked at the subatomic world they were looking down a mineshaft? What if the universe was really a bottomless pit and there was no floor beneath everyone's feet? If the world continued to retreat as physicists scaled down in size, how could they possibly know where it ended? How could they find out what the conditions were like down there?

The change in view that the Copernican revolution had presented to the world was still heralded as a move away from a merely anthropomorphic perspective toward a more objective, universal standpoint. Yet a similar anthropomorphism remained in everyone's perception of size. The world we lived in remained the standard of measure, and our thinking faltered as we approached the boundaries of this domain. The idea of more worlds lurking below the limits of observation, that as we probed into ever smaller spaces we found ever tinier particles and perhaps even entire systems of particles, the idea that they were not simple particles at all but perhaps infinitely complex entities in their own right, seemed out of the question. But why should our scale of size be the center of the universe? What if existence was centered elsewhere? What if entire universes dwelled within the smallest spaces, vast realms with limitless structures, all unknowable and unreachable yet everywhere all around us, lurking inside everything we touched? This conjecture reminded Jonathan of the physicist who had walked around his house wearing snowshoes because experiments had shown to him that atoms were mostly empty space and he was afraid of falling through the floor. Perhaps he needed a rope ladder instead, to pull himself back up when he fell into the interior of materiality and just kept falling.

Jonathan tried a different perspective. He reversed the roles and imagined that human beings were the inhabitants of the atomic sub-universe and that super large beings, part of a group of alien physicists, were attempting to observe us from outer space—what to them was the macroscopic world of their ordinary experience. They looked down with their microscopes and sophisticated instrumentation on, say, the island of Manhattan. They noted predictable patterns of activity. They detected a flow of tiny metallic capsules down concrete arteries reaching maximums at certain times of the day, moving in opposite directions on each side of a central axis, flanked by droves of protoplasmic corpuscles exhibiting a more random motion, entering and exiting vesicular spaces in concrete and stone structures, arrayed uniformly between the arteries. There were also patterns of electrical activity, including a luminescence peaking mainly during the first part of the dark cycle. The emission of carbon dioxide from the vesicles fluctuated and increased whenever the temperature dropped and various heating mechanisms were activated.

They formalized the relationships between these various phenomena and introduced mathematical concepts to account for the observed regularities, but being unable to zoom in further and take a closer look, there was no way for the giant observers to know what they were really looking at. They could never guess that these particles were actually people who had jobs and families, hopes and ambitions, thoughts and ideas, and that they had complex lives, that they were sentient beings who drove their cars by seeing the streets with their own eyes and walked down avenues on muscular appendages called legs, slept in their houses and apartments on beds and talked to each other via electronic devices. The observers would see only the overall patterns created by this microcosm of infinitely complex beings, individuals with personal agendas and free will, all the idiosyncrasies melding together into overarching group behaviors, averaging and canceling out their peculiar personalities and mannerisms, negating their unique destinies. The observers drew arrows of force over the throngs of pedestrians, wrote differential equations for the rates of change of traffic flow, and concluded that these behaviors were all bound by mathematical laws. The giant observers assumed the simplicity of the constituents from the simplicity of the schematic representations that they had fabricated in their minds. Jonathan could not blame them, because without more information, it seemed like the logical thing to do.

Everyone thought that they knew what proteins were--configurations of carbon, hydrogen and oxygen atoms--because they'd been handed this 'topographic map' of them and told that this was it, this was the reality of these entities, even though the complex behaviors of these molecules within living cells seemed to contradict such a simplistic characterization. Students were presented with childish stick-figure drawings of these molecules, circles connected by straight lines, similar to the pictorial families pasted on the rear windows of minivans. Whenever Jonathan was following one of these minivans, he tried to imagine what the kids really looked like, the oldest daughter and the youngest boy, what kind of dog they had and how well it got along with the cat. None of this information was conveyed by the diagram. The circles and lines were symbols referring to concepts, much as the x's and y's of an equation, and these entities were not really what was out there in the world. The biologist could not see a protein in any kind of detail because it was way too small, but he might be surprised if he could.

Perhaps no matter how deep physicists dug, they never unearthed the foundations of existence. Everyone was essentially suspended over an infinite abyss. People had once thought that there was nothing beyond the ocean, but in fact there was a new world waiting to be discovered. Perhaps these depths could not be plumed and must remain forever concealed, just as the expanse of the universe above our heads was tantalizingly presented to us but forever out of reach. The remote regions underlying the universe were purely matters of speculation, and the conjectures could easily be dismissed on these grounds, but the inferences were not nearly as far-fetched as wormholes, time travel and teleportation—the mainstays of the currently sanctioned scientific fantasies.

Just as the astronomical universe wasn't really above us, the hidden world of the infinitely small wasn't really below us. It was inside of every point, every particle and every empty space. The direction that we moved in order to reach it wasn't aligned up and down, but rather inside and out. Every point was a rabbit hole, the entrance to another room, and everything in that room, every point, led through another rabbit hole to another room. We kept traveling toward the center of the universe—well, not the universe, but existence itself—always pressing inward, only there was an invariance in what we found there, a uniform expansion of detail to counteract and balance the contraction in size. The complexity of reality never went away. Intricate patterns blossomed at each new level, in essence taking us back to where we had started—but there was no circularity here. We were always in a new place and the place didn't look like anywhere we'd been before, but it was a place nonetheless, a complex world in its own right, a small universe populated by a wealth of diverse entities with their own codes of conduct.

We were not moving back and forth in time and we hadn't entered another dimension. Somewhere out in the larger part of the universe was the world we had left behind, the place where we lived. Similarly all the pulsars, black holes, nebula and galaxies might just be the molecules of a much larger world, or a small volume of cytoplasm within a single cell that was part of an unfathomable organism. The flows of dark energy, the flickering of exploding supernovae, might just be the neurological activities of some giant brain, the slightest trivia in a mass of living tissue within the cerebral cortex of a cosmic consciousness. Or they might be the atomic interactions within a facet of material that was part of a structure of unbelievable proportions. Either there was an explanation for the bountifulness of astronomical events, or we must concede that it was all meaningless and the glowing gases, birthing stars, and colliding galaxies with their tremendous flows of energies and intense bursts of radiation existed for no reason. The creation of the universe with its staggering proportions and limitless inhabitants seemed like a lot of trouble to go through for no reason, yet it might be as pointless as the howling of coyotes on a moonless night.

If atoms and molecules were further examples of physical objects in the world, then why wouldn't they be similar to all the other physical objects in the world and share with them the same character of infinite particularity. Mechanics couldn't possibly reveal this particularity to us. From an astronomical point of view, the earth was simply a particle with mass and velocity, yet the earth was infinitely complex. The earth was a composite of smaller objects each of which in turn was a composite of still smaller objects, and each of them was made up of objects that were smaller still. The individual contributions of these smaller objects could not be detected because of the change in scale, so mechanics could not resolve the earth into these components. In physics, only the items that could be detected were real. Physicists let the instruments determine reality for them. Instruments turned atoms and molecules into simple entities made up of numerical quantities when in fact the assumption that they were infinite particularities was more reasonable. Why wouldn't they be? Everything else that we knew, every object of our world had this character, so why would these atomic and subatomic objects be exceptions? Being small was not a valid excuse, because small was a relative term. We were small in comparison to galaxies, but that didn't make us simple beings or relieve us of our burden of complexity, a quality that allowed us to be resolved into an infinite number of properties and quantities.

If this were the case, then there was much more to atoms and molecules than simple mechanics—infinitely more in fact. But as with the earth when considered from an astronomical perspective, the details were beyond our reach. More was going on in the submicroscopic world than physicists could ever imagine, yet they continued to plot imaginary orbits and hypothetical trajectories and calculate masses and angular momentums as if such things were valuable for an understanding of these entities. They were slaves to the technologies they had created and they let the machines dictate what was important to them and decide what was real.

No one could look at the earth from a great distance through a telescope and see Jonathan sitting under a tree. They could not possibly deduce that this speck of organic matter was thinking about the universe. No one could possibly guess the truth. Jonathan suspected that the same situation existed for the atomic world. Jonathan wanted to imagine the true state of affairs and picture to himself the tiny worlds that lay beyond the weblike facade of numbers and quantities that had been created by mathematical imaginations, to peek into the bustling societies that flourished beyond the reach of electronic devices and consider the possibility that one of these infinite worlds might possess a seemingly insignificant entity tucked under a canopy of the tiniest corpuscles, an infinitesimal being residing in a microscopic forest of unnamed and unknowable objects, an entity who at this very moment was speculating about the possibility of something such as Jonathan existing in a world that was utterly beyond its reach.

The physicist objected to Jonathan's reasoning at this point. If subatomic particles were really comparable to the earth in terms of their complexity, then this complexity would provide a much greater range of behaviors than what physicists observed in a laboratory. If relatively simple concepts could account for everything that physicists saw, then physicists had no reason to assume anything else, but moreover, they couldn't write equations for things they couldn't observe. But the relative simplicity of the particles had been deduced from the relative simplicity of the laboratory setups. The nice, neat mathematical arrangements produced nice, neat mathematical phenomena. Physicists thereby envisioned a tidy little world of empty abstractions and walked away with smiles on their faces, content with their illusions and satisfied with their efforts, only to confront a universe that was radically different from what they had observed in the laboratory. Outside lurked the mysteries of living organisms, the enigmas of inexplicable coincidences, the kaleidoscopes of meteorological events, a diverse and incongruous world that was forever spinning out of control, yet they had declared with straight faces that non-reproducible phenomena were comprised entirely of reproducible phenomena, existence was based on nonexistence, and a strict order of physical laws produced nothing but disorder. They decided that they didn't need to deal with any of that and they could simply ignore it all—a rather bizarre conclusion to say the least.

Complexity didn't add weight. It didn't change the mechanics of the object, although it might provide opportunities for the object to break apart or develop internal motions. Some objects might crumble and disintegrate, but most were quite resilient. After 150 years of major league baseball showcasing powerful sluggers crushing countless high-speed fastballs, Jonathan had never heard of a batter tearing the jacket off a baseball. Bats, on the other hand, didn't fare so well. The problem with mechanics was that it didn't reveal anything about the object—outside of its mechanical properties—and that was why mechanics couldn't provide physicists with an authentic knowledge of the world. More complex objects didn't necessarily have more complex mechanical behaviors. Yet even the self-imposed requirement that physicists limit themselves to geometry hadn't worked out for them. Even within this highly restricted domain, the equations had already blossomed and multiplied into increasingly unmanageable forms. Even within the context of these relatively simple mathematical arrangements, the mathematics had become bewildering, yet it still fell ridiculously short of the real world.

Jonathan continued climbing upward until he reached a crosswise fence where he found the remainder of the canyon denuded of vegetation by a herd of cows. His heart sank and he saw no reason to proceed onto this active range, so instead of crossing the fence he scaled the top of the ridge and dropped down into the next canyon in order to return by this parallel route.

As he walked along, he entered a long stretch of riverbed that had been swept clean. Over the years the summer floods had continually removed the sand that had been deposited by the winds. The rushing water lifted up the grains and hurried them along to their new homes. The bedrock became exposed once there was no more sand left to remove, yet the waters kept coming, washing and wearing away the underlying sheet of stone, then meticulously polishing its surface smooth. Dense rows of tall grasses now curbed the slick pavement and Jonathan found himself strolling along a narrow, winding road beautifully landscaped with trees and boulders. The magnificence of the place overwhelmed him. He felt that he was off to see the wizard and he might run into a dancing troupe of munchkins at any moment. The pleasantly civilized walkway finished with a series of rough, irregular terraces leading down to a nasty pile of rubble.

After a mile or so the canyon opened up into a shallow basin. Emma had found something up ahead and was making a racket, so Jonathan stiffened his pace. He drew close to the scene of the commotion and discovered that Emma had cornered a buck. The deer had taken refuge in the midst of a dense thicket where it had found a hollow, but Emma experienced little trouble penetrating its defensive stronghold. She assailed the animal through a narrow opening in the otherwise unbroken wall of thorny branches. The buck lowered its impressive rack and aimed the pointed weapon at its rowdy antagonist, lurching forward in spurts of locomotion. Jonathan noticed that the deer had a broken hind leg, snapped clean at the knee, now attached by nothing more than the leather of its hide. The lower portion of the leg swung freely in a wide circle, propelled every which way by the erratic movements of its host. Jonathan winced at the sight of the dangling appendage and the revolting injury. The animal could not properly flee, so it stood its ground and faced Emma's challenge. The leg displayed various surface wounds, indicating that the buck had gotten it caught in the nearby fence and snapped it off desperately trying to extricate itself from the barbed wire.

Emma was absolutely hysterical over the buck's behavior. The buck was acting weird. She had met deer before and they were never like this. She knew something was very wrong and the incongruity drove her mad. She couldn't understand why the buck was threatening her instead of just running away or simply standing its ground. The buck wanted to fight and Emma was game. Every time Jonathan moved in to capture her, she circled around and continued the pointless assault. He walked directly in front of the buck following a path that led him straight to Emma hoping to intercept her when the buck unexpectedly turned on him. It caught him off guard and he almost fell to the ground, but luckily he was able to maintain his balance and pivot on one foot. He took off running in the opposite direction. He could feel the buck's antlers poking him in the back and he could hear the buck stumbling and faltering as it tried to run, the broken leg banging against the rocks and dragging on the ground, but he had no time to glance over his shoulder at his pursuer. Instead he surveyed the desert in front of him where everything looked perfectly normal, all the rocks and trees perfectly calm and quiet. Jonathan laughed out loud as he churned his legs frantically running through an absolutely serene landscape, seemingly panic-stricken for no apparent reason. The reason was right behind him, even though he couldn't see it at the moment.

The buck was somehow able to keep up with him. It did not relent and the jabs continued as Jonathan fled down the slope. Finally satisfied with its effort, the buck let him go and hobbled back to its hideout, leaving him standing near a large juniper doubled over and panting heavily. After witnessing the buck's crazy behavior, so unlike anything Emma had ever seen before, she was now in a frenzy. She intercepted the buck and attacked it once again, growling fiercely and leaping up trying to bite it, now in an effort to protect Jonathan from further assaults. Although this was understandable from her point of view, Emma's excessively aggressive behavior was completely unacceptable. What made her think that she could pull off this kind of stunt and get away with it? Night was falling and once the skies grew dark, Jonathan could no longer manage the situation. Plus the night was going to be cold and he hadn't brought enough warm clothes with him. It broke his heart to have to do this, but Emma left him no choice. He needed to end this now. He walked over to his pack and got his gun.

Emma was barking loudly and fighting the buck through the opening in its stronghold. Neither would surrender the offensive and both remained fiercely aggressive. Jonathan loaded a shell into the chamber and joined the combatants. He gave Emma one last chance to redeem herself. He fired a shot into the rocks on the other side of the canyon, thinking that the concussion would drive Emma away. Emma paused for a moment and looked around, following the echo with her head as the sound bounced off the hills and disappeared down the canyon, but then she immediately dismissed the event as nothing of importance and quickly resumed the confrontation. Emma was a gun dog and she had been bred not to pay attention to gunfire. Jonathan should have known better than to waste his time with such a tactic.

Jonathan was out of options. He would have gladly left the deer to the gruesome fate of a slow and painful death, but Emma wouldn't allow that. He turned toward Emma and shouted "Why do you make me do this? Why are you always so stubborn? Why can't you just obey me?" Emma paid no attention. She thought that he was yelling at the buck and encouraging her to attack it.

Jonathan felt weary. He thought about the first time he took Sarah to meet his parents back in Chicago. His father was a gambler, so they all went together to the race track for some friendly wagering and a "get to know each other" session. Sarah was raising draft horses back in New Mexico, a profession that was no longer profitable because almost no one worked the land with horses anymore. She loved horses, so it seemed like the right thing to do. Unfortunately, two horses broke their legs in separate races while streaking toward the finish line, the terrible mishaps occurring right in front of the grandstands. Everyone had an excellent view of the grisly proceedings, each time the trainers struggling to load the frantic animal into a large van while the snapped leg flailed in every direction for the benefit of the audience, most of whom had already left to place their bets on the next race. Sarah was outraged and raved about banning horse racing altogether. Jonathan always imagined that they took the animals out behind the stables and shot them. He wasn't sure if killing such a large beast with a lethal injection was more humane or a better way of disposing of it, but that must've been what they did because he listened, and he never heard a gunshot coming from the direction of the horse stables the entire evening.

The buck had no chance with that leg. He turned toward the animal and slowly approached it. The buck stopped fighting and stared directly into Jonathan's eyes. It had a look of dignity—and resignation. It wanted this to be over and knew that Jonathan had come to end it. Emma's barking faded away in his consciousness and he found himself alone with the deer. He raised the firearm and pointed it directly at the buck's face. The buck watched him with a solemn expression, the eyes pleading for a merciful conclusion to this dreadful escapade. Jonathan stomped his foot and yelled causing Emma to jump back, then pulled the trigger twice. The deer's legs gave out and it collapsed gracefully on the ground as if it were going to sleep. Emma immediately stopped in her tracks. After a long period of frantic animation, the three of them were suddenly and strangely silent. Jonathan glanced over at Emma and she stared back with a look of relief on her face, knowing that the confrontation had finally ended. They all remained stock-still, frozen in their respective places, the deer inert on the ground, Emma poised with her legs splayed, her motion abruptly arrested in the midst of a mad dash, and Jonathan standing with his arm raised, the gun no longer pointing at anything. A pool of blood spread slowly out from beneath the deer's head in an ever widening circle. The deer was lying on its side staring with open eyes into the void of the canyon. Jonathan took two steps forward and fired two more shots into the base of its skull.

The sky was quickly growing dark and they were still two miles from the truck. Jonathan disarmed the pistol and returned it to his pack. Emma laid down nearby. She was thirsty following her intense battle so he broke out her dish and filled it. She greedily lapped up the cool water without bothering to stand up. He swung the pack over his shoulder and stepped into the emptiness of the endless desert, a tiny, insignificant speck released into an arena that was well beyond the scope of his understanding. He seemed to dissolve into nothingness as he walked away from the bloody scene, his spirit slowly vanishing into the vacuum of space. The blackening night assimilated him into its blackness and he was hopelessly lost in an infinite universe, his body gradually sinking into the gloomy depths of a vast ocean of murky air.

The cloudy, moonless night had no stars. The only light was the glow of his GPS unit as he checked his position from time to time. He didn't bother to search for a headlamp since he much preferred the darkness of the night. Emma was too far away to keep track of her with a headlamp anyway.

Jonathan emerged from the mouth of the canyon and faced a world that was opaque and silent. The truck was nowhere to be found, still too far away to be seen, even if it had been light outside. Emma was running around somewhere in this inky mess, her actions unbeknown to him, only the occasional note of panting or the gruff rumble of snorting reaching his ears, the faint undertones carried in his direction by the erratic currents of the cool night air. He could not protect her from the unseen dangers and could only hope that she would make it. Emma was carefree and didn't understand that her life might be in jeopardy, unaware that rattlesnakes and mountain lions also prowled the desert at night. Jonathan used to be like her, untroubled by the precariousness of his existence, but he had changed.

—12—

What did the mercury in a capillary tube have to do with two dissimilar metals fused together and shaped into a spiral? Not much—nothing really. Both were types of thermometers that anyone could purchase at a local hardware store. Jonathan could correlate the two movements by saying that, for example, a one inch expansion in the column of fluid equaled a six degree rotation of the coil. The two events, which previously had no connection to one another, were now coupled together by a mathematical formula. He could then tie these phenomena to other phenomena, such as the bubbles of steam in his teapot at sunrise or the ice crystals growing on his window pane on a dark winter's night, representing the freezing and boiling points of water. All of these unrelated things now had something in common: they shared "temperature."

Jonathan could calibrate every phenomenon in the universe in a similar fashion without regard for what it was or what physical processes were involved, thus establishing the universality of temperature. He spun his spiderweb of correspondences by extending a new silken thread to every new object he encountered until the entire universe was trapped under a vast network of gossamery interconnections. Temperature became all-encompassing. Everything in the universe was related through temperature, but only because he'd found a way to assign everything a temperature, comparing each item to one of his thermometers and ultimately to each other.

All quantities in physics were based on such comparisons. Not only could every two events and objects in the universe be compared to one another, they could be compared in an infinite number of ways, so everyone had to decide how they were going to carry out this operation. The procedure they used was entirely up to them. The physicist might think that he had the ability to see what was important in the world and what was not important, but more likely than not he didn't even consider the world in this regard. Everything revolved around him and his personal ambitions—pretty much the same attitude that everyone on the planet shared.

Jonathan couldn't trust the physicist to make these kinds of decisions because he knew that the physicist had a hidden agenda. The physicist wanted to turn the world into a mathematical system, so all of his judgements would be geared toward that end. Jonathan, on the other hand, had no preconceived plan for making such comparisons so he had to turn to nature or some higher authority for guidance. Jonathan wondered: what would Jesus do? Jesus had made lots of comparisons during his lifetime here on earth—a rich man entering heaven and a camel passing through the eye of a needle, the benefits of a man gaining the whole world against him suffering the loss of his soul. But those were not the types of comparisons that physicists would proffer many centuries later, a particle of matter and the solar system, a ray of light and the ripples on a pond. Jonathan didn't think that Jesus ever mentioned the similarity between a capillary tube and a bimetallic coil. Jesus certainly would have talked at great length about the supreme importance of capillary tubes and bimetallic coils had he known about them, and undoubtedly he would have made them a central principle of his teachings. Of course that wasn't true and he never would have wasted his time meditating over such trivial concerns. He understood that these mundane contrivances played no roll at all in the great scheme of things. His way of thinking did not grant these gadgets the same significance that the physicist bestowed upon them. Jesus instead bartered in the vague and unscientific concepts of morality and righteousness, good and evil, sin and redemption, love and trust. Would physicists have found Jesus' ideas more interesting if Jesus had taken the trouble to quantify them?

Let's say God actually was a mathematician, just as the early physicists and philosophers had portrayed him. God spoke to Jesus in a vision and told him to just calculate good and evil, giving him the formulas and the necessary definitions for the relevant quantities. Through Jesus, the mediator between God and men, God gave to the world a numerical scale of goodness. Jesus went around measuring everyone and telling them that this was God's plan for human salvation.

The data came in and the numbers were rather unsettling. On a scale of 10, Jonathan's neighbor had scored a 6.90, while Jonathan had registered only a 6.82. God had provided Jesus with a mathematical picture of the world that divided humanity into two mutually exclusive sets, one with the property of eternal damnation and the other with the property of eternal bliss. Jesus had only to draw the line between them, so he reached down into the depths of his soul, analyzed all the theoretical considerations, and came up with the number 6.85. As a result of what appeared to be a rather arbitrary decision, Jonathan didn't make the cut but his neighbor achieved salvation by a hair, even though there was very little difference between them. Jonathan imagined Jesus putting his hand on Jonathan's shoulder and consoling him: "Tough luck, soldier. I never said that the afterlife was fair. My disciples wrote down every word I ever said. You can check it out for yourself."

Jonathan didn't think that this was about fairness. It was about the absurdity of quantification. A person couldn't understand good and evil more clearly after they were put on a numerical scale, and mathematizing them didn't make the concepts more real. If Jesus had simply confronted Jonathan with his sins, Jonathan would probably have seen his point, but going to hell over the value of a number was insane. Nevertheless the physicist considered the number to be more real than the objects and events of the world. The reasoning behind this strange conclusion seemed to be that the quantity took him closer to mathematics and mathematics was obviously the ultimate reality of the universe, but on closer inspection, the opposite appeared to be true.

Most people pictured the Supreme Judge of Souls to be a wise man, a person who had the experience of old age and through his many trials and tribulations had learned a lot about the meaning of life. Over the years he had met many good people and many bad people, and these examples conferred upon him a deep understanding of the subtleties of virtue and wickedness. Jonathan pictured the venerable old sage stroking his long beard as he solemnly meditated on the merits of an individual case. Calculations, on the other hand, just cranked out numbers. The outcomes were determined by placing these quantities in formal relationships, equations that tied the magnitudes together with concepts that had nothing at all to do with the real world. The calculator multiplied the numbers and divided them, added them together and subtracted them from one another, then played all sorts of games with them, deluding himself into thinking that he was still studying the nature of sin.

If someone handed Jonathan the idea of pushing a camel's body through the eye of a sewing needle and the idea of a man entering a world whose location could not be specified, and asked him in what way these two events were similar, he would be at a loss for words. Comparisons when taken at face value were incomprehensible. A conceptual system had to first be put into place, and it was through this system that the comparison made sense. So what did blondes and the Bermuda triangle have in common? Not much —nothing really. The comparison was just a joke. A person had to understand what a joke was, in this case a play on words, before the answer made sense.

Still Jonathan was curious about associating two things that didn't seem to be related. What reason was there for making a comparison between blonde women and a region of the Caribbean? Well, he could think of several reasons for comparing them to each other. A blonde's hair matched the color of the sandy beaches in the Caribbean. Blondes had more fun and the Caribbean was a fun place to visit. Blondes were beautiful and mysterious causing their friends and acquaintances to disappear into their depths much as ships had been known to disappear into the Bermuda Triangle. While these statements might all be true, the correct answer was, of course, that they had both swallowed a lot of se(a)men. How could Jonathan have possibly known that this was the real connection between them? Who would have ever guessed? So what did a capillary tube and a bimetallic coil have in common? The answer was that they both had a temperature. No, this wasn't another joke. The physicist was deadly serious about temperature.

The connection between a capillary tube and a bimetallic coil only made sense because the physicist had brought along with him a bunch of mathematical ideas and agendas and in this context the similarity became apparent. At face value, the two thermometers had no more in common than blondes and the Bermuda triangle—or more correctly, the capillary tube and the bimetallic coil had no more in common than blondes and the Bermuda Triangle because the words 'two thermometers' already implied a set and an act of counting both of which were mathematical concepts that united the two objects together. Before the discussion of temperature ever began in earnest with the introduction of analytic geometry and a scale of measurement, mathematics had already slipped in through the back door and made itself comfortable in the living room. Mathematics shaped everyone's thinking from the very beginning, long before the concept of temperature had been firmly established.

The mathematical comparisons essential to the concept of temperature were no more real than straight lines, coordinate systems, sets of objects and the host of other constructions the physicist used to fabricate his fantasyland. They all came from a way of thinking and not from the world. Moreover, they came from a very peculiar way of thinking that no one had ever imagined before physicists began applying their mathematics to anything and everything they could get their hands on. Physicists were only drawing imaginary lines between things and then trying to convince everyone that these were the real connections between the actual objects. Of course the whole scheme worked from a logical and mathematical standpoint. It had been set up that way by using the concepts of logic and mathematics in the first place.

The physicist jumped in at this point and objected to Jonathan's characterization of temperature. Comparing the readings of thermometers was not the same as comparing blonds and the Bermuda triangle because these comparisons were carried out differently. If the capillary tube and the bimetallic coil were both placed in the same environment, then they would both produce the same number, so they were tied together through this common environment, an environment that was a physical reality and not merely a concept. Blondes and the Bermuda Triangle, on the other hand, were connected only through ideas, but the comparisons in physics had real physical connections.

Physicists put devices in specific locations to see how they acted. They then tied the behaviors together, calibrating one to the other, so that when the first apparatus did one thing and the second apparatus did another thing, they could say that the two behaviors were related, not in substance and not even in general outline, but in an artificial and highly abstract way. The physical processes weren't actually related to one another. Each one was self determined and the connection between them was imaginary, but the two processes were coupled through the environment by occupying the same place. The environment influenced each one differently, but temperature gauged the overall magnitude of this influence in a rather vague way.

Physicists bulldozed the situation over with monolithic concepts. So much was happening in the environment, all kinds of physical processes all occurring at once, influences transmitted in every direction in a maze of minute interconnections. A similar web of interrelatedness existed inside the apparatus. The combined actions worked not harmoniously, but often fought each other, competing for the attention of numerous elements within the device, pulling them one way or another. These influences played out in manners that were entirely different in each device, so where was temperature in all of this? The environment was not a unity and neither was the device. The ways that prolific elements in the environment changed the dynamic structures of the device were manifold, and the transactions involved in one device bore little or no resemblance to the transactions occurring in the second device.

How could a collection of objects as rich and varied as an "environment" actually unify anything? How could such a loosely connected array of diverse items tie anything together? The true state of affairs to which the word referred could never possibly do so, but the idea itself certainly could, so the physicist allowed the idea of an 'environment" perform this function and place the two thermometers in a harmonious union presided over by the overarching authority of a categorical idea.

The environment was brought in and called upon to help the physicist out, to play a specified role in his mathematical world. In this way, the physicist's correlation of temperatures depended on the existence of objects in the world, but the comparison between blondes and the Bermuda Triangle also relied on the existence of objects in the world. Blondes and the Bermuda Triangle certainly both existed and they were not just ideas. The comedian could take these two realities and use them to make a joke, but the joke was something he had thought up in his mind. The comedian fabricated the material and that was why he was given credit for the joke and it could be understood as his intellectual property. What made a comedian funny was that he started out with the realities of our lives and then built upon them, connecting things together in novel and unforeseen ways. The jokester created a world of humorous insights and spontaneous laughter much in the same way that the physicist created a world of mathematical precision and exact proportions.

But what right did the jokester have to turn everything into a joke, often at the expense of his subjects? Well, he liked jokes. He liked to make fun of people and everything they did. Were the jokes really funny? Well, that depended on your sense of humor. What right did the physicist have to turn everything into geometrical constructions and algebraic formulas, often at the expense of his subjects. Were the mathematical relationships really valid? Well, that depended on your perspective. Sometimes the physicist drew lines that were inappropriate. Sometimes the comedian told jokes that were off the mark. When a joke failed to evoke laughter, perhaps it was the audience that didn't understand the comedian's way of thinking. Everyone said that the comic had a weird sense of humor, but could they say for sure that there was no humor in his routine. Could they say that he was simply wrong? Lot's of operas and symphonies had been panned when they were first written, but later became accepted and even cherished by later audiences. Perhaps the comedian was merely an anachronism and he was out of sync with the rest of society.

When Jonathan had previously said that blondes and the Bermuda Triangle were both realities of the world, he was just kidding. The problem with the pun was that neither blondes nor the Bermuda Triangle existed, any more than the environment the physicist used to tie his thermometers together. "Blonde" was an abstraction, a general category, an idea, a mathematical set that could only be represented by an idealized and hypothetical blonde woman used as an illustrative example. Given all the natural shades of hair color complicated by the routine application of artificial dyes and the additions of accents, the general term "blonde" became a broad spectrum of yellow, flaxen, golden, platinum, ash and strawberry, ranging from towhead to brown-haired. The definition of a blonde was anything but hard and fast, but a woman who was placed in that category by the definition could then have all sorts of personalities and traits. Telling Jonathan that a woman was blonde didn't grant him an accurate picture of that woman, so she remained an unknown person. Jonathan could not go out and find the Bermuda Triangle because there was no such place. He could only look at your map depicting your idea of where the Bermuda Triangle was located. Someone else could draw a Square of Darkness or a Divine Circle over the region because none of this was real and the sizes and shapes of these areas were nothing more than arbitrary definitions. Thus the joke was based entirely on the relationship between ideas rather than the harsh realities of an indeterminate universe, and had only token references to the complexities of the world that everyone lived in. Neither the intricate and loosely-connected assemblies known as thermometers nor the raggedy environments in which they were placed were easily marshaled, so temperature was similarly just the relationship between ideas. The conclusions drawn by the physicist depended on an odd way of thinking, what Jonathan called "the physicist's sense of humor." A critic would find it just as easy to argue that the physicist's mathematical relationships weren't convincing as he would argue that the comedian's jokes weren't funny. The audience had to buy into the premise of the comedic skit before the laughter could ring out, just as everyone had to buy into the legitimacy of mathematical thinking before numerical relationships could be postulated as the laws of nature.

The emptiness of mathematics allowed many things to be connected together while the fullness of being prevented all such associations from being made. Mathematics connected things that weren't connected and that made the mathematics false—at least from the perspective of the world. The physicist was still able to compare forms by measuring temperatures, clocking events, or using any number of other procedures and then say that the objects and events were linked to one another, but if they were only related through the mathematics, then the relationships depended on the validity of mathematical thinking, not as a consistent set of logical propositions but as a way of accurately depicting physical reality. The physicist had to assume that such thinking revealed a reality beyond the mathematics itself, but if mathematics was alien to the world then mathematics would necessarily and exclusively lead to misrepresentations and blatant errors by drawing lines that were not really there. In order for the mathematics to be valid, the mathematics would have to be written into the world instead of being added to it in a separate step, the way that physics actually worked right now. The physicist could go ahead and draw pictures, only they were not pictures of what existed, but rather pictures of his own ideas. The mathematics did not copy anything in the real world since that was not where the mathematics originated.

The physicist had to first create a game board and then enlist nature in his game. The game board was necessary because that was where the mathematics came from. Mathematics wasn't everywhere in the universe, but the possibilities for fabricating all sorts of game boards were virtually endless and this created the illusion that mathematics was everywhere. The emptiness of mathematical forms gave the physicist the freedom to put the mathematics wherever he wanted it to be. Nature consented to play along only because nature did not have a mind of its own and could not protest the physicist's whimsical rules and definitions. Nature couldn't say: "I don't want to play that game." The physicist didn't bother to ask because in truth there was no one to ask. The physicist could do whatever he wanted and what he wanted more than anything else in the whole world was to have mathematics everywhere.

The physicist assumed that mathematical comparisons were meaningful much in the same way that the Christian assumed that biblical comparisons were meaningful, but what if Jesus hadn't really been the son of God? Then gaining the world had nothing to do with losing one's soul and the camel had no connection to the rich man entering heaven. And what if mathematics wasn't the reality of the world? In that case, one temperature had nothing to do with another temperature. The comparisons were misleading—or as Jonathan would say, irrelevant.

Temperature could not possibly be something that tied everything together. What tied everything together was the idea that every single object in the universe could be placed alongside every other object and through this comparison be put on a single numerical scale, and this idea was something that physicists had made up in their heads using ideas they had stolen from mathematics. Lumping every conceivable sin and transgression together without sorting them out didn't work either. Perhaps someone needed to step forward and tell God and Jesus that they were full of shit—and while they were at it, they needed to tell the physicists too. Physicists assumed that mathematics was the key to understanding the universe, but they had to create laboratories in order for this to happen, places where mathematical comparisons took on special meanings. Out in the desert, mathematical comparisons were irrelevant because nothing in the desert was mathematical. Physicists could still go out and measure distances and angles, velocities and momentums, and then relate the quantities to each other in formulas, but they wound up with an inadvertent mathematics that was similar to the mechanics of books.

The mechanics of books only became meaningful when the book was thrown at the television set in a fit of rage, when the book became a projectile with a target, but the book only became such an object inadvertently. The book could also become a paperweight or a doorstop. The mathematical equations of gravitational force then became pertinent, but no one was talking about the book as a book, but the book as something else. The book was also a nondescript material object in addition to being a book. As a doorstop, the book might spend its days getting kicked around on the floor, sliding over the smooth ceramic tiles, helplessly lying around collecting dirt. There it rested, a great work of art, a literary masterpiece with the pages left unturned, the noble ideas unthought, the powerful emotions unfelt, the triumphs of the human spirit unappreciated. The storehouse of wisdom and knowledge contained within its pages was utterly wasted. The eloquent words and carefully crafted sentences only held the door open. Mechanics reduced everything to its basest level, stripping objects of their most essential qualities.

Perhaps atoms and molecules were also great works of art that often performed menial tasks and this made everyone blind to their infinite potentials, duped everyone into thinking atoms were far less than what they were. Locked inside each of these tiny entities was a repository of knowledge and wisdom with unimaginable potential, not the massive release of blind, destructive energy that the physicists had elicited from them, but an intelligence, either conscious or artificial, capable of doing great things, like building a fantastic universe of infinite complexity. Physicists sold atoms short, reducing them to probability waves and the electromagnetic forces between particles, models generated out of their imaginations. But what were atoms really? The physicists could not tell us. They could only bounce them off one another and blast them apart, following their trajectories with sophisticated instrumentation as the pieces flew away.

In the same way Jonathan could speculate about the vast knowledge and intelligence hidden within a hockey puck, and talk about its infinite potential to do things. The problem was, a hockey puck was nothing more than a mathematical form and from the perspective of the world such a configuration was pointless. The molecules had never wanted to be a hockey puck and they never would have done so on their own, but they were forced to become a fabricated mathematical object against their normal inclinations, a proprietary mixture of poly-isoprene, antioxidants and bonding agents shaped into a uniform one-inch by three-inch cylinder by an equipment manufacturer. The hockey puck was the helpless victim of ideology and had only the intelligence of a fixed set of ideas deliberately imparted to it by its designer. Despite its limited origin and background, teams of men in brightly colored sweaters chased this object around to the great concern of huge crowds of spectators and large audiences of home viewers because the puck had been invested with a special meaning that an ordinary piece of vulcanized rubber didn't have. During a game, the puck was the most important thing in the whole world and all eyes focused on it, but only because people had given it this exceptional status. Everyone agreed that the trajectories of this innocuous black disk meant something very important, even though no matter how many times the players put the puck into the back of the net the actions never really amounted to anything, other than producing a number that was prominently displayed on a scoreboard suspended high above the ice. Physicists similarly invested objects with unnatural significance and tracked the behaviors of these objects with astute diligence while playing games with them in the laboratory, the elaborate setups being the scientific correlates of hockey rinks. Physicists observed the outcomes of the plays and kept score with computers, tabulating the results and heralding the final numbers as the ultimate realities of the universe, but the game was what made the plays important and the game was a purely human invention. A hockey puck that had been left laying out in the sand baking in the hot desert sun alongside rocks and lizards had no meaning. It was just a worthless piece of rubber wasting away in the elements and its positions and motions were totally irrelevant.

The ultimate capabilities of atoms and molecules didn't matter to physicists working in their laboratories. The infinite details of the atomic interactions that gave rise to the readings of temperature could be completely left out of the picture. These facts would certainly be different for different types of thermometers. The relationship between a given device and its lively, bustling environment would necessarily be highly complex, a maze of individual connections between a wide array of molecules. Various aspects came into play in distinct situations, but that was unimportant since everything would be based on the numbers alone. The physicist centered his ideas of the physical world around the readings he got, leaving the underlying atomic world to fend for itself, the entities working out the details on their own time. The ways that these molecules behaved and interacted with one another varied from one reading to the next as the device was moved around between various locations, placed in gases, liquids, and solids with different chemical compositions that were under a wide range of pressures and undergoing a variety of processes and transformations, and these interactions affected each type of device in unique manners so that what the physicist was measuring was different in each case, yet all these factors were consolidated into the measurements. The physicist needn't bother with the question "What exactly am I measuring in this particular case?" because the act of measurement was the only reality he had to deal with. As long as the readings were determined in some way by the unknown realities of the atomic world, he was not required to cite any specifics. He might say that this was an interesting question and seek to answer it by introducing and correlating the numbers produced by other kinds of instrumentation, but all of that was supplemental and absolutely unnecessary. He could pull out his tool bag of stock concepts and assemble an explanation that made sense to him in terms of the theories currently in vogue, and if these theories later fell out of favor with physicists he could reformulate his account, downplaying the fact that he had been wrong the whole time. Or he could simply dismiss the entire issue with a couple of words, a reference to molecular vibrations or kinetic energy and leave it at that, essentially saying "Who cares?" He'd gotten his hands on some mathematics and that was the important point. The fact that temperature was a chimera was of no concern to him.

Jonathan could tie objects together with concepts in an infinite number of ways according to how he chose to think about them, but that didn't bring the objects closer together or lend a reality to the concepts he had used to make his comparisons. Jonathan's jacket and a creosote bush were both examples of green objects, yet they really had nothing in common when viewed as concrete objects. The jacket consisted of threads of nylon woven into cloth, cut into panels which were then sewn together to form a body, two sleeves and a hood, with pockets and zippers. The bush consisted of sugars and proteins encased by cellulose walls, arranged into tubules and epidermal layers, which then grew by complex metabolic processes into leaves, roots, and stems. In the context of the infinite particularity of physical objects, the fact that the two items were shades of green was insignificant because that quality was lost within a limitless wealth of details. The infinite number of other attributes overshadowed the importance of being green. Each attribute of an object was an infinitesimal minutiae when placed against the backdrop of the fullness of being. The jacket and the bush were worlds apart and the color green could not overcome this fact and unite them in any meaningful way.

But what if Jonathan stopped thinking about the jacket and the bush as real objects and simply forgot about the infinity of existence? He made the jacket an object "x" with the property "green" and the bush an object "y" with the property "green." With this clever move he'd transformed the real world into something very different. He'd created stick figures out of the objects, taking away everything that made the jacket what it was and everything that made the bush what it was. Now "x" and "y" could be tied together through the property "green" because "x" and "y" were nothing in themselves. The jacket and the bush became members of the set of green objects and took their identity from this fact rather than from their existence in the world. Jonathan had radically altered the identities of the objects and changed the meaning of existence.

The physicist argued that Jonathan could not say that the two objects had nothing in common. They had "green" in common. But he had to let go of the objects and put them down in order to isolate this aspect and focus on it. Only in this way could he elevate this relationship to a central position. When he held the two objects in his hands and witnessed them in the fullness of their being, he saw that they had absolutely nothing in common. Only when he thought of them from an analytical standpoint did they become identified as two examples of a certain category.

Jonathan clarified the situation, not by answering the question "What is a jacket?" which referred to the idea of a jacket, but the question "What is my jacket?" which referred to the concrete object that was his jacket. Jonathan's jacket was the totality of details and therefore it was infinite. His jacket was the small rip in the fabric by the pocket, the loose threads by the drawstring, the frayed cuffs, the smudge of grease on the backside from crawling under the truck to check for damage to the skid plates and undercarriage, the coffee stains on the right front panel from awkwardly descending the stairs of the camper, the slight discolorations on the shoulders from long exposures to the sun, the musty odor from days spent crammed into his backpack, the way the zipper snagged on the 86th tooth. The particulars of his jacket were overwhelming, but they were what made his jacket what it was. Jonathan's jacket was not an "x" with the property "green." That jacket was a different jacket altogether. It was not a facsimile of his jacket because in no way did it capture the reality of that jacket.

The physicist could ignore the details of material existence and treat an object as if it were something else in the same way that he could substitute one term in an equation for another one. In mathematical reasoning, he could simply disregard inconvenient factors and annoying influences by dropping entire expressions for no valid reason other than that they were comparatively small. Jonathan had always sat in class with a pocketful of pencils and every time his professor committed a breach of existential etiquette he would snap the pencil in half. He wasn't about to let the error pass unnoticed. He bought pencils by the case at the local office supply store and the store manager appreciated his steady business. His classes were punctuated with the sounds of splintering pieces of wood. Where did physicists get the right to do these things? Expediency. It was all just a game anyway, and if the numbers came out right in the end, who cared about the number of violations of existential etiquette committed along the way?

The physicist found nothing wrong with analyzing an object by turning it into a set of properties. After all, each property corresponded to an actual attribute of the object. The object connected these properties to one another by bringing them together, causing them to coexist in the same place, and this function was carried out in a mathematical sense by envisioning all the properties to be members of a single set. The problem was that a property did not reflect the true situation. For example, none of the cellular structures in the creosote leaves—including the numerous membranes, ribosomes, nuclei, and a host of other organelles—were green. Even most of the pigments in the chloroplasts were not green. The color was the characteristic of a single substance called chlorophyll. The creosote leaves contained a network of yellowish veins, they were enveloped in a transparent waxy coating, and the surface was a profusion of fine white hairs. The property "green" was not just a blanket statement that glossed over the particulars, but a misrepresentation of the actual state of affairs. The leaf was a complex of substructures that were all overlapping and intermingling, not a clearcut and monolithic property. Most of the leaf was not even green at all, so saying that the leaf was green was clearly wrong, yet Jonathan must still put the leaf into this category whenever he began a logical analysis of the leaf.

Similarly, the nylon fibers constituting the bulk of his jacket were not themselves green, only the chemical dye used to impregnate the fibers was green. The fabric was then coated with a transparent plastic to make the jacket water resistant. The zipper was metallic silver and reflective white pin striping had been sewn along the seams. Out of all the materials that made up the jacket, only one minor component—the dye—was actually green, yet everyone said that the jacket was green, even though most of it was not.

Turning an object into a symbol with properties did something very special for the physicist. The transformation caused the properties to become the actual parts of the object, and as parts they too existed in the same manner as the object itself. The properties could now be detached from the object, then manipulated, compared to each other—as well as to everything else in the universe—and put into relationships. The atoms and molecules had just been stuck together, latching onto one another by means that were not fully understood, acquiring shapes and patterns and properties inadvertently and inconsequentially, but now the shapes and patterns and properties were real and they existed all by themselves, thereby going beyond the materiality of the object. This was an essential first step to bringing physics into the world: converting physical objects into sets of logical propositions, the starting point for the wealth of concepts that would follow and later become the true realities of the world.

An infinite particularity did not have properties, and specifically it did not have quantities. As with properties, quantities didn't capture the particularity of the object, something that was essential for maintaining its reality intact. Quantities bulldozed over the details and buried them beneath a suffocating monolithic concept, ignored the bewildering arrays of interactions and interconnections, but in physics that was perfectly alright because the details were considered to be unimportant anyway, when in fact they were the only things of any importance, the only things that actually existed. The object was made up of nothing but details. Jonathan's jacket wasn't really green. It only looked that way to a casual observer. The ocean wasn't really spherical. It only appeared that way from a distance. The tree didn't really have a volume because it was not bounded by a simple, continuous surface. Properties and quantities were illusions, mere appearances, misrepresentations of the concrete facts of existence, ghostly apparitions, but instead of castigating these empty and meaningless forms for their obvious shortcomings and failures, the physicist glorified them, and elevated them to the status of the underlying realities of the object. The physicist believed that nothing was lost in the resolution of the object into properties and quantities, and that in fact the object had merely been stripped of its obnoxious facade so it could emerge in all of its mathematical splendor. That was what everybody in the scientific community wanted to believe—so what was the problem?

Any means of generating mathematics was automatically considered to be good and it was not something to be questioned—particularly on philosophical grounds.

No one could deny that the jacket looked green, the oceans appeared to be spherical, and the tree seemed to have a volume, so why not start there, with the certainties of these direct observations? Why weren't these statements an adequate foundation for a knowledge of the universe? Because none of these objects were what they appeared to be. If an investigator delved into these objects and examined them more closely, the properties and quantities that he had previously established dissolved and vanished into thin air. The common belief that since the particulars had caused the forms to appear, and since the particulars were unquestionably real, then the forms were equally real, was simply not true, because no concept could be extracted from an infinity of particulars. An unfathomable mass of disparate elements implied absolutely nothing. A concept, especially a mathematical one, could only issue from a way of thinking, a perspective that was not incorporated into the particulars but rather added to them, built on top of them, externally imposed on them. The object was not bound by the whims of the theoretician, a person who had gotten his ideas from somewhere else, namely his imagination.

Jonathan realized that he'd once again lost track of Emma. She'd been running around nearby just a few minutes ago. He began scanning the terrain, starting with the places where he would most expect to find her, namely the gaps between the rocks on the nearby hillside, or the hollows between the bushes and clumps of grass in the arroyo just up ahead. Suddenly he spotted her in another direction altogether. The only part of her that was visible was her tail protruding above the surrounding vegetation.

Emma not only held her tail straight up in the air as she ran between the bushes, but she waved it back and forth as if saying "Here I am! Look at me! I'm over here!" The technique worked so well that Jonathan had adopted a similar strategy himself. When Emma lost sight of him and she was obviously trying to track him down in the area where she'd last seen him, he waved his arms over his head to attract her attention. His arm motion was his version of wagging his tail.

The behavior of Emma's tail was an effective evolutionary adaptation in terms of dogs hunting together with their human counterparts, but if Emma were alone or part of a pack of wild dogs and she were prey to other species, then this behavioral quirk would be a great disadvantage, allowing a predator to more easily locate its next victim. Thinking back to his many encounters with coyotes, Jonathan didn't recall ever seeing a coyote raise its tail in this fashion, so the tail behavior must have been a trait that was acquired after the domestication of dogs. Hunters were well aware of the benefits of such a behavior, and perhaps for breeding purposes they selected only those animals that held their tails up. Or perhaps the tail wagging was something that just happened on its own. Maybe in the process of selecting for other traits, the raised tail appeared along with them as part of a cluster of traits tied together by some mysterious underlying genetic connections.

The other end of Emma's body was equally fascinating. Emma was highly scent oriented. All dogs were to some extent, but Emma excelled in this capacity to an extraordinary degree. Emma was essentially led around through life by her nose. In terms of early evolutionary history, Jonathan wondered, did a canine's body exist first and then decide to put a nose at the front, and then make the nose the entire point of the body's existence? That made no sense. The nose must have been created first and the body added later as a vehicle to carry the nose around. Originally the primordial nose could only sniff scents that happened along its way, carried aloft by the capricious breezes of the atmosphere. One day the nose decided that it was not satisfied with this kind of existence.

It wanted to pursue those scents and find their origins, so a part of the nose began to evolve and gradually turned into a set of legs and toes.

Or perhaps the nose and the body, as biologists say, evolved together, starting out small, but after eons of hard work eventually building a grand multicellular empire. Beginning with nothing more than an olfactory cell attached to a neuron, the duo collaborated on the further project and enlisted the help of a third muscle cell, all the while dreaming of a glorious future together, a happy confederation of disparate souls. They had the vision and foresight to know that one day a large sense organ would be connected to a complex brain connected to an intricate network of muscles, the organs all moving together toward the common goal of tracking scents across the desert as one. The three cells had met by chance in the primordial soup and saw the potential of working together as they imbibed a steady stream of nutrient cocktails flowing past their tiny niche, the group situated in the midst of a huge crowd of other cells who were not invited to participate in the project.

Emma ran toward Jonathan and threw herself down on the ground, then rubbed her backside on a particular spot, as if she were scratching her itchy back. She leapt back to her feet and ran off. He didn't see anything on the ground other than some marks in the sand. She had obviously picked up a scent, most likely the urine of some desert creature, probably an antelope. He didn't know why Emma wanted to smell like antelope piss or what purpose that could possibly serve. Perhaps she was masking her presence. The desert must reek of antelope piss. Since the smell was everywhere, the antelope wouldn't know she was coming toward them until it was too late. Or maybe the odor repelled fleas and it was an evolutionary adaptation to avoid flea-borne diseases. Or perhaps she thought that antelope piss had a wonderful aroma and she just wanted to smell nice. In that case, she was doing it for no reason other than her personal vanity.

When musicians got together to play music, everyone said that they kept time, meaning of course that they kept time with each other. They synchronized their individual rhythms into a common beat, much as physicists matched intervals in their laboratories, synchronizing their instruments with certain aspects of the phenomena. The physicist could adjust the baseline clock of his oscilloscope to match the frequency of a radio wave and suddenly the pattern on the screen stabilized. He'd coordinated two events that were not connected to one another: the output of the circuitboard oscillator with the fluctuations of a particular electromagnetic radiation. He'd harmonized the two beats into a single melody. Neither the musicians nor the physicists participated in a universal system of time. Both created time on the spot, the musicians on a stage during a performance, and the physicists in a laboratory during an experiment.

The physicist constructed time in the same manner that he had constructed temperature. A swinging pendulum, the shadow cast on the face of a sundial, the rotation of the earth, the vibrations of a quartz crystal, could all be compared to one another and calibrated, then placed onto a single mathematical scale, thus establishing the universality of time. Time could now stand on its own. As a thing-in-itself, time could become a fundamental reality of the universe that existed independently of every physical process.

The physicist could now start timing everything, comparing the timing of each new event to the timing of other events that he had arbitrarily chosen as the standards of comparison, thus extending the web of interconnections to include every phenomenon. He could correlate the flight of a projectile to the hands on a dial by saying that the projectile's movement from point A to point B equaled six clicks of the mechanical clock. There was no real connection between the projectile and the clock. He just matched them together.

The raw materials of nature did not prevent physicists from formulating a neat and precise concept of time in their laboratories. Clock time was a manufactured artifact, not an extrapolation of natural rhythms or a refinement of the loose and haphazard occurrences of nature. Nature had provided them with countless irregular cycles, endless variations superimposed over other variations, regular and irregular cycles all crisscrossing and overlapping each other. Mechanical time was fundamentally different. Clock time was clearcut, linear and straightforward, precise and unequivocal. Just because physicists employed natural processes in the manufacture of time did not mean that time was an integral part of the natural processes. Physicists hadn't found mechanical time lying around waiting to be discovered. They created it out of mathematical ideas and then projected it back onto the world. Nobody needed clocks in order to live. The fact that modern society was set up around clocks only created an artificial need for them. Numerical time was nothing more than institutionalized mathematics, merely a social convention. If Jonathan took all the clocks away, numerical time ceased to exist, just as pulling up all the survey stakes caused the geometrical grid to vanish from the face of the earth.

When Jonathan was having fun, time seemed to go by quickly, but when he was bored, time seemed to almost stop. He remembered sitting at his desk in class and watching the second hand make each little jump. He'd turn away and stare at the blackboard for what seemed like an eternity, then turn his head back to the clock on the wall only to find that the seconds had barely advanced. Problems arose whenever people compared mechanical clocks to their mental processes. Some people couldn't figure out where the time went while others thought that time was frozen. These contractions and dilations of time occurred in consciousness, a domain that was beyond the reach of scientific methods. Physicists didn't care about anyone's perception of time anyway, even though this was ultimately a reflection of the physical processes occurring within their brains.

Clocks in schools and office buildings were synchronized with the signals produced by a rather intricate device known as a master clock, usually located in the principal's office or sometimes in a nearby administrative office. Every hour, each clock in the building was brought in line with the master clock. Only in this way would the second hands all reach twelve at exactly the same moment that the bell rang. This complicated arrangement was necessary because each clock moved at its own pace, as did the brains of the humans inside the building. Left to themselves, each person and each machine would be on its own time and each one would be different from the others. In fact, every physical process in the universe was on its own time, only physicists artificially synchronized them by correlating one to the other, performing a function similar to the master clock in a large institution.

One of the main concerns Jonathan had each day on his hikes was that he got back to the truck before dark. He'd learned to do this quite well by watching the sun in the sky and noting its height above the horizon. He found that he couldn't easily use clock time to accomplish this task since the length of the day and the time of the sunset were constantly changing. He'd have to play with numbers and do calculations in his head. For him, each day was a whole that had a measure of one. Although the lengths of the days varied considerably throughout the year, this was unimportant because he was comparing the sun directly to the day without knowing what the length of the day was. He didn't need to calculate anything because he could simply match the two intervals directly against one another.

The tilt of the ecliptic varied throughout the year and this greatly affected the rate at which the sun rose and set. In winter, the sun was slow to climb into the sky and lazily drifted along the eastern horizon, moving more laterally than vertically, but then in the evening the sun dropped like a rock, straight down, and quickly disappeared below the horizon. The situation was reversed in the summer. The sun shot straight up in the morning as if it had been a rocket launched from the earth and the day quickly got off to a flying start. In the evening the sun drifted horizontally, approaching the horizon gradually, much as an airplane landing on a runway. This trajectory also resulted in a protracted twilight because the sun continued to linger just below the horizon long after it had disappeared from the sky, granting Jonathan ample time for last minute delays— encountering unexpected obstacles or waiting for Emma to return from a sudden foray. He took this fact into consideration when he selected his route home, a speedy return during certain months of the year, and a more leisurely route home during other times of the year, knowing that he had an ample cushion.

Since both the rates at which the sun rose and set and the lengths of the days were constantly changing, they were at odds with mechanical time. Mechanical time was based on constant rates and that was not what Jonathan had here, so he was trying to match constant rates to variable rates. The relationships between them were difficult to get a handle on because he was trying to translate the reality of the sun's motion into something that was utterly foreign to it.

Physicists chose a fixed rate to represent time because that was what physicists wanted time to be—uniform and constant. They formulated this concept first and then went around and looked for physical processes that supported their preconceived idea of time. Physicists set the standards of time for themselves, picking and choosing from a roster of possible candidates, searching for the appropriate physical processes which they could elect to represent the vast populations of changing events. The rate at which the rain fell, the wind blew, the hills eroded, the plants grew, the heat built in the rocks and then dissipated, were all irregular and often spasmodic. In terms of this overwhelming majority, the physical processes that were uniform and constant were the real outcasts in nature, the anomalies, the elitist minority. The bulk of the universe was built from unstable and constantly fluctuating rates, but physicists chose to disregard all of that and take comparatively rare and until recently unknown processes as the absolute standards of time, rather obscure events such as the vibrations of crystals and atoms. This newly formed aristocracy of neat and precise events became the supreme rulers of the ragged and unruly mobs, standing in stark contrast to the vagaries of the general population. A steady, uniform rate became the model for time itself, just as a perfectly straight line became the model for space itself, however, mechanical time was as out of place in the universe as was the straight line. Both were introduced solely for the purpose of making calculations and not because they represented any kind of reality.

The regularity and uniformity of waves were impossible to deny. Electromagnetic radiation gave the physicist what he wanted, a model for the constant and uniform flow of time, just as previously it had given him a model for the straight line structure of space. So building on this foundation, physicists suddenly began seeing cycles and waves everywhere. Everything was flowing rhythmically and uniformly and this made it appear that the concept of time came from the world and not from mathematics. The world itself seemed to be suggesting the constant flow of time. None of this was apparent from the macroscopic world, but this fact only corroborated the physicists' cosmological view that an underlying mathematical reality lay behind appearances, exactly what they had been saying all along.

Although the reasoning was very compelling, electromagnetic radiations still didn't support a concept of time. Although these rhythms and flows were everywhere, the problem was that they could not be consolidated. They were all independent of one another. Each one could contradict the others, fall our of sync and even interfere with them, or disregard the others entirely. Although each one represented a rate, a flow of time, it did not represent the flow of time. The rates could be compared to one another, but that did not thereby connect them together. The comparison only established a mathematical relationship and mathematics did not have the ability to actually connect things together. This was where physics went wrong. Physicists erected a monolithic scale of time when they had no legitimate right to do so. Placing rates side by side made it appear as if they were related when in fact they weren't. If time had actually existed in the world, everything would be connected through time. In that case Jonathan could, at least in principle, manipulate all the rates in the world by manipulating time itself, but he couldn't do that, just as he couldn't change all the green objects in the world by changing green itself. He couldn't possibly get his hands on time—or green for that matter—because they were only concepts. He had no access to time because there was no such thing as time. There was no such thing as green either, only a plurality of green objects that had no real connection to one another. Everyone falsely assumed that they had the same green in each case, but it was always a different green, a second green, a green unto itself. A person could associate all of these colors in his mind if he chose to do this, or he could set them apart if he'd rather do that, because the connections depended entirely on his way of thinking. Physicists assumed that everyone was compelled to view everything in the same terms that they did. Physicists had chosen to think about objects and events in certain ways by placing them within the context of an encyclopedia of quixotic mathematical concepts, deliberately turning their backs on all the problems that were associated with such a decision.

New Mexico was mostly desert and it had few natural lakes. Jonathan had visited a number of them and they were all rather enchanting, although typically quite small. The state did have a good selection of man-made reservoirs located on several of its largest rivers. These were the perfect places for Emma to engage in all of her favorite summertime activities. She could run around chasing rabbits and geckoes until she got overheated, then plunge into the cool waters for a nice swim. Jonathan often made a point of stopping at one of these lakes whenever he had the chance, particularly during full moons when he could sit for hours and watch the moonlight glinting on the rippled waters.

Jonathan had been hiking in the southern mountains for over a week now and he felt that it was time to head home. His route took him past one of his favorite lakes, so seeing that he had enough supplies to last a few more days, he decided to detour to the lake. As soon as he arrived he found an ideal spot in a primitive camping area located right on the water, with plenty of off-leash hiking all around, although after dark he kept Emma tied up nearby so that she didn't harass the skunks and porcupines and whatever else was roaming the desert after sunset. This was one of the few times he built a campfire in the evening. Watching the flames dance against the backdrop of the peaceful lake was especially pleasing to the eye. The orange glow of the firelight was a nice contrast to the fading blue light of the sky overhead, challenging the authority of the lake to dominate the scenery and capture everyone's attention.

Jonathan and Emma settled in together for a very relaxing evening. A gentle wind carried the sounds of music and voices across the lake, adding a further dimension to the monotony of the sloshing waves, with intermittent melodic rifts lilting above the drone of nature's rhythmic background. It was another Saturday night and the lake was a popular place for people to come and enjoy the outdoors. Jonathan had a good idea what the revelers were doing around the scattered campfires, the flickering lamps illuminating small segments of the shoreline, but what were the stars doing tonight? The scene in the sky didn't reveal what was going on right now. The reason he couldn't know what the stars were doing right now was that he was not connected to them by anything. He was not tied to them by time or space, by light or darkness, by strings either imaginary or tangible.

When Jonathan looked up, he didn't see the universe as it was right now. He saw points of light representing the stars at various points in their distant pasts. He was looking at snapshots of solar systems that had long since grown up and changed, and perhaps even died away. The night sky was a photo album of the extended family of the universe, the way they all used to be once upon a time long ago, their forgotten childhoods and adolescences captured in the images of the stars tonight, their births and deaths all recorded by the light rays, the ancient stars embedded in the illumination much as insects embedded in amber. The stars in the sky were really pictures that had been taken in vastly different settings, dating all the way back to different millennia and different eons. Where was the universal moment in all of this? Where was the all-pervading "now" in the image of the Milky Way that Jonathan saw above his head?

If light had traveled with infinite velocity, Jonathan would be united with the stars by this light and he would be one with the heavens. The universal moment would be available to him. But because of the great distances involved and the comparatively slow messengers bringing him the news, light was completely unable to fulfill this function. To make matters worse, he had nothing else to fall back on, nowhere else to turn. If the universe could not be unified in a practical sense, Jonathan asked himself, then what right did physicists have to unify it in a conceptual sense with their notions of a monolithic scale of time and an all-pervading coordinate system?

The universal moment was just a calculation, a mathematical construction, a shaky speculation, and not the reality of what he was now witnessing. The current "now" of the stars was unknowable, a mental fiction, unimaginable and unreachable both in principle and practice, yet everyone believed that their "now" must exist simultaneously with Jonathan's "now" and the stars' "now", that a universal "now" pervaded the entire universe and represented the true state of affairs. Jonathan could say that at this very moment in a galaxy far away a star had become a supernova and the moment of its explosion coincided exactly with the moment of his utterance of those words.

But the problem didn't reside solely in the heavens. Sarah was many miles away and Jonathan was not connected to her either. He didn't know what she was doing right now and he couldn't even be sure that she was still alive. He could call her on the phone and talk to her. The cellular network could bring them together in a way that had previously been impossible. The technology had the potential to bridge the impenetrable gulf and overcome their detachment and isolation. The phone was a human invention that could unify their separate existences, at least for a few moments while they were using it, but its power was limited and the scheme only worked temporarily over comparatively short distances.

Physicists could fantasize about a connected universe all they wanted, but if they couldn't find a way to actually connect it in a real sense, then the idea of a connected universe remained just a pleasant daydream. They needed a connected universe in order for their mathematical ideas to have a footing in reality—or they could just pretend. Pretending was fun and it was easy to do, so naturally they settled on that. They drew pictures of a connected universe and consoled themselves with these paltry sketches, telling themselves that the universe was in fact connected in exactly the way they pictured it in their minds.

As with drawing lines and counting objects, the mathematics of time and space connected things that weren't connected and painted a false picture of a thoroughly disconnected universe. The interconnections that did exist in the universe were non-mathematical in nature, imperfect and provisional, a medley of forces, radiations and entanglements lacking the omnipotence of the all-encompassing mathematical constructions that had been devised by physicists.

Jonathan had gotten the idea of simultaneity from his daily life. He told himself that at this very moment, Sarah was probably lying on the couch watching a Swedish crime drama on public television, Tom was at the liquor store standing in the checkout line with several people in front of him, while Jonathan was stoking the campfire at the lake with Emma watching him from the shadows. But when he extrapolated the simultaneity of local events into the astronomical universe, he ran into insurmountable problems. Jonathan could blame the speed of light and not his conception of time. He still believed that the universal 'now' existed, only it ran away and hid behind an impenetrable curtain and disappeared from view. Jonathan realized that the universal 'now' was just a figment of his imagination. Simultaneity was really just the relationship between him and his immediate surroundings. He was here and these other things were here with him. Simultaneity did not incorporate the world at large—he had just extended his personal experience to include everything in the universe.

The physicist began lecturing Jonathan at this point: "You know, I can prove the existence of the 'universal now' by starting with what you just said. The idea of the universal 'now' arose from your connection to your immediate surroundings, but those surroundings could similarly be connected to their immediate surroundings, and then those surroundings could be connected to further surroundings, as so on and so forth. The relationship of you to your environment could be multiplied ad infinitum, meaning that all places were connected by such direct comparisons and therefore must share the same moment." The physicist was quite convinced by this argument. Jonathan replied: "You must assume that the relationship between me and my environment was that of absolute identity. Not just a close approximation but a perfect unity, because the slightest discrepancy would be magnified by the infinite multiplications and result in a separation, a drift, a disconnection. Once again, you've substituted a very small number with zero in order to get what you wanted. Identity was a mathematical concept that had no place in the real world. Mathematical thinking only led to errors."

The question "What were the stars doing right now?" could not be answered, even though the question seemed to make sense in terms of Jonathan's ordinary experience and his usual way of thinking, that is, when he was only referring to local events. However, he was not talking about that anymore. Physicists continued to complicate their mathematics in the vain attempt to conform the ideas of time and space to a reality that was utterly beyond mathematics, a reality that could not even be mathematized to any significant extent. Physical existence was not the problem—it was the mathematics that was the problem. Mathematics could not account for the universe because the universe was not a mathematical system, but rather something that defied all mathematical thinking. Mathematics and physical reality were ultimately irreconcilable.

As Jonathan sat mesmerized by the dancing flames of the campfire, a full moon began to rise above the mountains and he remembered a day when he had camped here last year. He had arrived just after sunset. The evening was so spectacular that he stopped the truck about halfway to the campsite, shut the motor off, opened the door and stepped outside. His memories of that moment were so vivid and he recalled the event with such fondness that he became obsessed with the idea of recapturing it, not merely experiencing it a second time, but actually going back in time and reliving it all over again.

He had left the truck near a dumpster which stood alongside a juncture in the road. He probably parked about ten feet away. Today if he parked merely an inch to either side, his action would not be exactly the same as it had been on that day. His tires were brand new back then and the tracks they had made in the sand were quite distinctive. The tires had since worn considerably and they no longer made the same imprints in the sand. He would have to correct that. He'd have to drive all the way back up to Albuquerque and buy new tires, even though the truck didn't need them at the moment. Tire manufacturers were continually changing tread designs so it was quite possible that he could no longer find the exact pattern that he had purchased last year. He had changed the engine oil over the winter, so he'd have to locate that oil and put it back in the crankcase, but the oil had been recycled and it was no longer available. He had discarded his shoes last summer after they had completely worn out, so he'd have to go to the dump and locate them, but after spending a year buried in heaps of trash, they were no longer the same shoes he had worn that day. All the leaves on the sage had fallen off over the winter and been replaced by new leaves in the spring, but these new leaves were different from the old leaves. He'd have to find the old leaves and put them back on the plants, but they had all decomposed and disintegrated. He'd then have to trim the branches so that they returned to the previous year's lengths, and completely erase the cuts the clippers made so that the bushes didn't look funny. The winds had blown the grains of sand all around, changing the shapes of the mounds and drifts, so he'd have to not only recapture last year's forms, a problem because he hadn't noticed and recorded what those forms were, but find each grain and put it back where it had been.

The water in the lake that day had long since been released down the river and been used to flood fields and irrigate crops over the course of the summer. It then percolated into the ground and evaporated into the air, yet somehow he'd have to retrieve every last drop of it and put it all back in the lake so that the lake could be exactly the same as it was before. He'd collected some empty aluminum cans and other trash on his way back to the truck that day and put them in the dumpster. He wondered where that trash was now, probably somewhere in the dump along with his shoes. Jonathan didn't remember exactly where he'd found all those items, so he couldn't put them back where they had been, yet he would have to do this in order to go back in time. Many people had camped at this campsite since that night. They had disturbed the ground, kicked stones, spilt beer, dropped bottle caps and toothpicks, flung cigarette butts into the bushes, and left their marks in innumerable ways. They had burned logs in the fire pit, and now Jonathan was going to have to take the resulting ash, since blown around by the wind and washed into the soil by the monsoon rains that had come in late summer, and reassemble the logs, logs he had never seen, by turning a dirty gray powder into incredibly complicated arrays of dead cells. He'd then have to figure out where the campers had gotten those logs so he could put them back in their proper places, but he had no idea who had camped here since then, probably a large number of people who were now scattered across the country, and perhaps well beyond its borders. Tracking them down would be a problem.

Jonathan could not locate the oxygen molecules that had been in the air that day and not only put them back in their original positions, but push them along in the same directions they had been traveling at that moment. They'd since been scattered throughout the atmosphere and they were now utterly lost, but if Jonathan didn't identify each one and move it back into place, instead using a different oxygen molecule as a substitute, then he would only have a facsimile of the past and not the actual past. The job required that Jonathan find himself a pair of tweezers, not just an ordinary pair of tweezers commonly found in medicine cabinets, but a miraculous pair that could pluck a single molecule out from boundless swarms and firmly hold onto it while he moved the invisible mote a great distance. He'd then have to keep the molecule in place while he began retrieving the others, no mean task considering that these entities wiggled and jumped around incessantly. Even if he could get all the molecules properly situated, the moment he let go of them they would all fly off in different directions, not even remotely copying the motions they had taken on that long lost day.

The planets had since changed their positions, the galaxies had turned on their axes and new stars had been born. The oceans had heaved, fish had been spawned while others had died or were caught by fishermen, and here in the desert snakes had shed their skins, coyotes had lost their winter coats, bees had followed their queens to new hives. There was no going back. Time travel was a nonsensical idea because time didn't exist—certainly not the way that physicists envisioned it in their minds, a universal mathematical structure assembled out of mathematical concepts that likewise didn't exist.

These considerations made one thing clear. No one could go back in time because no one could recreate the past. Even God couldn't recreate the past. It was more than just an infinite task. It was utterly impossible, but fortunately physicists did not have to put everything back in its place because the concept of time would do that for them. They didn't have to ask themselves what time was or how time could possibly perform such a feat because they didn't live in the real world where such questions had meanings. They lived in a blank world of abstractions and it wasn't necessary to think in concrete terms, thus they weren't able to see that the calculation of time wasn't a reality of the universe and that it had no power to do anything. Mathematical time was just a convenient fiction, a marker in a game, a game they played everyday in laboratories all over the world.

This was true for all the abstractions that had been created by physicists. No reality corresponded to a mathematical concept—unless of course a material object had been especially created to embody that concept. Thus the mathematical formulation of time applied only to mathematical objects and it had no significance outside of that context. These objects were found not only in laboratories, but also in all the networks of devices and equipment that were manufactured along the same lines as laboratory apparatuses and functioned in the same capacity, synchronizing events that were not otherwise synchronized.

If the universe were truly a mathematical system, then a mathematical concept would have a place where it belonged, a means by which it was tied to all the other mathematical concepts in the universe. It would not only be rooted in a coherent way of thinking where all the ideas fit together, but also in the ways that material objects were actually connected to one another. The physicist was imagining a conceptual universe that was utterly unlike the physical universe and he couldn't figure out why this conceptual universe did not transfer its properties over to the physical universe in the same way that it did for him in the laboratory. The laboratory, unlike the world, was in fact a conceptual system, and the concepts behind the objects gave the objects special powers, that is, it granted them the ability to produce more mathematics. No amount of labor and industry in the laboratory could ever turn mathematical time into a reality of the universe because the universe had no mathematical objects of its own and thus it could not produce the mathematics that was required of it. The physicist didn't care. He would go ahead and do the job himself.

But even if it had been possible to put everything back in its place, events would not have proceeded this time as they had done in the past. The world would launch into a new series of events that would not resemble what had originally transpired. The world was not a determinate system but generated events on the spot, so despite all his efforts, Jonathan was still not back in the past as he'd hoped. The moment he'd remembered had been so utterly unique that it could not be recreated, just as no one could recreate last week's thunderstorm or yesterday's sunrise. Physicists had this crazy idea that there was something called 'time' and if they could manipulate this entity they could manipulate all the endless details of existence. They imagined that this abstract idea had the power to bring the dead back to life, summon what had been irrevocably lost, move all of creation in a coordinated fashion. Physicists had a profound faith in abstractions. Abstractions were not only real, but they were all-powerful. How did they ever come to this ridiculous conclusion? They should have been able to see that abstractions were nothing in themselves and thus they were totally incapable of doing anything. Sure, things changed, but that didn't imply the existence of time. Time was just a concept, and to make matters worse, it was a mathematical concept.

Jonathan stood up. A faint blue light still lingered above the western horizon. He turned away from the lake and surveyed the surrounding area. He wondered what this place had looked like in times past. He had always assumed that it had been grassland prior to the arrival of Europeans. Then for several hundred years, sheep were driven north along the Camino Real and traded to the colonists for food and supplies. Sheep barons emerged and competed with one another for ownership of the largest flocks. The land was pushed to the limit and the ever increasing numbers of sheep chewed the native grasses down to nothing. Not enough of the grasses remained for these hardy plants to survive the inevitable periods of drought. Once the grasses died out, the door was opened for other plants to take their places. Creosote was the first species to avail itself of the vacant land and consequently the entire area was overrun with creosote.

All of this was merely speculation. No one had bothered to write down an accurate description of the land so there was no way to be sure what it had looked like. Sometimes Jonathan wished that he could walk that land again, go back and see it for himself, see with his own eyes what the first Spanish settlers had seen with theirs. In order to go back in time, a past universe would have to be reconstituted and for that to happen, a template would have to exist—not just a mere outline, but an exhaustively detailed recording of the past moment. Furthermore, this template would have to have somehow been preserved and carried on down through the ages. Not a template exactly, but an exact replica of the original. Not a general description or a formal representation, not a map or an atlas of maps, not a picture or an entire library of photo albums, not a sketch or a diagram, but a complete, exhaustive account of the entire universe down to the minutest details. The infinite particularity of the universe was irreducible. Nothing could serve as a substitute for this past universe, nothing could take its place. Each detail had to be specified independently of the other details and this meant that the universe could not be simplified. It could not be summarized, condensed into an abbreviated form, replaced with a more basic and less elaborate entity, filed away and stored as a bunch of mathematical formulas.

Where would such a replica be kept anyway, and how could anything that existed today reference this blueprint, this immense catalogue of details, and how could an actual physical universe then be generated out of the data? The amount of information was beyond comprehension. The individual blades of grass would each have to be measured, specifying their heights, widths and precise locations, but unfortunately this wouldn't do because the measurements would not include the infinite amount of other relevant data, the endless trivialities of cellular and chemical compositions that gave rise to the appearances and textures of the various blades of grass. Quantities and measurements could never capture the infinite particularities of the world and these particulars were what would be needed for someone to reconstruct the originals.

In order to recreate the past, the past would have to still exist, not just in some vague sense, but precisely as it once did, a complete universe of material bodies all standing in the exact same relationships as before. A person had nothing else to use as a guide, no way to determine what had once existed. If time were real, then each moment would be associated with the infinite particularly of the universe at that moment. If time were actually something in its own right, then time would unite the universe under a single banner and every past universe could be referenced by its corresponding moment. If time existed, then time would essentially contain every past within itself. But in truth, time was nothing. It was a pure flight of fancy, an empty reference, a mere concept, and the past was irrevocably gone. The past was unknowable to Jonathan because nothing had kept it alive —no human invention such as writing, no technology such as photography. The past was unknowable to the universe in a similar fashion because no force or energy or other mathematical structure could possibly serve in this capacity. Nothing existed that could actually connect the present to the past.

Physicists still believed that mathematics could provide just such a connection. They believed that the mathematics of the cosmic background radiation held within its form the mathematics of the Big Bang. They could discern in certain aspects of the current universe various aspects of the universe as it had once existed when it was first created. This nascent universe was conveniently exposed through the long chains of causality that were necessitated by the laws of physics, and the sequence of events that had led up to the current state of affairs could be mentally reconstructed. Given the abstract nature of the mathematical laws, this could only be done in the broadest of outlines. Physicists had to reduce the present universe to a system of ideas, and from these ideas, the ideas underlying the origin of the universe could be surmised by working backward, that is, by assuming the causality of ideas. But Jonathan wondered: was the origin of the universe really just nothing more than an interplay of ideas? The early universe had not been a system of ideas any more than the land of the first European settlers had been a system of ideas. The equations of physics could never arrive at the world of the Mexican sheep herders any more than the equations of physics could arrive at the early universe. Physicists were just playing games with numbers, trapped in a world of ideas that had little or nothing to do with the richness of an infinitely complex past.

Still the numbers in these equations had to come out right, so the numbers corresponding to the past had to agree with the numbers measured today. The details could all be ignored. Jonathan mused to himself: was that because they played no role in any of this, or was that because the details revealed that the system of mathematical formulas was not an accurate account of what really happened? Jonathan had repeatedly

275

demonstrated to himself that quantities did not incorporate the infinite details of the world within them, but rather excluded all of these nuances and intricacies, and thus the quantities were a new reality with unique properties and characteristics of their own, a reality that was very different from the reality of the world.

Astronomical events occurred on a scale that was beyond human experience, so in essence, no details could be supplied. The languages that had evolved in human societies did not possess the proper adjectives to describe a phenomenon such as the Big Bang, and numbers were, well, just numbers. The Big Bang could only be grasped as an abstraction, and conveniently this was what physics provided. The Big Bang was nothing but the interaction of mass and force and energy—terms that did not evoke a clear picture in one's mind of what had actually transpired. When the physicist talked about the Big Bang, he spoke only about the ideas of physics. No one could have possibly observed the Big Bang and that was fortunate, because none of the physicist's concepts related in any way to the particulars. In the physicist's mind, mass and force and energy all did a cosmic dance on an otherwise dark and empty stage, the abstract characters engaged in stilted, lifeless dialogues, blindly following rigid sequences of mechanical gestures and motions.

The long history of the Rio Grande valley had been very complicated, but Jonathan could summarize it all with a few simple statements: originally there had been grasses. Sheep were introduced. The sheep caused the land to become barren. The barren land turned into an expanse of creosote. Jonathan knew that the truth had been so much more than that. New events had transpired each day and brought with them unique configurations. Each year was different from other years and each locality had its own story to tell.

The statement that "the land had once possessed grasses" was only a relationship between ideas. It had nothing to do with the actual landscapes that had once existed. At each moment, all kinds of grasses and plants had been scattered about in all sorts of patterns and densities forming highly variable colonies and hybrid communities all in various states of vigor and distress. Jonathan's simple statement did not refer to any of that, and thus the statement was not about reality. Neither were the physicist's statements about the Big Bang. Talking about reality wasn't as easy as it had first appeared. In fact, it was the hardest thing in the world to do. Actually, it was utterly impossible.

Since the world was comprised entirely of details, any statement about the world would have to be the specification of a detail. In describing the desert of the first settlers, Jonathan would have to say something like: "A blue gramma grass once sat at this exact spot." A detail like that didn't tell Jonathan very much, but elaborating was difficult. He couldn't say that the plant had 189 leaves back then, because he was replacing the actual leaves with generic leaves and the plant did not have generic leaves. Each leaf was unique and it was this uniqueness that he needed to specify. He couldn't say that the plant was 18 inches tall and 12 inches in diameter because the plant did not have geometrical properties. He was only talking about his own definitions. Each aspect of the plant was itself an infinity of details and there was no way to even begin to enumerate them all. At some point, he'd have to switch over to ideas and start distancing himself from reality, start surrounding himself with his own thoughts.

The notion that the results of experiments applied to the real world made physics seem a lot more interesting than it really was. The calculations of time in the laboratory were imaginatively extrapolated to the everyday world and this gave everyone a sense that time was more than just a calculation. Mathematical time was understood as having a much greater meaning than merely the formal relationships between symbols. It was more than simply the ratios between quantities, numbers that only had significance in laboratory settings. Physicists felt that they weren't just studying their own mathematics, and they believed that the results of measurements were in truth statements about the world that everyone knew and loved and interacted with every day, however, when Jonathan attempted to insert the physicist's empty abstractions into the concrete world, the concreteness of existence destroyed them and they became laughable nonsense. Everyone could now see that mathematical time was a joke, the cornerstone of a nonexistent conceptual universe.

In order for time to be real, the universe would have to be cohesive, the individual objects and events would have to add up to something—that is, something other than a number. Everything would have to be connected together, otherwise the moment corresponded to nothing. Jonathan could tie the current moment to this particular rock or to that particular tree, but he couldn't tie the moment to the entire universe because like the forest, the universe was just a concept in his head. In truth, there was no universe, only an infinite number of individuals each going its own way. The universal moment fell apart into an unfathomable wealth of separate moments, each with its own time. They could all be tied together using mathematical concepts, but ultimately these formulations accomplished nothing. The equations didn't actually tie everything together—they only made it appear as if everything were tied together. Physicists could spin the illusion of time and fantasize about the universal moment, but they couldn't make their dreams come true.

Since time was just a calculation, nothing in the universe corresponded to time, and thus time could not be manipulated or altered. Mathematical time referred to nothing concrete and tangible, thus no one could put their hands on it: no one could change it, control it, or direct it, because this concept of time didn't imply a real entity. Time was strictly the product of mathematical reasoning—a bunch of vacant ideas and uninspired visualizations, an abstract picture drawn with points and lines, curves and surfaces, numbers and operators, symbols and placeholders, a fantasy about unsubstantial and implausible structures that could be superficially related to everything but had no real connection to anything. Time could not be a part of anything and no properties could be attributed to it. Time could not be connected to space because there was nothing to connect. Space-time was just a crazy idea produced by an overactive imagination. Everybody was on their own time, acting out their lives independently of one another, finding a path that was uniquely their own. The world did not move as one and everyone was not a part of a single overarching plan. Physicists created the illusion of unity, tying everything together with imaginary strings, instituting a system of creative mathematical relationships that did not copy the realities of the physical world.

If the musicians had thought about time in the same way that physicists did, then they could have made the universe a member of the band. In that way it could play along with the rest of the group, synchronizing its rhythms to match the ones the musicians had created. The universe did not listen to the song and join in because it couldn't. Playing in a band required a consciousness that paid attention to what the others were doing. The universe would first have to hear the music in order to add to it, understand what was going on in order to contribute, tap its foot to the exact same tempo as everyone else by observing and deliberately reproducing their metronomic cadence. The musicians were too smart to ever fall for such a puerile fantasy, but the physicists' imaginations were not so similarly restrained.

The physicist believed in Time much in the same way that the preacher believed in God. Their respective faiths were both so strong that neither felt obligated to provide any specifics regarding the supreme powers of these entities. Jonathan had always wondered: how did God actually create the universe? Did he wave a magic wand? Did green sparks and lightning bolts emanate from his fingertips just as it did from the fingertips of superheroes in comic-book movies? Was God able to transform his thoughts directly into material objects without having to use his hands to fabricate these objects? If God actually did have hands, then shouldn't we be afraid that he might knock the earth out of its orbit as he was reaching for something else and send us flying into the sun?

Maybe time couldn't do everything in the same sense that God could do everything, but in physics time was still capable of mysterious and unexplained feats. Time could fold back on itself and provide an avenue to a universe that no longer existed, bringing that universe back to life, recreating it in much the same manner as God had created the original universe. Every past moment was an entire universe in itself, a universe that had long since vanished, but here time faced a task that was even more daunting than God's project of creating a universe out of nothing. Time had to not only create a whole universe, but to assemble that universe out of the same materials that had once existed back then, and many of these elements had since broken down, dissolved, evaporated, melted, combusted, tarnished, and degraded. Time would have to locate every grain of sand, polish it up and put it back where it had once been, but no one could do that—not even God. The task was not just extremely difficult, but utterly impossible. Yet the physicist talked about time moving forward and backward, a prospect that made absolutely no sense.

The physicist claimed that the direction of time could be inferred from the relationship between cause and effect. Time always moved from the former to the latter. A rock hitting the ground must occur after it was let go by the person who dropped it. Many simple actions were like this. Jonathan pushed a button and his computer screen lit up. If the screen lit up and then Jonathan pushed the button, time would be going backward. If the rock hit the ground before he let go of it, time would be moving backward. Since all actions had consequences, physicists could determine the flow of time through these sequences. Certainly one thing led to another—but not necessarily so. The computer might not light up this time. The wind might surge and blow the rock slightly off course hitting a dead branch nearby instead of the ground.

Sometimes Jonathan tripped over rocks. If he fell forward and then afterward snagged his toe on a rock, would that event mean that time had moved backward. If his GPS unit came off his belt first and then he pressed the release lever on the carabiner, would that imply time was now streaming in the other direction. If he wearily returned to the truck before he'd enthusiastically left it behind, exited the truck before he'd opened the door, locked the vehicle before he'd inserted the key, would these all be statements about the flow of time? Saying that time only moved in one direction was like saying that everything was what it was. The real world did not provide examples of time reversing itself. This phenomenon was something that had to be fabricated in a laboratory with the introduction of mathematical objects. It was a relationship between abstractions, a new reality created by the abstractions themselves. Time could not possibly gush one way or the other in the real world thus changing the order of events. It was a completely nonsensical idea. In fact, the concreteness of existence refuted the very idea of time itself.

Here Jonathan was taking certain events out of context and setting them apart as if nothing else was happening. Time couldn't be moving backward in one string of events while still moving forward in all the other events surrounding it. More specifically, time couldn't be moving backward in an experiment while still moving forward in the room

278

where the experimenters were idly standing by watching the events unfold. Time would be simultaneously moving forward and backward in the same area, quite an acrobatic feat to say the least. When physicists talked about time twisting and contorting itself, Jonathan could now see what they meant. The laboratory phenomena suddenly turned around and parted ways with the experimenters who continued traveling in the same direction as before. If time actually had some sort of geometrical form, then maybe they'd meet again someday. Unfortunately time was just a calculation. It could come out one way in a certain laboratory phenomenon and another way in other parts of the laboratory.

Change alone did not imply the existence of time. Time was ordered change, change that was structured. Time implied that all changes were related to one another, that all changes were united through a geometrical construction of time, an all-encompassing mathematical framework. If Time were a reality, then there was only one Change, the unrelenting march of Time, the changing of the Universe as a whole.

The physicist retorted: "So what if time travel was utterly impossible and it was a nonsensical idea. I never said it was anything other than that." Jonathan was quick to reply: "You created the idea of time travel with your way of thinking, by believing in the reality of abstractions, complete fictions you created in the laboratory with the arbitrary introduction of mathematical forms. Time travel was a natural extension of your concept of time as a thing-in-itself. But if mathematical concepts led to such inanities, didn't that make you suspicious of them and force you to question their validity? Clocks were just ticking away—they weren't measuring anything."

The questions posed by relativity theory were usually cast in terms of imaginary vehicles, typically either spaceships or trains. Jonathan was old-fashioned and he much preferred trains. Jonathan imagined two identical twins standing side by side on a railway platform. One twin boarded the train. The train left the station and accelerated to a speed that approached the speed of light. Experiments seemed to indicate that the twins would no longer be identical. Physicists were convinced that the twin on the train would age less than the twin on the platform because the time on the train had slowed, but the truth was that during his journey the railway twin's liver had aged much more than his heart, a condition that could be blamed on a major drinking problem, while his brain was essentially the same age as before, showing no detectable signs of wear and tear or the ravages of old age, mostly due to an obsessive addiction to solving math and crossword puzzles every day. The notion that all his organs changed at the same rate because they were all tied together by a common denominator called the flow of time, an entity that externally controlled their individual processes, seemed completely unfounded. Time was just an assemblage of mathematical ideas entertained by a certain class of eccentric minds, the formal relationships between empty abstractions, variables in equations, symbols signifying mere concepts. Nothing in the world could possibly correspond to them. Time was a ghost that haunted the vivid, childish imaginations of physicists who were still naive enough to believe in such things. In their own minds they had proved the existence of time through logical reasonings and scientific observations, but in reality they had constructed time using fanciful ideas, resorting to mathematical entities that didn't exist in the world.

Discussions of relativity theory were always full of clocks and yardsticks because that was what time and space were to physicists—the numbers generated by these devices. But try to define time without reference to a clock. Non-numerical time might be something important—but what was it? Without a clock, two in the afternoon didn't exist because there was no way to identify it. Without a clock, Jonathan couldn't say that he'd be back in an hour. It was a meaningless expression. There was no way to know when the hour had passed unless he had a clock in his possession. Still things changed, events occurred, and the moments passed away.

Thus the passengers always had their clocks and yardsticks handy, not so that they could observe the phenomena of time dilation and space contraction themselves, but so other people could observe these effects, people standing alongside the tracks or riding in other trains traveling in different directions. Of course, these other people could not possibly see the clocks and yardsticks held by the passengers as they zoomed past. Physicists were talking about something else entirely, the mathematics of mathematical objects, the forms and arrangements of laboratory experiments, and these did not translate over into the real world of concrete objects, so the metaphor broke down and it was not required to make literal sense. It was not clear that the passengers would notice anything unusual at all. The dilation of time and the contraction of space were afflictions of the instruments and not the people who were holding the instruments. If Jonathan's pocket watch ran faster or slower, what difference would it make? In fact, his watch had started running slow some time ago and he'd put it in his dresser drawer hoping to take it to a jeweler someday and have it fixed. He was always misplacing his cellphone and he often couldn't find it, so he didn't much care what numbers appeared on its face.

Still let's say that the acceleration of the train had somehow taken a toll on the internal mechanisms of the passengers' pocket watches causing them to no longer keep the correct time. But what was the correct time? Even here on the surface of the earth, all the clocks were running at different rates and had to be be constantly brought back into alignment. This was formerly a huge problem, but now people had a global cellular network that did the job for them, so the illusion of universal time became thoroughly entrenched in everyone's mind. Everyone knew that the person on the other end of the line saw the same numbers on his touch screen as they saw on theirs—well, after they converted time zones— because all times came from a single source, a federation of corporate giants who owned all the phones and equipment. This network of towers and satellites embodied the concept of time as the synchronization of objects that had nothing in common, establishing the necessary correspondences through the deliberate actions of electronic devices that were in constant communication with one another. The concept of time commonly employed by physicists also envisioned time as coming from a single source, in this case believed to be time itself.

If time altered the functioning of the passengers' watches to make the dials read differently, exactly how was this feat accomplished? What specific changes had been made to the clocks' mechanisms? And if time additionally altered the organs of the passengers' bodies, each one a compound of biological processes bearing no relationship to the mechanical processes of the clocks, how could one and the same entity—or force, or agent, or mechanism, or whatever it was—accomplish such diverse tasks, and moreover do so in a unified and coordinated way? How could time affect everything at once and elicit changes in each object and event separately and independently, change each rate accordingly, even though each rate was peculiar to the specific processes in question, actually a series of rates that had to be be balanced to one another in order for the organ or the mechanism to work in the first place? Such a force was quite baffling, yet the physicist did not explain it, or detail the unbelievable mastery and total command of this strange force, or account for the mechanism of its agency. What kind of entity must time be in order to have these stupendous powers? By what physical means did time accomplish tasks that were so varied?

Physicists would say that time changed the clock's mechanism by changing the mass and length of its component parts, but if time increased the mass of the pendulum weight, where did time get this additional material and how did it attach it to the existing weight? Did time externally plaster this substance—presumably composed of the same chemical elements as the original weight—to the side of the weight, or did it infuse the molecules uniformly throughout the interior of the weight? In order to make the clock run slower, did time lengthen the chain of the pendulum by adding additional links and thus increasing the total number of links, or by distorting each individual link, narrowing and stretching them all to the same degree? Or did the measurements of abstract quantities just make it appear as if the weight had increased and the dimensions had been altered? Perhaps the whole thing was nothing more than a conceptual facade, a shift in the masquerade of symbols written in equations, the behaviors of quantities that only represented a false world of ideas, the patterns of simple numerical values that did not correspond into any actual rebuilding or reconfiguration of the mechanism.

Being a simple device, playing with the mechanics of the clock to alter its rate was easy enough to do, but what about the human heart? Did time mess with the cardiovascular nerves that triggered the beats of this organ, or did it relax the heart muscles causing them to flex more leisurely? Cellular processes could be slowed by decreasing the ambient temperature and causing the chemical reactions to proceed more slowly, but how could time do this directly, telling each organ and each cell within that organ to adjust its metabolism and coordinate its activities with all the other cells and organs? They would all have to be affected simultaneously to the same degree, that is, their activities would have to be dictated by Time itself, an entity that existed everywhere and had supreme power over everything. Jonathan couldn't help but marvel at the physicists' creations and all the dazzling and stupendous things that these mathematical constructions could do. Physicists endowed these imaginary entities with fantastic powers which the mathematical constructions then wielded with reckless abandon, forcing every creature and every object to obey. Such was the nature of mathematical law—at least in the minds of physicists.

These days Jonathan had many more opportunities to go hiking and camping than he'd had in the past, although he'd always tried to get out as often as possible. During busy stretches in his life, sometimes he would sit on the couch after a long, hard day and promise the dogs that he would take them camping next week. They looked back at him with anxious expressions letting him know that they were bored and eager to get out of town. Then one day he figured it out.

When he was a philosophy student in college, Jonathan imagined that he would die on his eightieth birthday after attending a party with all his philosopher friends, just as Plato had done, but he was in good health and with good fortune he could easily make it to his eighty-fourth birthday. Jonathan could reasonably expect Emma to live for twelve years, or exactly one seventh of his lifespan. Thus one day in Jonathan's life corresponded to one week in Emma's life. The latter part of next week was two months away as far as Emma was concerned, and that was a long time to sit around and wait for Jonathan to take her camping. If Emma was going to do a lot of camping in her lifetime, then she didn't have that kind of time to waste. If he took her hiking tomorrow, then that was actually her next week, so when he said next week to Emma, he really meant tomorrow, and when he said two months from now, he really meant later next week. Time wasn't the same for both of them and that was a problem.

Jonathan considered some of the other inhabitants of the universe, ranging from the tiniest atoms to largest galaxies, and examined the circumstances surrounding each one's existence. Given the great differences in the events and motions, time could not possibly be the same in each case. The life cycle of a galaxy, for example, was unimaginably long and slow, spanning countless human generations. The human race would have come and gone long before a galaxy made noticeable progress in its rotation. Yet this timing was perfectly normal to a galaxy and humans were the ones who were incomprehensibly fast and short-lived. Perhaps each scale of size had its own time, so as objects ascended the scale of magnitude, time readjusted itself to match the new size and the object always had the same corresponding level of time. A larger size simply meant more time. Since every comparison could be expressed as a proportion, combining all the numbers and putting them together on a single scale didn't make time the same for everyone. Ratios did not establish any real connection between objects, yet in physics, time was universal and all objects shared the same time.

No one could deny that the universe was in a constant sate of flux. What seemed equally clear was that everything was not all flowing together as one. The fact that objects interacted with one another and were influenced by their surroundings did not prove that all these diverse motions were united, that every object participated in the same time by being parts of a single mathematical entity. Physicists compared motions and outlines and felt that these pure forms were enough to make inferences, but appearances were misleading. Sometimes the clouds along the horizon appeared to be a mountain range and Jonathan could easily believe that the rocks and clouds shared something—but what was it? They really had nothing in common and the similarity was just an illusion.

Once time became a universal aspect of all things, physicists could connect time to other abstractions such as space. They connected space and time together in much the same manner that Escher connected the top of his staircase to the bottom of his staircase in a painting. Physicists were only drawing pictures of reality, just as Escher was only drawing pictures of staircases. Physicists were connecting the various parts of their diagrams together by drawing lines, that is, expressing the formal relationships within equations. Escher could only have done what he did because he didn't have a real staircase, but merely an artistic representation of one. The mathematics of time and space weren't really time and space, but only schematic renditions of concrete realities, systems of empty abstractions portraying a world that was infinitely more than a few lines on a piece of paper or the basic relationships between numbers in an equation. The physicist wasn't actually connecting anything together, but like Escher, only making it appear that he had done so. The mathematics allowed the physicist to freely draw lines all over the place, putting them wherever he wanted them to be in order to create whatever effect he desired—on paper, at least.

Jonathan roused himself from his deep reflective state and found that the campfire had languished. The once numerous flames had receded beneath a thick layer of whitish ash and the smoking coals no longer projected much in the way of either heat or light. The moon had since risen high above the black silhouettes of the mountains. It seemed hopelessly lost and wandered gloomily through a wasteland of shiftless clouds, vainly searching for a way out but finding only blackness and haze in dismal combinations. Its dull and confused light cast a pall over the desert landscape, affecting each object in the same manner, transforming everything equally, changing the character of every rock, limb, and bush simultaneously. This was what physicists wanted time to do, to touch every object in exactly the same way and impart the same temporal features to each one, but where was the luminous center of time broadcasting its influence far and wide and causing what Jonathan saw all around him to assume a peculiar nature? In what sky did the glowing satellite of temporality trace its course? There was no such sky over the universe—unless Jonathan allowed for the one that had been mentally fabricated by physicists.

Jonathan stiffly rose from his chair and turned his back on the dark and lifeless fire. Emma raised her head as he approached and looked at him with sleepy eyes, ready to turn in after a long day of hiking and traveling. He clipped a leash to her collar and led her out into the desert so that she could relieve herself one last time before bed. As he walked down the road in the moonlight, he continued the thread he had started when he first sat down by the campfire.

Considered abstractly, teleportation, like time travel, seemed to be a workable idea. All one had to do was build a machine that could scan the human body, the machine using some sort of electromagnetic radiation to determine the precise positions and orientations of all the molecules, and then record this information on a computer hard drive. This numerical data could then be put into service recreating the original patterns. When expressed in these terms, the scheme seemed simple enough. The problems were all technical in nature: understanding the geometries of diffraction and reflection, categorizing the interactions of the radiations with matter, identifying the individual molecules involved and determining the atomic parameters that needed to be copied, specifying and plotting all the spatial relationships between molecules—the kinds of problems that could be solved using geometry and mathematics.

But once again, in the concrete world it was a different story. The human body was not an apparatus in a laboratory but rather an object in the real world and thus the laws of physics were utterly irrelevant for understanding what it was. The human body was completely beyond mathematics and could not be reduced to a formula or a set of formulas. If biological processes were governed by something other than mathematical relationships, then how would the details of the body be specified? If the body could not be understood in mathematical terms, then technological devices could not determine what was important because they were all based on mathematics. In the real world, all mathematical descriptions were superfluous and misleading because it was the concreteness of the world that was central to its being, the one thing that mattered most.

Saying that the human body was a spatial distribution of molecules was like saying that the earth was a sphere. Certainly ways of thinking could be developed wherein each statement assumed a certain element of truth, but from a concrete point of view, each characterization failed miserably to capture the actual nature of the object. The human body was infinitely more than a spatial distribution of molecules, just as the earth was infinitely more than a sphere. Once the spatial patterns had been copied in a second location, these molecules would then have to assume the roles they had previously adopted and resume the work they had already begun. Merely placing new molecules in certain positions would not automatically recreate these relationships because these connections had been established over a long period of time through a history of previous associations. The machine would also have to impart this storied background, the training and conditioning that had originally molded the behaviors of the molecules, going all the way back to the very inception of the organism.

Synthesizing all the proteins, nucleic acids, lipids and other molecules in the human body from basic elements or even precursor compounds was a daunting project considering the complexity of the chemical reactions involved, so naturally transporting the ones that had already been synthesized by the body would be much easier. Teleportation didn't need to create a second human body, but merely dissolve the original one and send its constituent parts along, somewhat like putting fruit in a blender then pouring the contents down a funnel, in this case a channel of light or some other form of radiation, whereupon the pureed contents would be meticulously reassembled molecule by molecule, the blueprint apparently having being sent in coded form along with the molecules themselves. But what if a certain percentage of the molecules were destroyed by the intense radiation of outer space or collided with the swarms of micro-particles sailing along on the solar winds and were knocked off course, never arriving at their destination? The reconstituted person would then become like a chunk of Swiss cheese that was full of holes. Given the large number of teleportations required for sustaining a long career of space travel, after a while the missing parts would most likely constitute a serious health problem.

Science fiction writers, using the same logic and reasoning that was commonly employed by physicists, assumed that once the molecules were put back in their proper places they would remember their former lives and simply pick up where they had left off, continuing their prior efforts as if nothing had happened, perhaps feeling a bit groggy after their ordeal but quickly shaking off the haze and confusion. As far as the person being transported was concerned, the time spent in transit was not merely an NDE, or a certifiable death followed by a miraculous resurrection, but the total annihilation of the body followed by a pure act of creation, building a completely new body from scratch. The old body not only became lifeless, but completely disappeared for a while. Where was the person during this time? Was he hovering in the general area as a disembodied spirit waiting for his body to arrive so he could inhabit it once again?

Undergoing a major surgery in a hospital was a terrifying enough experience, particularly when the patient truly thought about it, losing consciousness for a few hours while a doctor cut the patient's tissues with razor-sharp scalpels and poked and fiddled with the patient's organs using stainless-steel forceps, but in teleportation, some kind of high-energy hatchets thoroughly chopped up every organ, completely trashing these delicate structures by mincing them into a molecular slurry, disregarding the fact that the person was completely dependent on these organs to be alive. The extreme measures went way beyond the minor alterations involved in an ordinary surgery. In contrast to the common practice of hospital wards sedating the patients prior to the operation, Jonathan never saw anyone in the movies administer tranquilizers to the teleports to calm their nerves. Tearing the body apart like that should at least require some sort of anesthesia to deaden the pain, which must be excruciating. Still the tissues took no time to heal and the patients required no rehabilitation or physical therapy in order to return to a normal lifestyle, a strange situation given the extreme severity of the wounds that had been inflicted upon them.

Typical of mathematical thinking, teleportation was visualized as an abstraction, a set of rather vacant ideas and meaningless expressions that ignored the grim realities of the actual events taking place. In truth, the procedure was very scary and the prospect of an equipment malfunction was an ever-present danger, something that was largely downplayed in sci-fi movies. Being trapped inside a tin can flying around in an infinite void was reckless enough, but the crew members were soldiers and death was a reality they faced everyday. The characters in the movies never showed any fear and they nonchalantly shrugged off the pressing danger as they confidently stepped onto the platforms and then dramatically turned toward the camera, striking defiant poses reminiscent of the idealistic revolutionaries in communist propaganda posters. "Go ahead. Violently hack me to pieces with your sophisticated butcher knives and scatter the tiny tidbits of meat scraps across the vacuum of space where all kinds of hazards lurked, then gather up all the bloody entrails and dead matter, collect all the biological materials and corral all the organic substances, and stitch me back together again, just as I was before. No big deal, and certainly no reason to be alarmed."

After all, look at the technological miracles that we took for granted today, yet everyone knew from experience just how unreliable technologies were and in view of this fact the unshakable faith of the space travelers was rather touching. Jesus would have been highly envious. In fact, he might have been a bit miffed that more people hadn't trusted him that way. But trusting in a machine was nothing at all like trusting in a person's word. Scientists were different because they definitely knew what they were talking about and their proclamations could be taken as absolute truth—unlike most people who just talked carelessly, saying things that they could not really substantiate. Sorry, Jesus, but we can't be sure you got that story about salvation right. Well, we can't be sure the machine will work right either. Everyone had to believe in something.

The teleport device could do nothing more than map the various aspects of the human body. Just as a map of the landscape could never generate the actual state of affairs that existed out in the desert, this map could never actually generate a living organism. Every concrete object went way beyond mathematics. Every object in the real world was infinitely more than a simple mathematical rendition of it.

The absurdities of time travel and teleportation did not center around the anticipation of some limit to the abilities of physicists and engineers to mold materials into mathematical forms, but rather on a misunderstanding of mathematics and its role in the universe. In the real world, mathematics did not exercise command over material bodies as it did in the laboratory, a power it wielded only because those bodies had previously been bestowed with precise mathematical forms. Mathematics did not rule the universe, but rather was powerless in the face of brute existence. When confronted by the works of nature, mathematics quickly became a matter of no importance. Electrical engineers could make a television set only because a television set was a mathematical object. The human body was another story altogether.

Jonathan suddenly found himself standing at the door of the trailer. Emma had understood the purpose of their walk and diligently done her duty. He opened the latch to let her inside and took one last look at the moon soaring overhead as he entered the shadowy enclosure of his roadworthy residence, the gleam of the silver palace now greatly subdued in the pale light of a hazy moon.

—13—

If Galileo had climbed the stairs of the Leaning Tower of Pisa and launched paper airplanes from the balcony, then faced the gathered audience below, a mishmash of street urchins, learned scholars, and curious pedestrians, and exclaimed in a commanding voice, "Look, this is what you're up against! This is the way the world works! Calculate this you fools!" and then off flew another paper airplane careening into the air, Jonathan would have admired him for that, because he would've been standing up for the world and defending its sovereignty. Galileo would have been displaying a keen insight into the nature of things. Instead Galileo chose to negate the world by devising an experiment with an uncharacteristic result, an orchestration designed ahead of time to be reproducible. He cleverly selected—of all things—a cannonball and a musket ball, so that each time he dropped the two spherical objects over the edge, the result was the same. But the world was more like the paper airplane. The world took off in different directions. It lurched madly, first one way and then another, and its future course could not be predicted.

But from now on scientists were only going to concern themselves with reproducible events. The fact that all events in nature were unique would disqualify them as subjects of investigation. The first men of science realized that science had to be based on nihilism. The erratic occurrences of everyday life were all meaningless and such things had to be swept aside to make room for a new world order, an ultramodern society built entirely from mathematics. Political extremists blew up police stations and government buildings in order to create room for their personal vision of utopia, but scientists merely had to turn their backs on nature and focus their attention exclusively on the artificial and highly contrived phenomena of experiments.

Reproducibility was a very strange request. The memorable experiences Jonathan had accumulated over the years were all once-in-a-lifetime events. He understood that some people found comfort and security in a structured and well-ordered existence, but instead he welcomed the unexpected. He sought adventures and surprises. He did't want today to be a copy of yesterday, or this afternoon's hike to be a replica of some other hike he'd taken in the past.

Why was reproducibility so important to physicists? Well, if physicists seriously wanted to connect objects and events together with mathematical relationships, then that goal required a constancy in the outcomes of events. Mathematics was eternal and unchanging so the world had to follow suit. But that meant casting the bulk of natural occurrences aside, along with most of human experience. Physicists had to put the pieces of paper back in the desk drawers and the frisbees back on the closet shelves and limit themselves to exactly reproducible phenomena. Basically they had to fabricate these phenomena themselves because nature didn't provide enough material to work with. The program was absolutely astounding, as radical as any manifesto concocted by a lunatic fringe of idealistic rebels seeking to destroy society as we know it. Physicists threw the world out the window because it didn't meet their demands. In its place they erected another world, the world of the laboratory, where they made everything conform to their standards. The early scientists realized that they'd have to erase the world in order to rebuild it.

Galileo probably never threw anything off the Leaning Tower of Pisa. The story is now considered to be a fabrication, one of the many charming anecdotes that pepper the romantic history of science and make it more interesting. But he could have done so and the demonstration would have played out in the manner described in textbooks—or would it? Jonathan wondered where his bowling ball was right now. Since he'd smashed his hand, his fingers no longer fit into the custom drilled holes so he had no use for it anymore and he'd never found anyone who wanted it. He could make it his cannonball. And a few weeks ago he had found some some golf balls when he and Emma were walking the links at the University of New Mexico course in Albuquerque, a popular after-hour dog walking destination for many people where the dogs could run off leash on the lush, well-manicured grass of the fairways. These could serve as substitutes for the musket balls. He needed to locate a high cliff somewhere. Jonathan suddenly came to his senses. He didn't have to go through all that trouble because the result wasn't important. The idea that physicists could express non-reproducible events solely in terms of reproducible events was utterly insane. The outcome of Galileo's experiment didn't matter because the experiment itself was meaningless. The world didn't need that kind of nonsense.

What the world needed was a champion, someone who would not stand by idly while its integrity was trampled and its sovereignty refuted. The world got slandered and ridiculed at every opportunity. Nobody upheld the world anymore. In the bitter conflict between *res cogitans* and *res extensa*, everyone took the side of ideas. Jonathan wondered: who were these people who thought that they could overthrow the entire universe with a simpleminded experiment or capture its essence with something as trivial as a mathematical truth?

Rather than being a truth about the world, the Leaning Tower of Pisa experiment was a blatant lie. The physicist was exasperated that anyone would be dumb enough to make such a statement. He said that no one in their right mind would ever deny that the two objects fell at the same rate. Jonathan was forced to explain. The objects did not represent the world or how it normally functioned. The world was not composed of perfect spheres with smooth surfaces. Physicists pictured atoms as spheres, but they had no reason to believe that this was what they truly were. A sphere was just the symbol they used to represent atoms. Galileo could have easily dropped rocks off the balcony and gotten a comparable result, but he understood that physics wasn't about the works of nature or the objects of nature. It was about mathematical objects. Natural objects commonly had irregular outlines that often included flat or concave areas and these would create turbulence that would push the objects around, diverting them from straight-line trajectories. Lighter objects would be subject to the resistance offered by the air, essentially putting them at the mercy of the wind. Objects seldom had a clear path and rarely got very far before running into something, greatly complicating their behaviors. This was the real world, the way that virtually everything worked. Thus the experiment was not characteristic of the world and did not capture of the truth about material existence. Still physicists could make the experiment an archetype, a model for how they wanted the world to work, and proceed to generate similar examples in their laboratories, essentially replicating the experiment in an endless stream of forms and configurations. Now they could say that this was the way the world actually worked. Well, not the real world, but the world they had created through experiments based on this paradigm.

Reproducibility enabled the physicist to formulate mathematical laws, but from the point of view of the world, rather than being an enabler, reproducibility was actually a stone around the physicist's neck. Reproducibility hindered every attempt to understand a universe that was almost exclusively irreproducible. The vast majority of irreproducible events could not simply be swept under the rug or demoted and pushed into the background. The physicist's strict adherence to the criterion of reproducibility weighted down his explanations and became a load he could not carry. The irreproducible events in nature could not possibly be understood in terms of the reproducible events in a laboratory. Physicists needed to discover the essence of irreproducibility and find a way to deal with it on its own terms instead of trying to transform it into its opposite for the sake of mathematical convenience.

But physicists had a much deeper problem with reproducibility. Reproducibility was not just an inappropriate concept for understanding the world, but it failed to function in the manner that they had intended it to function, namely to provide them with an objective knowledge of the universe. Jonathan's eyes did not present him with reality, but only with a vision, an image that did not reveal the vast intricacies of the microscopic and molecular worlds. What he saw was relative to the way his eyeballs were structured and therefore his vision was subjective. Jonathan's eyeballs were instruments in the same sense that the apparatuses in a laboratory were instruments, and the views these devices presented were subjective in that they were determined by the structure of the instruments, in this case by the mathematics embedded in the designs. Instruments produced a mathematical result only because all objects in the laboratory were already mathematical forms. When objects acted on their own as they did out in nature, they produced no mathematics and could not be mathematized to any significant degree, even with the most ardent efforts. Natural phenomena not only resisted mathematization, they made it impossible. The world was so thoroughly non-mathematical that mathematicians and physicists could make no significant inroads or even begin to predict the outcomes of events.

If a law of physics could only be observed by someone who was already in possession of mathematical objects, then that law was subjective. Jonathan's perception of a thunderstorm was also subjective since his observations were relative to his possessing eyes with which to see. This was a universal problem and ultimately the thunderstorm could not be observed by anyone. The infinite processes going on within the clouds were not available to Jonathan and he saw only images of something he could not fully understand. What Jonathan saw was not the actual thunderstorm, but only a picture of it. Similarly, a person who was in possession of mathematical objects was only looking at other images, viewing the thunderstorm through a filter of preconceived ideas, observing the events from a certain ideological vantage point, a system of concepts that centered around the geometries and algebras of the apparatuses and the designs of the instruments, including the patterns they produced. Jonathan's understanding of the thunderstorm was only his point of view, and the designs that mathematical objects drew were only from the point of view of the mathematics that had been built into them. Jonathan could never observe the thunderstorm in its entirety and neither could the physicist. Both had only limited views of it—Jonathan by his restricted sense of sight, the physicist by the limited ability of empty forms to capture concrete realities. The laws of physics were just narrow perspectives created by a short-sighted way of thinking called mathematics, and it was a mathematics that had been forced upon the world. Artificially fabricated mathematical objects generated sketches, graphs and drawings of a world that ultimately could not be grasped through these skeletal outlines. Jonathan's eyeballs were fabricating patterns that did not copy what was truly going on within the clouds, and the physicist's apparatuses only fabricated other patterns, another sort of vision that was equally limited and deceptive.

In a universe where the phenomena were incalculable and the realities that resided behind these phenomena were unknowable, there could be no objectivity. A universe of mathematical objects, on the other hand, would be fixed and definite, structured and perfectly ordered, but an infinite particularly had a comprehensive indeterminateness to it and this prevented the theoretician from getting his hands on it. The real world was elusive and it continually slipped through his fingers. Every conceptual framework fell short upon further examination, the ideas and premises began to generate anomalies and enigmas, the exceptions and mismatches piled up and became impossible to ignore. Every statement was a subjective interpretation of something that could not be grasped.

Morden science had originally emerged from a world where superstition, witchcraft and magic were commonly held opinions. Early scientists had sought to counter these unfounded mythical beliefs with direct observations, but little did they know that the concept of observation was itself unfounded. Already fully immersed in mathematical thinking, they considered observation abstractly, which would have been appropriate had they lived in a mathematical universe where everything was conceptual and abstract. They never could have anticipated what followed. It turned out that the reality of the world was unobservable.

This simple truth still appears to not have dawned on many people. Physicists continue to grapple with the concept of observation in their experiments and generate endless nonsense about the strange connections between the observer and the observed, when that was not the issue. The issue was that physical reality, by its very nature, could not be observed. This was a strange conclusion, but it said a lot about physical reality. It's not what anybody thought it was. No one seems to have come to terms with what they were dealing with here. Physical reality was set up in such a way that nothing could be observed in its entirety, leaving everyone with only partial views, limited and illusory perspectives of a world that could not be fully embraced. An observer could only glimpse certain aspects of the infinite particularities taking place. No observation was thorough and comprehensive, and none were definitive. The peeks and glances that were available to physicists could not be put together mentally because they were not tied together by any conceptual system. Not just the endless details but the boundless diversity of details could not possibly be marshaled, making an exhaustive understanding of anything impossible.

Observation was a completely different matter in the laboratory because here physicists were only observing the mathematical forms of their mathematical objects. Physicists had discovered that they didn't have to observe the real world. They didn't need to know the details of existence. Instead of trying to unravel a tangle of misshapen objects with no apparent order, they compared the marks on a yardstick or the hands on a clock to phenomena that had clearcut edges, uniform shapes and precise timings. They could turn their backs on the real world and replace it with an artificial world of abstract concepts, turning it into an unrecognizable mathematical analogue that was much more to their liking.

Physicists sidestepped the problem of objectivity by focusing instead on reproducibility. They said: "If another person stood where Jonathan stood and looked in the same direction that he looked, then they would see what Jonathan saw." Any person could do this, so physicists said that this made Jonathan's view objective. Beyond being relative to the structures of his sense organs, what Jonathan saw was also a matter of what he thought he saw, how he interpreted the patterns that were presented to him, thus the way he thought about the world determined what he saw, so even if another person stood in his shoes, their vision of the world might not be exactly the same as his. Ultimately all knowledge was subjective and to expect otherwise was unrealistic. Placing an absurd standard on knowledge accomplished nothing, so philosophers and physicists both decided to back away from this ultimatum and not concern themselves with the unattainable. They both had to admit that absolute knowledge was impossible, yet that didn't mean that they didn't know anything. Compared with complete ignorance, what they had was definitely a knowledge of sorts. Comparing oneself to the highest standard only served to expose one's inadequacies. When stood alongside Eric Clapton, Jonathan couldn't play the guitar at all, but when compared to someone who knew nothing about music, Jonathan was a true musician.

By specifying what another person had to do in order to see what he saw, Jonathan was merely objectifying his subjectivity. In order to recreate the results of an experiment, physicists would explain in great detail what they had done in the first place to get those results, but that didn't make the results objective—it only made the results reproducible. A person who had made a mistake could explain the circumstances that had led up to the mistake, and someone else could then go ahead and make the same mistake—but it was still a mistake. Reproducibility did not guarantee that the results were correct.

The presumed objectivity of physics was premised on the neutrality of mathematics. Mathematics didn't change the reality of the situation or add anything to it, but merely reported what was already there, contradicting the fact that converting concrete objects into abstractions created new realities. Mathematizing something that wasn't itself mathematical wasn't being objective. It was the most extreme prejudice. If mathematical imaginations were fantasies, then the laboratory wasn't the real world and the imposition of law and order in an experiment was tyranny.

Physicists had no claim to objectivity from the outset, continually adding their mathematical designs to everything and forcing disorderly material objects to conform to their mathematical molds. Jonathan did not allow himself to get caught up in such a program of falsification. Unlike physicists, he did not turn a blind eye to the obvious truths of the world around him. Jonathan wasn't convinced that the desert was inhabited by spirits and conscious entities—he was just open to the possibility that certain aspects of the desert required explanations that went beyond the simple mechanical actions of laboratory phenomena. The role that consciousness played in all of this was uncertain and at this point merely a suggestion, but if consciousness did figure into anything, eventually he would uncover its presence and bring it to light. Physicists tried to explain everything in the universe with mechanical actions, but they were only talking about their own experiments. They couldn't even begin to explain anything out in the real world, but that didn't seem to bother them.

In order to prove the existence of consciousness in the natural world, Jonathan would have to show that consciousness was a necessary assumption—but in the context of what way of thinking? Certainly not from a mathematical point of view since consciousness had no mechanical aspects and thus by its very nature it was non-mathematical. If Jonathan thought about the world differently, then perhaps consciousness might become a necessary assumption. The physicist's claim that Euclidean geometry was a necessary assumption only had meaning in a mathematical universe, and because Jonathan was trying to think about things differently, he wasn't driven towards that conclusion. In the desert, Euclidean geometry was unimportant and therefore unnecessary.

Instead of giving free rein to mathematical innovation and carte blanche to a symbolic way of thinking, doggedly striving to envision imaginary geometrical constructs everywhere in the universe, Jonathan chose a different path and placed consciousness at the center of his world, thereby picturing a living universe. Jonathan had never related to the barren abstractions of physics, the mindless forces and the blank quantities, because that was not the world he saw around him, a world that was so much more than simple mathematical relationships. He was never able to incorporate the vision that had so animated physicists throughout the centuries—the vision of a lifeless world. The mechanical aspects of the nature were all trivial concerns to him and he felt that no one in their right mind would ever waste their time trying to understand these things because there was nothing much to understand about them. However, just as a computer seemed to come alive through the endless sequences of basic binary operations, physicists sought to create a living universe through the endless sequences of simple mechanical actions. Why not just imagine a living universe instead? Why not just picture an elemental consciousness that was as ubiquitous as the mechanical forces postulated by physicists?

Earlier societies had already envisioned the world in terms of spirits and deities and Jonathan wasn't interested in resurrecting these ancient myths or returning to such naive cosmologies, but perhaps these people basically had the right idea and their attempts to understand the universe through consciousness were in some ways correct. The only problem was they generated these mythologies entirely out of their imaginations and modeled them on human behaviors and social interactions, and instead they needed to ground these speculations in a metaphysics of concrete objects and natural events, the materials and living bodies of the physical universe. This approach had already been adopted by a certain subculture of open-minded investigators regarding both plants and microorganisms and even water molecules, but these findings had been roundly rejected by a mainstream scientific community that was still bent on perpetuating the illusion of a mathematical universe. Instead of a science based on mechanics, serious investigators needed a science based on the fact that everything in the universe acted as if it were alive and seemed to exhibit some degree of awareness, which imparted a sense of direction and purpose to their activities. Physicists orchestrated everything in the laboratory with a preset goal in mind because they sought to create another world and in doing so they ignored what nature had done when left to its own resources. People needed to understand not what physicists had created but what nature had created and that meant putting mathematics aside and focusing on what was important and essential to the natural impulses of all things, the drive to generate new forms and develop uniqueness and non-reproducibility everywhere, the endless motions that cultivated originality and non-conformism in each individual. New types of explanations would lead to new types of experiments and perhaps these new directions would finally lead investigators toward the real roots of existence.

Jonathan hoisted his pack on his shoulders and struck out through the tall grass of summer, leaving the Airstream behind, the shiny aluminum shell sitting alone in the middle of an open field across from the old adobe farmhouse. Emma had already gotten a good head start on him and had disappeared into the forest. The predawn sky had brightened, but the sun had not yet risen over the horizon. The early morning atmosphere was populated by a large number of individual puffs of black clouds, widely scattered across the emptiness, eerily similar to the detonation residuals of antiaircraft shells that Jonathan had seen portrayed in war movies. The scene over his head appeared to be the aftermath of an intense air battle, the attack fighters now heading back to base somewhere off in the distance, the battle zone once again quiet and serene.

The isolated clouds did not appear to have broken up from a single cloud, although they might have each had a correspondingly similar origin. Jonathan imagined invisible seeds having been sown across the fertile sky, each one germinating on its own. The seedlings grew at comparable rates forming a garden of identical cultivars. The morning was perfectly calm for otherwise the clouds would have been shaped by the winds, the movements of air continually distorting them as they grew, merging them into dense streams or fanning them out into feathery wisps. But often the sky was more complicated and very distinct species of clouds occupied the same region of the atmosphere.

Yesterday evening was a good example. An extraordinary cloud appeared that was comprised of parallel strings of golden beads glowing in the last rays of the sunset, all stretching in more or less the same direction, spread out over a sizable section of the sky. The strings were distorted, pushed upward in places and downward in other places in such a way as to create the illusion of a desert landscape suspended in the sky, a strange land complete with rippled sands and sweeping dunes. The effect was uncanny and Jonathan ended up staring at the fantasy for quite a long time. The lines conveyed a sense of depth and he felt as if he were peering into another realm.

Right next to it, a bank of indeterminate black splotches highlighted with vague radial streaks floated at what appeared to be the same altitude as the precise chains of well-defined gems. How could two clouds so different from one another be in the same place? They obviously had utterly different causes and seemed to have nothing in common with each other, yet there they were, together side by side, sharing the same space, acted upon by the same temperatures and pressures and humidities. How could the same environment, the same identical forces and energies, result in such diverse forms?

The other day the light of dawn had found Jonathan lying on his back staring up at the clouds. The sky was a mosaic of four diverse species, intermingling and coexisting, again at what appeared to be more or less the same altitude. There was no simple explanation for this. The atmosphere was not layered or simply ordered. Instead a vast, complex structure was suspended over his head, distinct regions communicating with each other via invisible networks, meticulously incorporating convoluted flows of materials and energies and particular arrangements of molecules. The atmosphere was not composed of pressures and temperatures and any attempt to construct the atmosphere out of these abstractions was misleading. These quantities only pointed to the unknown realities of a world of unfathomable intricacies, a detailed and highly organized entity that was constantly changing right before his eyes.

On some mornings flocks of fluffy sheep mingled with broad flows of glacial ice, strings of compact beads were adorned with vortices of wispy smoke, stark pinwheels with jagged edges pushed their way across fuzzy seas. The sky was often a menagerie of vaporous fauna that reminded Jonathan of the abundant game roaming through an African wildlife park. The migrations were always unique, peculiar to that given day, yet they originated from the same composition of air hovering over the same expanse of terrain heated by the same radiation from the sun. Why then did he not see the same clouds every day? Each day the relevant parameters were basically the same yet the sky was always unique, and not only that, it was so radically different that it bore no resemblance to any previous day. Where was the steady chain of ideas at work here, the simple relationships, the ironclad rules forcing the exact unfolding of events—or even a vague facsimile of such notions?

The patchwork of brown and green fields sank from view as Jonathan followed the old horse trail that paralleled the river and climbed out of the valley. No one ever used this trail anymore and over the years the once clear path had been completely washed out in places, frequently blocked by fallen trees, obliterated by dense stands of bushes and saplings, or buried under deep drifts of pine needles, making progress rather difficult at times. Jonathan had established convenient shortcuts and detours around some of the worst sections and he could still make good time along these alternate routes, but today he abandoned these trails and started climbing away from the river, winding his way through the irregular arrays of piñons and junipers, adopting a course that would eventually lead him to the top of the ridge.

He came upon a remarkable u-shaped channel about 15 feet across gouged into the side of the hill, formed long ago by unknown events. The rift gradually widened as it tumbled downward and eventually opened up into a small but steep canyon in the forest below. He couldn't see the bottom of the fissure as he approached it, so he couldn't accurately gauge its depth, but the obstacle appeared to be impassable. Emma nonchalantly strutted up to the rim and without altering her cadence in any way she abruptly disappeared from view, as if a trap door had opened and she had fallen into an underground vault.

Jonathan stood and watched, pondering his own solution and wondering where Emma had gone, when suddenly she reappeared, popping straight up into the air as if she had worn springs on her feet. He imagined that a restraint had been released by someone pressing a large button and this had caused her to jump up like a jack-in-the-box. She nimbly landed on the other side of the channel and immediately resumed her normal gait, surveying the rocks and bushes as if nothing unusual had happened. She didn't bother to turn around to see if Jonathan had been watching her. This was routine stuff for her. She performed outrageous stunts all the time and thought nothing of it.

Sadly, Jonathan could not do the same. He reached the edge of the channel and realized that Emma's feat had been more incredible than he had previously thought. He didn't understand how she could so easily accomplish a crossing that was, from what he could see, utterly impossible. Scrambling up and down such vertical walls was certainly out of the question for him, so he decided to head downstream to look for a crossing there. Emma must be proud of herself. Perhaps she had been showing off after all.

Jonathan came upon a falls and struck a diagonal toward the near corner of the rocky embankment, paying close attention to each precarious step. After a minute or two he reached the stone bridge spanning the opening. The channel dropped another eight feet straight down to his right. Large boulders were carelessly piled on top of one another clogging the bottom. Similar piles of rubble also prevented passage along the channel to his left, but the narrow path over the dam was flat and easy walking, and the only way through this mess.

Jonathan labored a couple hundred feet up to the ridge, then followed it back to camp. The day began to get hot, but the clouds were rapidly gathering momentum. Jonathan finally returned to the trailer around noon. Emma jumped inside the moment he opened the door and promptly took refuge on the bed while he fussed over his gear, refilling the empty water bottles and putting them in the fridge, recharging the drained batteries in his GPS and airing out his crumpled rain jacket. He took the jacket everywhere with him, even on sunny days, because he used as it a filler to keep the water bottles standing upright in his pack. Sometimes the caps loosened for no reason and the water spilled out if they were allowed to lay on their sides for any length of time. He started preparing lunch for himself. He put a bowl of chopped cabbage on the cutting board next to a skinned onion and began adding ingredients to the hot dutch oven on the stovetop. Soon the trailer stank of cooked meat and vegetables.

Emma jerked her head up and started barking with a level of alarm typically reserved for intense and immediate danger, directing her wrath at the wall right next to her head. Jonathan stopped working and remained quiet. He heard two people talking outside and realized that they were both standing directly alongside the trailer. He peeked out the window through the blinds. Two guys had pulled up in a late-model SUV bearing the insignia of the Taos County Assessor's Office on the doors. One of them looked like the Devil. Jonathan surmised that the other must be his cohort. He swung the curved hatch of the trailer open and stepped down. With a few hesitant steps he approached the occupants of the very official-looking vehicle.

The Devil turned and spoke to him: "Are you Jonathan Silver?"

He replied: "Yes, I am." Jonathan squinted and blinked at the harsh glare of the outdoors. He looked up at the sky and noticed that the clouds appeared unusually menacing. In fact, they had a frightening appearance. A slew of stinging raindrops began slapping him in the face as he walked toward the two men. He cringed and averted has eyes, then invited the men inside to escape the cloudburst. The Devil gladly accepted his invitation. His cohort obediently followed him.

Jonathan turned off the stove burner, embarrassed by the cooking odors that had accumulated in the cramped space, then turned to face the Devil. A jarring bolt of lightning lashed out from the sky at this point and struck the ground nearby, causing Jonathan to jolt, and for the first time he realized that this guy didn't just look like the Devil, but he was the Devil. A barrage of insulting raindrops, a terrifying bolt of lightning, all set against a backdrop of evil black clouds—that was exactly the sort of fanfare Jonathan would expect from the Devil whenever he introduced himself to someone for the first time. The Devil paid no attention to the remarkable events taking place outside. He nonchalantly brandished a manilla folder with several 8"x10" closeup photos of the old adobe house. Jonathan was taken aback by his threatening mannerisms. He thought to himself: where in the world did he get these photos? The Devil thrust the photos toward Jonathan, one at a time. The images made Jonathan feel uncomfortable.

The Devil questioned Jonathan harshly: "Do you know this house? Can you tell me where it is?" The Devil knew damn well that Jonathan knew where the house was. He had called Jonathan by name. He must have known that Jonathan was the owner of the property.

"Yeah, it's over that way." Jonathan made a motion with his head, indicating the general direction.

"Well, we're interested in this house. Can you tell me who lives there?"

Jonathan thought to himself: what did any of this have to do with assessing the value of the property? Monetary value was a matter of acreage and square footage and building type. It didn't matter who lived there. And what did 'interested' mean? The Devil had a strange way of putting things.

"My wife and I used to live there, but we moved out. The house was infested with snakes and rats and spiders. I was bitten by a brown recluse several years ago." Jonathan went into his story about his bacterial infection and his adverse reactions to the sulfa drugs the doctors had given him.

The Devil listened patiently to his tale, then tried to summarize: "So you're not living there right now?"

Jonathan told the Devil that he and his wife were thinking of possibly putting a pre-manufactured home on the property. They'd talked to a salesman a few weeks ago, but they just sold their house in Albuquerque and they hadn't received the money yet.

The Devil mumbled. "I'll have to check into this Albuquerque property." Jonathan wasn't sure if the Devil was addressing his cohort or simply talking to himself. He couldn't figure out why the Devil would want to do that. The property was located in another county. This whole conversation didn't make sense and Jonathan began to wonder: who were these guys? He started freaking out. Something was wrong here, but he couldn't put his finger on it. These guys were weird.

The Devil repeated himself: "So let me get this straight. You're not living there."

In exasperation, Jonathan tried to clarify: "The house was abandoned. Right now the rooms were stacked with furniture and boxes of junk, furnishings they had inherited from their parents after they died. They had nowhere else to put everything. If they ever settled on purchasing a pre-manufactured house, then they would take some of that stuff out and use it to furnish the rooms."

The Devil abruptly abandoned the line of questioning and changed the subject. "Do you do any agricultural work around here?" The question required a whole new set of explanations. Jonathan replied: "I farmed the land for many years with my '49 Ferguson tractor—not Massey-Ferguson, just Ferguson—but the motor finally gave out. I took it all apart, but it was a bad design. I didn't know that when I bought it. The engine block had castings between the cylinders that were too flimsy and I discovered that they were all cracked, along with several of the crankshaft bearing mounts." Even Jonathan found the details incredibly boring. He then launched into a rambling speech about the problems associated with life in the valley. At this point he was just chatting them up. They gave him curious looks.

The Devil faced his minion and said: "Come on, let's go." Another fantastic blast of lightning struck the ground somewhere in the immediate vicinity of the trailer. After this second blast Jonathan was afraid to go outside, but the Devil and his cohort showed no fear so he followed them anyway. As he walked back to their vehicle, Jonathan questioned them about the tax liability he might expect from his new pre-manufactured home, but they seemed to have no interest in the question. The Devil turned to his cohort and asked him if a pre-manufactured home was considered to be a mobile home. His partner, with a smile on his face, replied that it probably was. They didn't seem to know anything about tax codes. As a final statement, Jonathan said: "So does this mean that my taxes will be going up?" He was referring to their visit, the point of which he still didn't understand. The Devil looked him straight in the eye and said in an ominous tone: "Taxes always go up."

The rain was still coming down as the two strangers attempted to turn around, backing up into the unplowed field, wheels dipping into ruts and flattening molehills. Jonathan fled back to the safety of his cozy trailer. Emma had a worried look on her face. He told her: "I don't get what just happened." Then it dawned on him. Those guys were cops. The dialogue was straight out of a TV detective show. They questioned him as if he were a suspect or a person of interest, showing him pictures of the victim and asking him if he knew the victim and what his relationship to him was—only the pictures were of his old house.

The rain stopped and Jonathan went outside to start unloading the wall panels of a storage building that he'd just brought up from Albuquerque. Suddenly he realized that he'd made a grave mistake. He'd failed to mention that he'd been distilling well water in the kitchen of the old house. Jonathan had developed a terrible reaction to the water out of the faucet. The uranium or arsenic or whatever set off his nerves, and the condition got progressively worse as he continued drinking the water. He sold distilled water to some of his friends to help pay the electric bill for the distillers. He still bathed in a giant clawfoot, cast-iron bathtub in an attached greenhouse rather than suffer the tiny fiberglass tub in the Airstream. Years ago he had installed a solar hot water heater, but the system had broken down repeatedly. After a while he finally gave up on it and started using the electric heater element built into the overly large 80 gallon water tank. He also ran a freezer in the old house to keep a small stock of food on hand. When he had talked to the cops, he implied that the house was not being used for anything, which would make the monthly electric bills look suspicious.

By failing to make these matters clear, Jonathan had inadvertently created a bad situation. On top of all that, they now knew that he owned a gun. Earlier a neighbor had called him to report that last weekend a mountain lion had approached a man who was fishing with his two young children along the river and threatened them, so Jonathan started holstering his gun on his belt—more for Emma's protection than for his own. He'd placed the gun conspicuously on top of the DVD player across from the stove when he'd been cooking lunch. The cops obviously saw it. They were probably armed themselves. If they suspected him of running a grow house, then he'd just turned a casual investigation into a swat situation. Perhaps he should load up the camper and get away for a few weeks. The first thing the police did when they came to break down your door was shoot your dog.

Jonathan told himself to calm down. He might be overreacting. The two guys might have actually been from the Assessor's Office and they were just out re-evaluating everyone's property, perhaps as part of a new government program to update the department's information. The old house had a land line. He hurried over there and called a neighbor.

He broke right in: "Have you seen the Taos County Assessors lately? They just stopped by my place."

"Hmmm. That's strange. I wonder what they wanted. No, they never came over here." The response was not what Jonathan wanted to hear.

"Yeah. Can't figure it out either. Ok, thanks." He called another one of his neighbors. They had exactly the same reaction.

"That's weird. What in the world were they doing down here?" The neighbor paused and then added: "Maybe they were looking for Bertha. She's trying to sell her place."

"Wow! I didn't know that. Well yeah, maybe. I"ll talk to you later." No, they weren't looking for Bertha and they weren't from the County Assessor's Office. Thinking back to the moment when he first greeted them, he remembered that they never introduced themselves or at any point said they were from the Assessor's Office—he just assumed that by looking at the emblem on the door, as they knew he would.

He decided to pack up as quickly as possible and head out—just to be safe. He didn't want to get caught up in the middle of a police raid. They could look around the old house all they wanted. They'd only find a few dead rats, their heads snapped cleanly by the powerful springs of the traps, the sinister devices lurking in the dark corners behind piles of boxes and stacks of furniture.

People talked. They made up stories, spread rumors, speculated about the lives of others, told lies. Jonathan stayed out of it for the most part, but in the past he'd heard some crazy stuff that he knew wasn't true. Maybe some people had been talking about him and the police came over to check out the reports. Their visit might have been the result of an anonymous phone call to the police hotline by someone he didn't even know.

The reason behind their visit didn't really matter. The upsetting encounter gave him more than sufficient reason to leave. He could have the camper loaded and ready to go in a couple of hours since he was well practiced at the routine. Once he was safely out of the valley, being on the road would feel good. Sure, he had lots of work to do around the farm, but all of that could wait until he got back, after all of this blew over.

Jonathan was gone in short order, successfully escaping into the anonymity of the open road. He wrestled with the sudden and unexpected change in plans while he tackled the sharp curves on the way down the mountainside, the road twisting and turning as if it was uncertain which way it wanted to go next. The excitement in Emma's eyes made everything alright. She knew that they were going on an adventure together and what could possibly be wrong with that.

Instead of mindlessly listening to some old music on the CD player, Jonathan thought about what he was doing right now—driving a pickup truck down a winding road. The strange part was that he could drive the truck without knowing anything about auto mechanics. The only thing he really needed to know was how to operate the controls. He turned the steering wheel and the truck went left and right. He stepped on the gas and the truck accelerated. He stepped on the brakes and the truck slowed down. He didn't need to have the slightest comprehension of hydraulic pumps and master cylinders, gearboxes and brake calipers, tie rod ends and ball joints, fuel injectors and spark plugs. He could manipulate the behavior of the vehicle without having any idea of what he was actually doing.

Automobiles obeyed certain laws and Jonathan could specify what those were. He could measure the response he got when he applied pressure to the brake pedal and express the results in mathematical terms. He could plot the resistance of the wheels against the linear displacement of the pedal and come up with a formula. The equation became a law of automobile dynamics and he could say that he now understood the connection between the pedal and the wheels, but without a knowledge of the component parts of the brake system and an understanding of how each one worked, he didn't comprehend the real connection between the pedal and the wheels at all. The mathematics told him nothing about any of that because the equation was only a bare schematic and it was utterly devoid of content. The measurement of simple mechanical quantities could never lead him to the underlying physical truths behind his observations. The formula was completely superficial. The mathematics didn't cause anything to happen, nor it it tell Jonathan what was actually happening.

The relationship between the rate of change of the truck's velocity and the pedal displacement was not so simple because many other factors influenced this result: the condition of the road surface, the wind speed and direction, the slope of the road, the wear on the tires. Breaking the truck on a steep uphill grade while driving into a strong head wind on a pavement that was nice and dry with good tires would be a lot easier than having the truck pushed along by a tail wind while going downhill on an icy surface with bald tires. Still Jonathan could state these relationships using the language of mathematics and then claim that he fully understood the situation because certain variables were connected to other variables in an equation. What was actually happening appeared nowhere in any of this.

Let's say that Jonathan never found the small lever molded into the contours of the dashboard that opened the latch to the hood of the vehicle. Without being able to look inside the engine compartment, he had no way to discover the real connection between his actions in the driver's seat and his observations of the truck's responses. Without such access, all he could do was make up something in his head, explaining the connection in terms of concepts that came from somewhere other than the truck's internal mechanisms. But even if he could open the hood, he would still not be able to make sense of the hoses and boxes and canisters. The connection between his actions and the behaviors of the truck would still be as much of a mystery as before. The tangle of wires and hoses would be incomprehensible to him and of no use in forming an explanation. Wasn't this what physicists were up against when they looked at the behaviors of atoms and molecules and tried to explain them in terms of their internal mechanisms? Why were physicists satisfied with merely discovering the mathematics of observable patterns when this couldn't possibly tell them what was really going on?

Jonathan had always wanted an explanation of gravitational attraction but all he had ever gotten in school was Newton's law of universal gravitation, a formula stating that the force was directly proportional to the product of the masses and inversely proportional to the square of the distance between them. He later read Einstein's General Theory of Relativity and was amazed at how complicated the simple mathematics of Newton's formulation had become. What started out as a relatively simple problem had turned into an incredible array of abstract lines and symbols. He wanted to open the hood on gravity and see how the feat was accomplished, trace the pathways, follow the connections, discover the means by which a mass was turned into a force, but no one else seemed to care. Apparently he was supposed to be satisfied with just the formulas, even though the mathematical relationships told him nothing about the underlying reality of gravity. Gravity was weird. If it was action at a distance with nothing in between, then the phenomenon was utterly incomprehensible. If gravity distorted space, wasn't this just an alternate way of mathematically describing the observed behaviors of objects? How was he expected to be comfortable with gravity if his teachers couldn't explain it to him? They couldn't tell him what it was or what caused it to appear.

Physicists had learned to drive—not the car of nature, but another car, one that was built out of laboratory apparatuses—but they had no idea of how anything really worked. They then figured out how to pop the hood and discovered a mass of components and interconnections, the atomic and subatomic systems behind their observations, but these parts and mechanisms were as inscrutable as the phenomena themselves. None of this mattered because they were going for a ride. They knew how to handle the vehicle and had become very adept at the fine points of driving it. They could skid sideways into a tight parking space, yank the steering and lift one side off the ground then drive around on only two wheels, jump over a long line of police cars, fly off a cliff and land rightsize up or upside down depending on what they wanted. They could make the car perform all sorts of tricks and they could put on quite a show. When Jonathan asked them how they did it, they answered by saying: "I floored the gas, turned the wheel to the right at this point and stepped on the brakes five seconds later." The real explanation was much more complicated: small parts whose names were never mentioned even though they played crucial roles, important events that went unrecorded, actions and motions that were always hidden from view, but none of that affected the final outcome. Physicists were making donuts in the parking lot and having a grand old time of it. They'd figured it all out. God knows what they would be up to next, jumping over the Grand Canyon perhaps. The vehicle of the laboratory allowed them to wow audiences all over the world, but nobody really knew what gravity was or why molecules formed bonds with each other. It just happened.

The drive down the highway was smooth and uneventful. Jonathan made good time, having to contend with only light traffic along the way, so luckily he managed to beat the rush hour traffic into Albuquerque. Jonathan liked to take Emma to the pool hall whenever he had the chance. The place was dog friendly and all the bartenders and doormen loved Emma. The hall occupied the entire second floor of a large building downtown. His timing was perfect today. It was almost four o'clock and the establishment would be opening soon. He texted his friend Richard, a long-time snooker aficionado, while Jonathan was stuck in traffic waiting for a red light, and let Richard know that he was in town. Sure, Richard would like to join him for pool and perhaps a drink or two. Jonathan found a parking place at the edge of the downtown area where a few grass medians and small patches of lawn intervened between the ubiquitous sidewalks and paved streets, providing Emma with plenty of opportunities to relieve herself. Emma knew the way from there, down alleyways and side streets, and led Jonathan in the proper direction, tugging politely on the leash to encourage him to walk a little faster.

Jonathan rarely went out on the town anymore because he seldom drank alcohol and he never left Emma alone in the truck for any length of time. He restricted his urban activities to outdoor patios and dog-friendly establishments. Still he was looking forward to tonight. He spent most of his time out in the middle of nowhere. Meeting a friend downtown for pool reminded him of his college days.

He had broken up with his girlfriend during his freshman year and he was all alone, so he greeted everyone with open arms. He was sharing a large farmhouse outside of town with two other students, but they seldom saw each other because they were always working, attending classes, or out on the town. Jonathan was as bad as the others. He was willing to do just about anything, to participate in any kind of social activity or community function, just as long as he didn't have to go home. Home was empty and dark and unbearably lonely. He couldn't stand to be there for very long, so he would stay out all night doing whatever other people wanted to do just so that he could be with them. He even convinced himself that these were the things he also wanted to do. He chased after any warm body that would sit next to him and the more people he found to accompany him on his search the merrier. They were all lost souls like himself, trying to fill the voids of their solitary existences. They traveled in groups, hitting all the taverns, clubs, restaurants, and living rooms they could find. They usually met in the university commons in the afternoon after classes were over and from there it went on and on, sometimes until the sun came up the next morning. One summer Jonathan witnessed the sun rise on twelve consecutive mornings. They were unstoppable in their quest to satisfy their hunger, but their efforts were mostly useless because nothing worked. Nevertheless, the next night they would try it again with the same results. The months came and went, the semesters passed, and they were all still alone.

If only Jonathan had learned how to relate to a dog back then, he could have saved himself a lot of trouble—not to mention money and brain cells—however, his adventure with dogs would not begin until many years later. He now realized that the whole affair had been pointless. It was one of the few things in his life that he truly regretted, and he often wished that he could have those days back again.

During that time he sat around listening to everyone talk, both inside and outside of classrooms, expounding theories and philosophies from behind lecterns or from behind cups of coffee or mugs of beer, putting ideas together in novel ways. He realized early on that organizing ideas into systems meant nothing. Anyone could do it, and even convince everyone who was gathered at the table that the reasoning was absolutely impeccable. Sometimes they would all nod in assent, their judgement colored by the caffeine or the alcohol, their enthusiasm fueled by the good times they were having with their friends. Jonathan wanted to believe in the truth of ideas, but after a while he despaired. The futility of such pursuits became all too obvious to him. He began to see the expounders of doctrines and ideologies as nothing more than crackpots and charlatans, well-meaning individuals who had become self-delusional. They believed their own stories and that was the problem. They had spent too much time in books and classrooms, amusing themselves in a Disneyland of ideas, tossing abstractions back and forth in lengthy, erudite conversations. They had gradually lost touch with reality and began replacing it with incoherent nonsense. They ended up with unsightly concoctions of petty notions all put together according to the whims of the designer. He couldn't read books anymore and after a few pages he set them down. He found his mind wandering as a self-defense mechanism, trying to block out the steady stream of meaningless connections. He was so glad that Emma couldn't talk and he didn't have to listen to it anymore. Emma was smart. She understood what she needed to know. What else was there in life?

Jonathan first learned calculus as a freshman in high school. He had seen the strange symbols of integrals and differentials somewhere in a book and with his youthful imagination he fancied these characters to be the cryptic runes revealing the hidden mysteries of the universe. He would be sorely disappointed when he later discovered the truth about them, but his curiosity was uncontrollable. He went to a local bookstore and found a 'programmed' textbook on calculus. This was a book that had been structured along the lines of a computer. Back then in the late '60's, lots of people were infatuated with the newly invented technology, even writing books that mimicked the algorithmic structure and logical flow of these devices.

The book had Jonathan read a section explaining an idea, then gave him a problem to solve which tested his ability to apply the idea. If he got the answer right, he moved on to the next section. If he got the answer wrong, he was directed to another section which explained the idea once again. Instead of reading the book sequentially, he jumped back and forth between the pages. Jonathan never again read a programmed textbook. He didn't believe that the fad lasted very long, probably dying out along with Nehru jackets and elephant bells.

The book worked and Jonathan grasped the rudiments of calculus, but he was forced to concentrate for long periods of time and the intense efforts gave him headaches. Two years later his high school offered a course in calculus for eligible seniors, but since he had gone to summer school each year and had earned enough credits to graduate as a junior, they made an exception for him and let him take the course.

He was then required to take calculus again as a freshman in college, so this was his third time around with it. The other students struggled with the new concepts, having had no previous exposure to them, while Jonathan cruised through the now hackneyed material. Eventually the other students got the idea. They would look at problems and follow the lead of the teacher. They would say to themselves: "This was what the teacher would do in this situation." Jonathan always wanted to do something else. The more people pushed him into a corner, the more he rebelled. He knew damn well what the teacher would do in this situation, but that would be the one thing he refused to do. As far as he was concerned, the teacher was always wrong. Jonathan wanted to be free and to escape the confinement of this prescribed way of thinking. He was determined to find another way of looking at things, so he argued with his calculus professor every day after class, pushing his quirky perspectives and philosophical criticisms to extremes. With raised voices and faces flushed red with anger, the two of them sparred up and down the now deserted hallways, the other students having fled to more congenial environments. The arguments got them nowhere and Jonathan knew that he could never be a mathematician.

Mathematicians and physicists needed everyone to be on the same page because they wanted an army of clones to participate in a universal mission, the erection of a single edifice of mathematical rules and fundamental truths about the universe. They dictated everything to the students: the meanings, the interpretations, the perspectives, the proper conclusions. You would draw the line here. You would measure this distance. They encouraged dissent only so they could quash it and show everyone that there was no other way. They had figured everything out in advance so they knew that there was no problem with either the methods or the results. This was simply the way it was. A student could question something only so that he could then be shown the correct way of thinking and be made to fall in line with the others, including the ones who had gone before and since moved on. Everyone thought about things in the same way so that they could all have conversations with each other and know that they were using the same expressions in the same contexts. Well, it didn't actually play out that way, but that was the general idea, the way physics was supposed to work. Everyone was in agreement. Everyone shared the same values, had the same visions, made the same mistakes.

Jonathan began to question many of the formulations of his physics professors as well, and not always with positive results. Jonathan's problem was that he understood the meanings of the symbols and knew how the physicists had arrived at the equations. He saw the equations for what they truly were. He could easily follow the logic and comprehend the meaninglessness of the games that physicists played. Since he had understood the mathematics from a rather young age, he quickly went beyond it, penetrated its facades and grasped the dark impulses behind all of the physicist's deceptions. The physicist didn't fool Jonathan in the least. He knew exactly what the physicist was doing. While the other students were trying to understand physics, Jonathan was trying to understand the physicist. He did not need an explanation for the arcane ideas of physical theories—he needed an explanation for the twisted psyches of physicists. He knew that the physicist had a hidden agenda and Jonathan wanted to expose him, uncover the lies, reveal the devious tricks and deceptions that the physicist routinely employed to get what he wanted, the illusion of another world, an imaginary place full of unwarranted assumptions and fake connections. Jonathan asked himself: "Why would anyone want to do the things that physicists did?" Gradually he came to understand the power they wielded. They could make the world into something it was not. He was confronting an age-old dream, a vain fantasy that had captivated the human imagination for centuries, the idea of turning something ordinary and commonplace into something special and valuable—lead into gold, water into wine, or as a struggling musician once put it, music into cash.

Physicists sought to turn a chaotic mess into a mathematical system. They had a plan. They realized that there was only one way they could accomplish this feat. They had to deny the reality of what they saw around them. The world wasn't actually a chaotic mess —it just looked that way. Underneath the facade of appearances lay a mathematical realm of precise designs and fixed relationships, a stable, predictable world. No one could see it and it gave no outward signs of its presence, yet they could prove it was there. The way they would go about this was strange. The realm could not be accessed through the appearances themselves, by breaking them down and analyzing them as they were, but could only be reached in a separate move that had no direct connection to the appearances themselves. Since the appearances were based on something else entirely, the appearances alone could not reveal this fact. They would have to put the appearances aside and come back to them later. The only way physicists could understand the appearances was to turn away from them and focus exclusively on mathematics. For this, they would have to create a mathematical world for themselves.

As with all mathematics, calculus was a one-size-fits-all approach. Since everything in the world was constantly changing, rates were everywhere. Calculus allowed the physicist to tie these rates together in equations, no matter what the rates were. The method was hollow because it remained on the surface of things. An understanding of an object should come directly from the object itself, flowing out of the unique nature and character of that object, but the mathematics had been developed with no consideration for the particular object at hand, because the physicist didn't start with the physical object. The theorems and proofs were well established long before the object was brought to his attention. The physicist then took the universal ideas of the mathematician and molded them to fit a particular situation, tailoring ideas that had nothing to do with anything to suit a given phenomenon. Instead of the ideas emerging from the particulars of an object, the ideas

301

were pared down and restricted so that the object became a special case of reasoning that was all-encompassing. Rather than starting with the object, the physicist finished with the object, thereby imposing mathematics on the object by coming at it from a place as far removed from the reality of the object as possible, from the far-away, never-never land of pure mathematical forms. Shouldn't the object take precedence over the mathematics and be the starting point of the investigation, the foundation of the physicist's understanding of the world?

Jonathan and Emma had almost reached the pool hall. They turned the last corner and discovered that a summer festival was taking place this weekend. People were already gathering along Central Avenue which had been completely blocked off to automobile traffic. Emma pulled up to a plain oak door with a slim vertical window wedged between an outdoor patio bar and a restaurant and waited for Jonathan. He opened the door and released her at the foot of the stairs. She bounded up the narrow hallway and began to turn the corner but immediately stopped in her tracks, her tail wagging frantically. She was greeting whoever was there today manning the entrance and checking ID's. She quickly disappeared and pranced over to the bar where she was enthusiastically welcomed by the staff. Everyone had smiles on their faces as Jonathan pulled up a stool, thanking him for bringing Emma along with him today.

Richard arrived carrying his pool cue after taking the elevator. He had a bad knee and needed to save himself for the strenuous effort of circling the pool table and executing shots. They made themselves comfortable at the snooker table, the last table at the end of the hall.

Jonathan and Richard rarely had the opportunity to play pool anymore, so their level of ineptitude with a game as difficult and demanding as snooker was roughly equal and their matches were highly competitive. Their calculations were routinely a bit off the mark and the table was unforgiving, therefore the games took much longer than usual to complete due to the many missed shots. They struggled to make the necessary adjustments and come up with the correct solutions to the problems posed by each layout of the table.

Classical mechanics could be used to describe the dynamics of everything from billiard balls to planets, so what Richard and Jonathan were doing today on the pool table was not unlike the work of physicists. Snooker required each player to mentally calculate angles and measure distances because the pool table was a mathematical object, however, no such arrangements existed in nature. In the chaotic world of atoms and molecules, the precise alignments of pool tables were totally missing. The atoms and molecules were bouncing around and flying freely in every direction, yet the physicist tried to see this mélange of haphazard trajectories in the same way that a pool player saw billiard balls on a pool table. The pool player manipulated the angles and momentums of the balls with the definite goal of potting them into pockets, making all of the actions as well as the calculations behind these actions deliberate, however, nature wasn't making shots and the mathematics behind the mechanics of individual atoms and molecules was not designed towards any end and thus had absolutely no significance. Physicists imagined geometrical constructions that were quite different from the rectangular pool tables with their regular arrays of pockets, yet they still made calculations that were not unlike the ones made by pool players, only this was not the point anymore, because nature wasn't playing games with the particles. Physicists could get nature to play all sorts of games in the laboratory and thus impart a deliberateness to the actions of the particles, but out in the real world nature didn't play those kinds of games and therefore the geometries couldn't possibly figure into the outcome of events. Nature couldn't manipulate the particles in the same way that pool players could manipulate the billiard balls because nature didn't have mathematical frameworks in which to operate and it didn't engage in mathematical thinking. Saying that geometrical constructions still existed and determined the courses of events in the same way that a pool table created the realities of the shots was too farfetched to be taken seriously. Pool tables were artificially fabricated arrangements and mathematical objects didn't occur naturally, and that was why the ideas of physics failed so miserably out in the world of clouds and trees.

The once spacious and empty hall steadily became more crowded as the evening progressed, with numerous groups of friends and acquaintances competing at the dozen or so tables. The booths and sitting tables in the lounge area also became packed with boisterous revelers. Emma often joined one of these tables, particularly if they were eating pizza, and stationed herself between two of the seated patrons. She looked like one of the gang, her head just above the level of the table, her intent gaze shifting from one person to the next as if she were following the conversation. Jonathan noticed that Emma had disappeared, so he scanned the crowd looking for her. He didn't see her at first because she blended in so well, then at second glance, he realized that at the table in the corner one of the people was not a person at all, but a dog, wedged in between two of the other customers.

After a while he walked over to the table and asked everyone if Emma was bothering them. A chorus of voices replied at once: "No, she's fine. She's not bothering us at all. I love having her here with us." He pressed the issue: "Are you sure?" Yes, they were sure. Jonathan made his way back through the crowd and returned to his pool table to resume his game. He would let Emma enjoy the company of her newfound comrades and check on her again in a little while. A short time later Emma came sauntering over. The people at the table must have finished eating, or perhaps the conversation had waned.

A man carrying a large box skirted their table and unlocked the vending machine next to Jonathan's chair. Emma knew that snacks were kept in there because she had seen people get them out before, so she ran over to the man thinking that she might entice him into sharing a tasty treat with her. The man knelt before the machine, unlocked the glass door and swung it out, then opened the cardboard box and pulled out cartons of cigarettes, which he piled on the floor in front of him. Emma was fascinated by all of this. She stood right next to him, tail wagging, her nose exploring the freshly exposed innards of the machine, sniffing the assorted packs of cigarettes arranged on the bottom racks. The man pulled out fresh packs from the cartons and stuffed them into the appropriate slots.

Emma snatched one of the empty cartons and pranced away with it to the delight of the vendor. Emma never took food off the table at home, but she often grabbed pieces of paper—sometimes very important pieces of paper—from the desk, the dining room table, or one of the end tables next to the sofa. Emma was an inveterate dissector, a quality she shared with her little brother Henry, Sarah's fifteen pound chihuahua-poodle mix. They would not rest until the squeaker had been extracted from the stuffed animal, the pages of the magazine torn out and shredded into thin strips, the stuffing pulled from the pillow, or the fibers of the rope tug-o-war toy permanently separated into individual strands. Emma sometimes tried to eat some of this material, but generally she ended up spitting it out on the floor.

Emma took the carton over to an open space between the pool tables and the bar and settled down for a nice, enjoyable dissection. She ripped small pieces of cardboard off and playfully flung them into the air, first to one side and then to the other. A circle of litter began to congregate around her on the carpet.

A woman seated at the bar had been watching this unfold with a smile on her face, but then decided that the mess was getting out of hand. She slipped gently off her barstool and approached Emma in a slow, deliberate motion, stepping softly so as not to make a sound, sneaking up on Emma as if she were stalking prey in the jungle. She began moving her limbs in exaggerated arcs as if she were walking through molasses or pantomiming the act of sneaking up on someone in front of an audience. This was going to be good. Jonathan put down his cue and suspended play for a few moments. Emma stopped ripping the carton and carefully studied the woman's motions. A twinkle appeared in Emma's eyes. If there was anything more fun than dissecting cardboard, it was a good game of keep away.

Emma stared directly into the woman's eyes, a broad grin on both of their faces. The woman was only one step away from obtaining the prize. Emma remained motionless, fascinated by the figure closing in on her, transfixed by all the mesmerizing sequences of strange t'ai chi moves. The woman stopped and stood at Emma's feet, then gradually bent at the knees in a comical fashion, slowly extending her hand toward the remaining portion of the carton as she lowered herself toward the floor. By all appearances, Emma was going to let her have it, but then in a blur of motion worthy of the best martial arts films, Emma snatched the carton away at the last possible second and sprang backward, clutching it in her mouth. The move was so swift and decisive that the woman made no last-ditch effort to grab it, but instead stared at the floor and the vanished carton as if she'd just witnessed a magic act. Emma was now several feet away dancing around with the carton, bucking like a little bronco, taunting the woman to come and try to get it.

The woman shrank back and sheepishly turned around, then crept over to her seat at the bar, the folly of her actions now quite apparent. The woman faced the other way and pretended to study the enormous selection of beers prominently displayed behind the glass doors of the coolers, oblivious to the triumphant look on Emma's face. Emma was disappointed that the game had ended so quickly and looked around the room for any other takers, but no one else seemed to be paying much attention.

Jonathan walked over to Emma and pulled out his bag of dog biscuits. Emma immediately dropped the carton on the floor and came toward him. He held out the treat and said "C'mon Emma, give me!" Emma turned around, fetched the carton and placed it in his hand. The woman at the bar was busy conversing with the bartender and unfortunately didn't see Emma's exemplary behavior. Jonathan deposited the carton into a nearby waste can and cleaned up the mess, meticulously collecting every shred of cardboard and leaving the carpet as pristine as it had been before, an artistic pattern of beer stains and interesting discolorations from a host of other sources, a few of which were rather puzzling.

Jonathan and Richard gradually tired of the frustration of repeatedly failing to make difficult shots and decided to call it quits. After one final conversation at the bar while the bartender calculated their bill, they left some money on the counter and parted ways. Central Avenue was hopping as Jonathan and Emma burst onto the scene, the broad thoroughfare now clogged by throngs of pedestrians. Together they wound their way between the tents of street vendors and the long lines of eager consumers, dodging waves of people flowing in various directions across the pavement. They approached the backside of a battery of bleachers, the metal structure spanning the entire street, and Jonathan took a seat at the end of the first row facing a dark and empty stage. He was soon approached by a man who asked if he could say hello to Emma. Jonathan told him that of course he could.

The man reached down and began ruffling Emma's fur and talking to her. A woman strolled past and saw them, then came over and stood behind the man. Several other people joined her and suddenly there was a line forming, all of them waiting for their chance to meet Emma. Emma received each one in turn, just as if she were the queen addressing her loyal subjects in the royal court. Jonathan listened patiently as each admirer proclaimed Emma's beauty, while Emma squirmed with delight at the touch of human hands. He couldn't help but smile at Emma's popularity. Finally the line dissipated and Jonathan rose, well rested, to resume his leisurely stroll.

He reached the edge of the festival zone and entered the city of shady convenience stores, ill-lit alleyways and closed shops. A homeless black man was sitting on the curb looking dejected. Apparently life had not been kind to him and he seemed lost, unable to join in the fun of the more well-to-do residents frantically searching for entertainment. Emma wanted to go over to him and started pulling on the leash, dragging Jonathan along with her. The man looked up, but said nothing, He seemed to welcome Emma's advance, although he didn't actually move or say anything, so Jonathan consented to the encounter.

Emma walked right up to the man and put a paw on each of his knees, then lifted herself up and slid both paws forward along the man's thighs, settling down with her front legs around his waist. She turned her head to one side as she did so and pressed the side of her face against the man's chest. The man was taken aback, not sure what to make of this strange behavior, then suddenly he understood what was going on. He lowered his head and rested it on top of Emma's head, then put his arms around her and returned the warm hug. The two sustained the embrace for a few moments, then Emma kissed him on the cheek and slowly backed away. The man was so moved by Emma's sincere affection that he seemed a bit flustered. Apparently no one had shown him such kindness in a long time and he had almost forgotten what it was like. A spark of happiness flickered across his face. He turned to Jonathan with a broad grin on his lips and a new light in his eyes and thanked him for allowing Emma to come over to him. Jonathan nodded. "You take care now" he told him as he and Emma receded into the darkness of the lonely, deserted street. Emma looked at Jonathan as they walked away, a smile of satisfaction on her face. Jonathan said to her: "How did you know? How could you tell that was the right thing to do?" Emma did not answer, but focused her attention on the long walk back to the truck, sniffing the ground, careful not to miss any of the fragrances. Her satisfaction grew with Jonathan's obvious approval. She was proud of herself and her eyes told him: "I knew I was right. I'm just glad that you understand it too."

Richard lived downtown, only a few blocks from where Jonathan had parked the truck. Jonathan and Emma would be spending the night in his driveway, launching their expedition into the wilderness with a bit of urban camping. Tomorrow they would awake to an angular landscape of condominium blocks and government offices, the panorama framed by the vinyl windows spaced regularly along the canvas shroud encircling camper, the bed comfortably elevated above the level of the street.

Richard's house was a member of one of the earliest residential neighborhoods in the city, built when Albuquerque was just a small town and the downtown that everyone knew today didn't exist. The first houses built were also the first to fall into disrepair, and when restoration became impractical or too expensive, the dwellings were torn down and replaced by parking lots and high-rise office buildings. A downtown emerged and overran the area, eradicating the once quiet and peaceful neighborhood. Richard was a holdout, only because no one had yet made him a tempting offer. The walls of his house were cracked and chunks of plaster had fallen from the ceilings. Richard suspended bedsheets under the disintegrating surfaces to catch the dust and dislodged fragments of mortar, similar to bandages placed over wounds. No one wanted the house anymore, but the land had considerable value. He had purchased the property 30 years ago as an investment, but so far the parlay hadn't panned out. Now he lived on a small island of suburban life surrounded by an inner-city reality.

Richard had a wide selection of vehicles and utility trailers parked around his house, but he had left a spot open for him, so Jonathan pulled the truck up to the weathered doors of the old garage taking advantage of the slope of the driveway to level the camper since the back end of the truck was higher off the ground due to the heavy-duty leaf springs in the rear. He raised the small, round access door on the side of the camper and pulled out the power cord, uncoiling the heavy cable one loop at a time. It gradually emerged from its secret quarters somewhere in the dark netherworld below the refrigerator. He plugged the end of the cord into an orange, heavy-duty extension cord protruding from under one of the garage doors. Emma watched him through the windshield, curious about this unexpected routine, an event incongruous with her past experiences. They'd visited Richard's house many times before, but they'd never camped in his driveway, although Richard had extended an invitation to Jonathan quite some time ago.

Jonathan swung the homemade wooden stairs down and away from the back door and set the bottom edge on the ground. He then circled the camper unbuckling the latches and entered the cabin to crank up the top. He lowered the shades on one side to block out the light from the streetlamp on the corner, a chore he never had to perform out in the wilds. He finally let Emma out of the truck and they walked around the block so he didn't have to rise out of bed in the middle of the night to take her outside to pee. Once they were both inside, their world was the same as always and they might as well have been anywhere.

The city finally went to sleep and the night became peaceful—but not silent. The city emitted a strange sound unlike anything Jonathan had ever heard before, a noise that came from no place in particular. The faint rumble was similar to the whir of heavy machinery in a textile factory, or the mingled voices in a large restaurant, or the murmur of an audience in an opera theater prior to the start of the performance. The sound was like the rush of the wind through the pine needles, or the hum of a compressor in an industrial air conditioner, or the roar of spectators in a distant stadium. The steady purr was punctuated occasionally by surprisingly clear shouts and loud voices, the whoosh of tires rolling on pavement, the growl of pistons straining under metal hoods, helicopter rotor blades punching the flaccid air, a string of muffled gunshots in an alley several blocks away, but the background noise was all around even though the streets were largely deserted and most people were cloistered in their houses, asleep in their beds. He couldn't really put his finger on it, the strange sound a city made at night. He listened to it for a long time, staring up at the vinyl sheeting peeling off the ceiling of the camper, trying to figure it all out, until the drone lulled him to sleep. He didn't know what happened after that.

—14—

Jonathan expanded his reclining chair and positioned the pivoting aluminum tubes on the irregular and crumbling sidewalk facing a no man's land of bustling traffic with numerous drivers looking for non-existent parking places. No one paid any attention to him as he relaxed and sipped coffee with Emma leashed to the armrest. She stately surveyed the scene before her, both of them equally enjoying the orange sunrise reflecting in the mirrored windows stacked high above their heads, even though they were incongruous and out of place here, neither of them having a lasting part in any of this. Jonathan was merely camping with his dog in his friend's driveway, settled in his chair as he would normally find himself on such a morning, but ordinarily in the middle of nowhere, lost in a desolate wilderness without a soul for many miles in any direction. Jonathan savored the absurdity of the current situation.

He drained his coffee mug and set it down on the broken concrete, then pulled out his laptop and used Richard's network to get online, hoping to catch up on the latest news and check on the weather, replaying a video of the morning weather forecast that had been broadcast earlier by a local tv station. A slim chance of rain was projected by the computer models but the forecaster, with a wry smirk on his face, pointed out that this was an idiosyncrasy in the program and the probability for rain today was way too small to be taken seriously.

Jonathan was glad to see that the weatherman was thinking on his own and not letting the computer do his job for him. A battle of the brains was taking place right before Jonathan's eyes, a matchup of wits pitting two different processes against each other: mechanical calculations vs intuitive insights. In what sense, Jonathan wondered, were the contestants both thinking about the weather. The question dated back over half a century. In 1950 a mathematician named Alan Turing made a strange claim. He claimed that there was a way to determine whether a computer was thinking or not. First he would direct a person to sit down in a room and type questions on the computer's keyboard, and then set up the machine to generate texts in response. The human interrogator would be given the opportunity to analyze the sentences and then use this information to decide whether these words were the responses of human being or a device. If the person wasn't certain which one it was, then the computer was thinking, meaning that it was intelligent.

If Jonathan had ever gotten the slightest inkling that his laptop was thinking he'd be a lot more suspicious of it. The device was aware of his presence only in the sense that it responded to his keystrokes, but in his daily life, every one of his actions was met by some sort of response from the world around him. A piece of wood responded to his hammer blows by splintering and his truck responded to his turning the steering wheel by rotating the front wheels. The computer wasn't aware of anything in terms of being sentient. In essence, Jonathan had his hand up the puppet's back, pushing its buttons and causing it to perform tasks. The idea that it could think was just a joke. The only intelligent person with him on the sidewalk this morning was Emma and she never said anything. She responded to quite a few words in English, but these were all words that Jonathan had taught her, so why did he think that she was intelligent? Scientists connected intelligence to language because the analysis of language was something they could understand, just as they connected physical phenomena to mathematics because mathematics was also something they could understand.

Nevertheless, Turing had invented a fun game. Jonathan wanted to play. He didn't have anyone to play with so he created a person out of his imagination. He said: "You be the interrogator and I'll be the computer programmer." First, Jonathan programed the computer to respond to the person's statements and questions as if the computer were some anonymous human being, someone this person had never met before. There were all kinds of people in the world and not all of them could speak correctly or even fashion a complete sentence, so he made the computer talk as if it were an inarticulate dunce, printing texts full of misspellings and improper word usages. Could he say that the computer was still intelligent, or was it now just a dim-witted fool? He then made the computer answer as if it were a crazy person, spinning incoherent tales and piecing together disjointed fragments of thoughts. Had the computer now lost its mind?

He then programed the computer to talk in the manner of someone this person knew by studying the peculiar speech patterns of that individual and copying them. The person told him "Yes, that sounds just like Tom." Jonathan could even get the computer to do this all by itself. He entered a record of Tom's conversations and had the computer play them back with variations on themes. The computer did this by defining various contexts and then inserting the response that fit the context. This involved nothing more than the logical analysis of grammar and the identification of certain characteristic associations of thoughts in Tom's mind. The computer gleaned patterns from a large pool of examples then repeated the patterns—it was the kind of thing that computers could do well, but it had nothing to do with being either intelligent or sentient. No one was talking about awareness or consciousness here, even though that was a prerequisite for people to think and act intelligently. The computer program made a formal duplication based solely on outward appearances—the essence of all mathematical relationships.

What if Jonathan wrote a program which caused the computer to generate text in such a way that it made the person think that the words were coming directly from God? The computer effectively created the illusion of God sitting in the next room and speaking directly to the person through a series of written messages. Let's say that Jonathan had done a good job and the machine convinced everyone who came along—or at least those who believed in the possibility of God—that they were actually witnessing God's words, just as if they were reading the bible. Had he therefore made the computer omniscient?

The imaginary player said he wasn't satisfied with just the sentences typed on a sheet of paper. He wanted to hear God speak. Jonathan wondered what God's voice would sound like. Did God have a smooth delivery, speaking as if he were trying to sell you something, perhaps convince you that salvation was a good deal and that you should take it? Or did God sound presidential, what you'd expect from the leader of the Christian world and the maker of the universe, perhaps with a touch of bravado and an air of smug confidence. Was God friendly, drawing you in as if he were your buddy and pal, or was he distant and authoritative, never letting you forget that he was vastly superior to you and this was merely a formal relationship? The phrases and sentences on the computer screen would take on different meanings depending on how they were spoken, so Jonathan understood why the player wanted to hear these words coming directly from God rather than basing his decision solely on vocabulary and grammar. Obviously the speaker had to not only say the right words but also sound convincing, constantly reassuring his listeners that he was genuinely the person they thought he was.

Jonathan continued his thought experiment. This time he had the computer write sentences that made it sound as if it were an alien from another galaxy. Since aliens came from technologically more advanced civilizations and had larger craniums than humans, had Jonathan raised the IQ of the computer and made it smarter than it was before? People were easily fooled, but in real life it wasn't so funny. Jonathan could spend a large part of his life living with someone and then discover that he didn't know that person at all.

308

He could wake up one morning and realize that the person next to him was a complete stranger, even after decades of conversations with that person. A person he had considered to be his best friend might turn out to be his worst enemy, simply because Jonathan had been duped by the person's specious words. What could Jonathan possibly deduce from the words of a conversation? Analyzing the grammar of speech was just a game and ultimately it led nowhere.

Jonathan could not deduce the inner thoughts of other people from their outward behaviors because these were two different things and the connections between them were uncertain—unknowable in fact. Sometimes people just said things and these utterances didn't reflect their true feelings. Yet the other person knew the connections quite well, because that person had access to his or her inner thoughts and therefore knew the truth.

If Jonathan could not deduce another person's private thoughts through the words of their conversations, then he could not deduce either intelligence or consciousness from the text of a computer printout. He still needed to know what was on the inside and mathematics could not help him here. Sure, the computer acted as if it were a human being, but sooner or later he must confront the fact that it was all mimicry and there was no person inside the metal box. Mathematics always stayed on the surface of things, relating the mere outlines of events to one another. Mathematics was limited to external behaviors and never got at the fullness of being or the underlying realities of material existence. But to the mathematical way of thinking, outward appearances were the only realities mathematicians, physicists and computer scientists could ever know, so if they could get the machines to act as if they were humans, down to the minutest details, their job was finished. Machines were intelligent, just like us. Jonathan was always amazed that mathematicians didn't see the fallacy in this reasoning, but then, he hadn't spent his entire life doing mathematics and it hadn't become a compulsory way of seeing the world, a knee jerk reaction to the analysis of every scenario. What he didn't understand was why mathematicians didn't come under more suspicion given that mathematical thinking led to such inanities.

The real problem was that the computer had no thoughts. Adding numbers together could never produce consciousness no matter how many computations were made and regardless of whether the numbers were expressed in base two or base ten. The result was still arithmetic and arithmetic could never generate anything other than sums, products, and quotients. Despite all the mathematics that had been written over the past few centuries, consciousness was as much a mystery as it had been before. A person might as well strive to mathematize love and solitude. The world was not about mathematics and mathematics was not about the world.

Some scientists were genuinely concerned about the problems associated with quantification, particularly in the case of something as complex and elusive as the brain, and struggled to create quantities that matched the cognitive and neurological phenomena as closely as possible—but they were wasting their time. Quantities could not possibly do the job. Consciousness was not the result of velocities and momentums, forces and energies, or the collision of molecules. The proper terms had not yet been invented and the language needed did not exist because no one knew what they were looking for and no one knew what they were talking about. The investigators must put their pencils down, move the icons for launching the modeling software to the trash bin, and unplug their computers, because the problem of consciousness could not be solved with algorithms and the logical flows of data. They must first answer the question: what is consciousness?

Well, they could begin by stating what consciousness was not: consciousness was not a theorem in Boolean Algebra.

The computer wasn't thinking at all. It was just juggling numbers. It was transferring numerical values coded in electrical potentials, sending electronic signals along copper wires and influencing the voltages in the transistors located in an accumulator and matching their values to one another, then switching other voltages in other locations called addresses based on the results of the comparisons. Physicists and most everyone else enjoyed talking about all of this in the glowing metaphorical terms of a living brain and comparing the processes to a thinking person because these analogies made the boring and humdrum events seem a lot more interesting than they really were. The computer was only operating according to the mathematics that scientists and programmers had put into it. A computer used mathematics to function because it had been built that way by a person who had understood mathematics in the first place, a person who then designed the circuitry in such a way that it functioned according to the theorems of Boolean algebra and the numerical ratios of Ohm's Law. The brain didn't utilize mathematics in this way. If it did, there would have had to have been an intellect responsible for putting the brain together. How else could the mathematics have gotten there? The mathematics just happened to appear by chance? Did the biological processes of the brain stumble blindly on the mathematical forms in the course of growing tissues? Did the brain tissues deliberately arrange their functions according to mathematical rules because they knew that the mathematical relationships were universally true and that these relationships could be recreated in physical processes?

When people thought about something they were not necessarily following any schematic or obeying any rules of logic. The associations people made in their minds could be just about anything, a continuation of the types of associations they had made in the past going all the way back to early childhood. The computer scientist needed only to outline the formal correspondences and specify the relationships between written symbols and not consider the biological realities behind them. Intelligence existed merely on an abstract plane, nothing more than an interplay of words and sentences, a self-referential routine disconnected from the actual existence of a sentient being.

In the Turing Test, duping people into believing something that wasn't true, namely that they were communicating with another person in another room, had been turned into a test for reality—but for what reality? The reality of a bunch of definitions and logical relationships. Physicists never had any intention of studying the reality of what existed, the reality of earth and sky, the reality of human consciousness. They played games because a game created a reality of its own. They could fabricate all the realities they wanted simply by inventing new games, then take these games to represent the reality of the world. All they had to do was manufacture mathematical objects and set up game boards. It was almost too easy. Unfortunately, the results were all counterfeit images of the true underlying realities of the world.

The computer was an impersonator that took on a persona without actually becoming that person. Much as an actor in a movie, the computer pretended to be someone it was not. In the cinema, everyone fell in love with the adorable character portrayed on the big screen and then confused this character with the human being who created the illusion for them. Jonathan could have cheated during the Turing test and opened the door to the computer room. No one was there. The lights blinked and the fans whirred exactly as before, the metal racks with countless printed circuit cards inserted into parallel slots stood impassively in neat rows, the bundles of color-coded wires all following the vertical and horizontal axes of the cabinetry. None of these things had changed in the slightest. It was the same computer with the same wiring diagrams performing the same menial tasks. All the different people he'd imagined, the dunce and the crazy fool, his friend Tom and the austere presence of God, the intelligent alien—had all been illusions. The computer had never been thinking about anything, but only going through the motions, blindly tracing the logical steps of algorithms. As Descartes so aptly put it, in order to think, the thinker had to be conscious. The thinker had to exist as a thinker. The astounding fallacy of artificial intelligence was that if programmers could make the illusion of consciousness seem real enough, then at some point consciousness would magically appear, as if the effects could create the causes, the mathematics create the underlying realities. Mathematics could never become these realities because the world was infinitely more than mathematics. At no point could mathematics even begin to capture the fullness of being.

At first glance computers seemed to be different from ordinary machines because they could do all sorts of things: play music, edit texts, customize photos, display graphics. But it was all done in the same way. A computer's structure was just as fixed as any other machine. The fear that these adding machines would multiply and take over the world once they became more intelligent than us was completely unfounded. We could always pull the plug on them. Or we could start feeding them erroneous data and the resulting sums would be incorrect. These options had been eliminated because we had deliberately made ourselves dependent on these devices, an act that we would surely one day regret.

No matter how many numbers the processors in a computer added together, these numbers could never become another computer—unless it was a virtual computer encoded in the algorithms of a host computer, a sub computer with a certain degree of functional independence. Otherwise it was not clear how computers would make babies and how these baby computers would grow up to be more powerful than their parents. Computers didn't evolve. Manufacturers just came out with new models every year so that last year's versions were obsolete and all the money people had spent on them was now wasted. The evolution of computers was nothing more than the endless cycles of consumerism, driven not by random mutations and natural selections, but by the insatiable greed of large corporations.

The dream of computer scientists was to manufacture a computer that was really a living organism, to find a way to invest the dead machine with the powers of life, namely, the ability to grow and replicate, the capacity for movement and free will, the intelligence and awareness of a conscious mind. Getting an adding machine to possess these traits was no mean trick. There wasn't a realistic plan at the moment, or even a clear idea of how this might be accomplished in the distant future, but computers could easily be made to seem as if they were alive, so let's just say that was good enough. We could make them talk and respond to our verbal commands, but on the inside it was all done by adding numbers together, lots of numbers processed at lightning fast rates. Of course, that was not the way that human beings did it, but it was the mathematical equivalent of these organic processes and to the mathematical mind, equivalence allowed for the substitution of one for the another. We could replace the machine with the real person because all of the external relationships had been satisfied and therefore the equation of life had been solved. The machine went through the motions and no one could tell the difference—well, mathematicians couldn't tell the difference because they were only thinking on one level, considering only the formal aspects. They were so well trained in this regard that the conclusions became conditioned reflexes which could not be broken. A equaled B and B equaled C, so therefore...

Jonathan had never been able to understand why mathematicians didn't come under ridicule since their thinking led to such fallacies. Shouldn't everyone at this point be saying: "Wait a minute! Something is very wrong here. Let's back up and consider another approach, something more responsible and a bit more substantial than mathematics." But instead a disconcerting number of people believed it to be an indisputable truth that computers were more intelligent than us and that they would one day take over the world and run us out, simply because they could be programmed to play games like chess and Jeopardy. We would not be able to compete with them in the one game that mattered, the game of life, and in terms of Darwinian evolution, that meant the extinction of the human race. But what resources would computers and humans be competing for? It certainly wasn't food and water. Was it that computers would take all the electricity for themselves and we would no longer be able to plug in our toasters and coffee makers? Without our appliances we would surely perish. What would computers do after we were gone? Learn how to drive trucks so they could repair and maintain the transmission towers and hydroelectric generators themselves? And since machines were so wonderful and perfect, of course they would never crash into one another and wreck their vehicles the way that human drivers so often did.

Any injury that tore away part of the skull and exposed the brain caused most people to recoil. Living tissue was fascinating but strange—and sometimes horrific. The brain was wet and mushy, countless sacks of protoplasm piled together and touching one another, all bathed in more fluids. Jonathan had never touched a living brain, but he imagined that the surface was silky and very delicate. The underlying electrical activity was all chemical in nature, electrical potentials due to varying concentrations of ions suspended in organic solvents. The number of chemical reactions occurring simultaneously within the cells was staggering, all directed by a class of proteins called enzymes. The brain functioned through metabolism, generating energy by turning oxygen into sugar. Tiny rivers of blood flowed around the islands of protoplasm bringing fresh supplies of hemoglobin, the molecule that carried the oxygen.

A computer shared none of these aspects. The physical processes in the brain bore absolutely no resemblance to the processes occurring in the computer. If the brain didn't function in the same way that the computer did, how then were their behaviors related? They weren't. A mathematical object and a non-mathematical object were fundamentally different from one another and any comparison between them was not just superficial, but illusory. The whole thing was a facade. The disparity went well beyond computers and higher organisms, because physics itself was also just a facade. The natural processes in the world bore no resemblance to the logical steps in mathematical proofs or the strange structures that resulted from them. Clearly the world didn't function in the same way that numbers did and the two behaviors weren't related at all.

A computer was so utterly different from a living brain in every detail of appearance, structure, and function that Jonathan couldn't understand how anyone would ever think of comparing them. As concrete objects, they had nothing in common. Therefore he was astounded when as a student in college he discovered that someone had written an entire book doing nothing but comparing a brain to a computer.

Obviously the book had been written by a mathematician, a man who was otherwise well-respected in his field, a man named Johnny von Neumann. In the fashion of all mathematical thinking, the content had been deleted and the author wasn't talking about real objects. The approach raised a crucial question: what could someone deduce from pure forms? The answer was not much—nothing really. Sure, a softball had the same geometry as a grapefruit, but in reality they had nothing in common. As concrete objects they were not related in any way, but mathematically they shared the property of being spheres of roughly the same diameter. Mathematics didn't penetrate into the core of existence but always remained on the surface, drawing superficial comparisons and remaining satisfied with mere appearances.

Jonathan asked himself: "So what if a softball had the same size and shape as a grapefruit?" The similarity was inconsequential. What was important was that he could eat the grapefruit, plant the seeds and grow more grapefruits, compost the rind and fertilize his garden. What was important was that he could hit the softball over the fence and win the game for his team, or throw a baserunner out at the plate. These were the realities of each object and they did not overlap. He could not deduce structure and design, or function and purpose, from pure forms. He didn't understand how anyone could throw away the realities of things that existed in the world and then claim that there was something leftover, and if that weren't outrageous enough, then make the emptiness the deeper reality. An understanding of these objects could only be achieved by clinging to the fullness of their being and never letting go of it.

Unlike a grapefruit and a softball, a computer and a brain didn't have similar shapes. The imagined similarity between them was hard to define. They both shared something called an "electrical impulse." The impulse didn't exist all by itself but rather it was a general pattern in the courses of two events that otherwise shared nothing in common: one event involved the concentrations of sodium and potassium ions in solution on either side of a lipid bilayer, and the other the free electrons floating in the netherworld between the atoms of a metallic crystal. Although the physical processes were complicated and the impulse was hidden somewhere within the intricacies of these dynamic structures, the comparison became clear when a mathematician plotted variables on a graph. Now for the first time the corresponding shapes became visible. The computer scientist drew lines on a blackboard, placed them alongside one another, and declared: "See! The computer is really like the brain!" The statement was similar to saying that a grapefruit really was like a softball based on a comparison of their bare forms. Where was the brain and the computer in any of this? Suddenly scientists were talking about pictures of mathematical entities and then trying to compare the realities of the objects themselves through these artistic inventions, which was clearly impossible.

If two objects had the same shape, what did that tell Jonathan? Not much—nothing really. The electrical impulse was a figure drawn on a piece of paper. Neither the line nor the graph which supported it existed in nature, but both were constructed by putting mathematical ideas together. In a similar way, the Monopoly board did not exist within the dice. The dice played a key but minor role in the complexities of the game, that is, of putting houses and hotels on specific streets, collecting rents from tenants, drawing opportunity cards, and getting thrown in jail. All of these events were constructions based on imaginative ideas. Since the dice determined the outcome of the individual plays, every action on the board could be traced back to the numbers that had been thrown on the dice. Scientific reasoning seemed to say that since the plays on the board were derived from the rolls of the dice, these plays were direct consequences and necessary conclusions and therefore they were just as real as the dice. The fact that the game didn't occur naturally made these plays an artificial reality that did not copy or capture the reality of the dice, but rather added to it. Human minds created the Monopoly game by making up rules and definitions—exactly the way that mathematics had been created—and that gave the game a life of its own. The dice could not make the plays on the board exist—

only people could do that. In the same way, the chemical reactions in the brain could not make the shapes of the impulses exist. These entities were created independently of the brain and they formed a separate reality residing outside the brain, something that existed in addition to the brain and went way beyond the realities of the brain's existence. The brain played only a token role in a mathematical system of ideas which could not be derived from the physical existence of the brain.

Sometimes a rock looked like a triangle, a dust devil looked like a human being, a rock wall looked like stonework, the clouds looked like a mountain range—or perhaps a human face—one graph looked like another graph, or a computer talked as if it were a person, but what could Jonathan conclude from mere appearances, from nothing more than the outlines of empty forms? He saw that the forms were not only insignificant, but misleading as well. They ushered him into numerous errors and deceptions, enticed him into drawing unsubstantial comparisons, and caused him to construct artificial ties between things that had nothing in common. Johnny von Neumann might as well have written a book comparing grapefruits to softballs. Of all the ills that plagued the human race and all the demons that preyed upon people's minds, there was no greater curse than abstract thinking.

Jonathan closed his laptop and looked up. The traffic on the street had subsided and the thoroughfare was now deserted. Everyone who had business downtown was safely ensconced in either an office or a cubicle, consumed by what Jonathan considered to be meaningless tasks. It was time for him and Emma to be on their way. He grabbed his coffee mug, turned off his computer, and folded his chair. Emma jumped up, glad to be finally doing something. He walked over to Richard's porch and expressed his gratitude for Richard's hospitality through the locked security door, unable to see Richard but hearing his voice emanating from the recliner chair positioned in the back corner of the living room. Richard apologized that his knee was too sore for him to get up. Jonathan broke camp and drove to the grocery store to stock up on supplies. The store had an elaborate buffet of cold salads and hot dishes so he included several of these selections in his shopping cart. Instead of trying to find a place to sit in the crowded eating area, he loaded his groceries into the camper and headed for a nearby park where he sat with Emma in the cool shade of a sprawling tree and leisurely enjoyed his lavish fare. He strolled twice around the park allowing Emma to thoroughly explore the well-worn pathway. Soon he and Emma were merging onto the interstate and heading out of town as a fierce downpour ravaged the city causing gutters to overflow and puddles to swell into small lakes.

Jonathan eventually exited the interstate and struck out across a flat and empty landscape, continuing onward until he reached a secondary turnoff. Emma had been pressing her head against his shoulder for quite some time now, telling him in her polite way that she needed to go to the bathroom. He also needed to stretch his legs. He hadn't seen any signs of civilization—other than the road itself—for the last ten miles, so he pulled the truck over onto some bare ground and stopped. Once outside the truck, the fresh air and rolling hills invited him to mingle with the rocks and grasses, so he grabbed his pack and started walking. Emma sprang into action, not expecting that they'd be heading out from what was obviously meant to be only a brief stop.

Jonathan couldn't help but notice that this particular locale had lots of rabbits. Rabbit populations varied greatly from place to place, but here the rabbits had lived up to their reputation. They had multiplied and the area was now overrun with them, their numbers more or less equally divided between black-tailed jackrabbits and desert cottontails. Often a rabbit ran away as Jonathan approached, bolting as he walked past its hiding place, his circuitous steps sending it scurrying into remote bushes. Strangely Emma paid no attention to any of them. She went about her business exploring the desert, sticking her nose into every dark corner and subterranean hollow that she came upon, acting as if she didn't see all the fancy footwork and neat acrobatics. On numerous occasions rabbits darted and disappeared over the low rises and merged into arroyos full of vegetation where they quickly became lost from view, but Emma didn't even turn her head to look at them. The rabbits were all getting away, sometimes fanning out in groups and crossing paths. Jonathan couldn't believe that she didn't notice them dashing between the clumps of grass and hopping madly over the stones and rocks. Something must be wrong with her. For some reason, she was out of touch with reality today. He'd heard that labs often developed eye defects. Perhaps she had started suffering from impaired vision, or maybe she had been stricken by some sort of neurological disorder. Hunting rabbits was her great passion in life and here she was completely surrounded by them yet she was missing countless opportunities to pursue them, oblivious to their presence. Despite Jonathan's often rosy estimations of Emma's mental prowess, perhaps she wasn't that bright after all.

After puzzling over this anomaly for some time, the answer finally dawned on him. Emma was anything but obtuse. Even though she never cast the slightest glance at any of the rushing rabbits or turned her attention away from the scents on the ground, Emma knew the rabbits were there, frantically running away from her in all directions. Emma was actually very clever. When a rabbit had gotten a head start on her, she had figured out that she couldn't possibly catch it, so she ignored it. She was smart enough to know that if she wasted her time with these impossible chases, she would miss out on the one rabbit that she could get, the one that she would take by surprise, the one that was cowering right in front of her nose. Forget about the others and let them go. This was the one she wanted, the one she was looking for, the one she would catch. She had the mental discipline not to let herself be distracted by the flurry of activity all around her, not to take her eyes off of the potential victim that was within reach, the rabbit that was still tucked under the branches, the one that had committed the fatal mistake of waiting too long and not taking the final opportunity to escape. She was ready to pounce on it in an instant, focused on the lightning quick reaction that would be required of her in order to spring one way or the other depending on which way the rabbit went. In the short span of four years she had acquired the skills and instincts of a true hunter. Emma was not some goofy dog who ran around blindly chasing everything that moved without ever thinking about it. She refrained from squandering her resources, deliberately saving herself for the one effort that mattered. Building on her considerable experience out in the field, she had developed a strategy that worked. She had caught twelve rabbits so far, and she was looking to make it thirteen.

Despite Emma's best efforts, the rabbits all got away. The afternoon matured and the rays of the sun began to weaken. Jonathan knew that he could find better camping further down the road so he decided to get going. Emma came running when she saw that he was leaving. She hopped back into the truck and Jonathan drove onward. He finally settled on a familiar place, a spot where he had camped before. He backed the truck onto a level patch of ground, unbuckled the hold down latches and cranked up the top, then unpacked the contents of the cabinets, raising the stove guard to return the teapot to its rightful place on the front burner. He lit the refrigerator by repeatedly pushing the ignitor button until he heard the wisp of gas burning in the background, then put the cooler alongside the rear wheel of the truck to sit quietly in the shade while the refrigerator cooled down. He dragged the reclining chair out and unfolded it in a good position, facing the black mesas.

Emma had been running around exploring the new territory the whole time, expecting that they'd take off on a good walk any moment now. Just as Jonathan was finishing setting up camp, a vehicle approached. Jonathan was parked on a seldom used two-track that looped around and rejoined a somewhat more substantial road a mile or two away. There was no reason for anyone to be driving on this road—unless they were coming to harass him.

Sure enough, the vehicle stopped right by his camp and he could see that the sole occupant was wearing a tribal police uniform. Jonathan wondered what he wanted. He walked over to the open window and offered a pleasant greeting. The guy looked around at all of Jonathan's stuff and was obviously trying to figure out what was going on here, searching in his mind for an explanation as to what Jonathan was doing.

He finally came to the inescapable conclusion: "You're….camping out here?"

"Yeah, I'm just out with my dog." Jonathan looked down. Emma was walking alongside the truck sniffing the tires. She seemed so vulnerable, so small and frail next to the gigantic tires of the massive three-quarter ton truck. Emma was lucky that she had Jonathan to watch over her and protect her.

"Well, you know you're trespassing on pueblo land."

That wasn't exactly true. He was actually on BLM land jointly administrated by the pueblo. There was a sign at the entrance stating as much, next to a metal plate affixed to the fence which read: "Excellence in Grazing." He'd talked to the local BLM office in Albuquerque when he first started coming out here several years ago. They'd received several complaints about the pueblo intimidating people in this area. Apparently the tribal police had pulled guns on some campers when they refused to leave. Jonathan was standing next to the driver's side door and he could see that this guy was wearing a firearm on his belt. He felt the weight of the Star 9mm in his back pocket.

"You can't camp here without a permit," the driver announced.

"So what do I have to do? Drive all the way to the governor's office?" The pueblo was another 30 miles down the interstate, then 10 miles down a two-lane highway.

The guy informed him that this was for his own protection. If Jonathan registered with the pueblo and paid a fee, then they would check on him and make sure that he was alright. This was a desolate place, and a long way from civilization.

"You need to be careful out here. Last year, someone got stuck in the mud right over there by that abandoned stock tank. Driving around on these roads can be very dangerous."

Jonathan looked over at the tank and thought about that day. It had been the last time he'd camped here. He figured that he'd better not tell this guy that the someone he was referring to was him. He didn't want the ranger to find out just how often he'd been coming here since apparently each time he had been breaking some tribal law.

Last summer torrential rains had washed out many of the roads in the area. The few people who lived in the vicinity of the interstate were suddenly surrounded by lakes of stagnant water and ended up being trapped in their homes for weeks. A state of emergency was declared. Jonathan, of course, loved it. The backroads were severed in numerous places, the substrata having been swept away by the raging waters, leaving sheer drop-offs. The deep fissures often came up on him unexpectedly if he wasn't paying close attention. The uniformity of the sand camouflaged the missing sections of roadway and he couldn't see the narrow gaps until he was right on top of them. The openness of the country caused his eyes to focus on the road farther up ahead, watching it disappear off into the distance, and he had to look down to see the immanent danger.

Backcountry travel was still possible because in such open country detours were not hard to find. By scouting on foot, Jonathan had been able to discover alternative routes, places nearby where he could maneuver the truck across the riverbeds without getting hung up on sandbars or steep inclines. He could then rejoin the road on the other side of the impasse. Jonathan ended up driving down roads that no one had driven since the disaster. He and Emma had the whole place to themselves, cut off from the rest of the world, alone with nature. Jonathan hiked for many miles, confident that they would not be disturbed in their revelry and that no one would invade their private sanctuary.

In the course of their adventures, Jonathan had turned onto this dirt track and started crossing the meadow with the tall grasses concealing the quagmire beneath it. The truck sank into the muck and became hopelessly stuck before he comprehended the gravity of the situation. He opened the door and found that he did not have to step down because the ground was level with the floor of the cab. The wheels were half submerged in the yielding sediment, and only the axles prevented further descent into the black, fragrant sludge.

He got out of the sunken truck and stood on a dense clump of grass, testing his footing. He gave Emma the ok to disembark. She sprang from the seat and bolted across the meadow to investigate the surrounding area, while he began assessing his plight. Proceeding further across the mud-hole was clearly impossible and that left him only one option, backing out using the ruts he had already plowed. He took a deep breath and smelled the fresh air splashing against his face. It was a beautiful day and there was no rush to go anywhere. He casually walked over to the tank and scavenged some boards that had fallen off the dilapidated corral. He pulled out his hi-boy jack from behind the seat and using a short, wide board as a platform for the jack, he lifted each wheel out of the mud, then wedged another board underneath the tire. He abutted additional boards behind these, placing them into the ruts and fashioning a long pair of ramps. He put the truck in reverse and backed down the makeshift bridge to the security of solid ground. The project had been fun. He never felt threatened or feared for his safety. People who drove the backcountry knew how to handle predicaments like this.

Jonathan's protracted silence made the ranger think that Jonathan was plotting to deceive him by just pretending to leave and then returning once the ranger had gone. The ranger sought to quash his devious plans.

"If you try to camp here, the patrols on the adjacent pueblo will see your lights and they will notify us."

Jonathan looked up at the faraway lands on the other side of the interstate, many miles to the north. Through an opening in the hills he could see a few trucks on the highway, tiny flecks of white, their motion almost imperceptible to the eye. They looked like spider mites crawling along a barely visible filament of webbing. Long ago he had installed a dimmer switch on the cabin lights and he always kept them on the lowest possible setting. Combined with the heavy tint on the camper windows, these lights were barely visible even from right outside the camper. The ranger apparently imagined that Jonathan would be sitting here tonight with a Coleman lantern blazing away, but Jonathan didn't even own one, and he almost never made a campfire anymore. He usually lounged in his reclining chair staring up at the stars, watching for the occasional meteorite to steak across the heavens, until he got cold and went to bed. For Jonathan, an important part of camping was the darkness of the night. The ranger's cohorts across the way would never see him nestled in between all the other amorphous black shapes, thoroughly camouflaged by a heavy blanket of darkness.

Disputing the ranger's authority might lead to a confrontation and Jonathan certainly didn't want the guns to come out. The result would be the same regardless of whether he shot the ranger or the ranger shot him: his life with Emma would be over, and that was the one thing he didn't want to happen, so he resolved to let the whole issue go. There wasn't really any other camping in the area, so he'd drive back to Albuquerque and hang out for another day, then head somewhere else. But Jonathan knew in his heart that his time here had come to an end, and this wonderful place would never be the same for him again.

Being alone and cut off had made Jonathan feel safe, but now he knew that his camp was subject to invasions by these roving inspectors who might show up at any time of the day or night. He did not need these people to protect him from the mud holes and washouts—he needed the desert to protect him from the insanity of modern society. He needed the endless miles of sand and gravel to intervene between him and civilization, to establish a barrier to shield him from the rude incursions of his fellow human beings.

Jonathan saw that the driver of the truck was actually Death himself, the great and all-powerful Death, convincingly disguised as a lowly human being, an ordinary man bearing the terrible news of his great loss. This wasn't Jonathan's place after all. Jonathan understood that death wasn't simply the loss of his own life or the loss of a friend or loved one. Death was everywhere, an everyday occurrence, an essential part of this transient existence of his. Things constantly passed out of his life never to return again: places he had found, feelings he had experienced, times he had known. In the end, everything would all go away.

"Ok. Give me time to pack up."

The ranger didn't wait for him, probably in a hurry to get home. Without looking at Jonathan or acknowledging his words, the ranger put his vehicle into drive and continued on his way, staring impassively out the windshield at the indistinct dirt tracks ahead, satisfied that he had accomplished his task of evicting Jonathan from his home. The ranger's truck disappeared behind a low rise and Jonathan was left alone with Emma. He explained to her that they weren't going hiking after all. Emma was baffled by Jonathan's strange behavior, her large brown eyes beckoning him to resume his usual routine, but instead he turned away from her and started collecting his gear. He put it all back in the camper and turned off the fridge. The drive back to the city was long and dreary. Emma curled up on the seat next to him and put her head on his lap and went to sleep, dreaming of running across the desert and chasing rabbits, while Jonathan jockeyed with long lines of semitrailers on the interstate, wishing that he could have stayed the night in that desolate place.

—15—

Jonathan and Sarah have been married for sixteen years now. They lived together for sixteen years prior to their marriage. For a long time they traveled across the desert together. They always camped in a tent because they were too poor to afford anything else. Sarah was never fond of hiking, but Jonathan struck out with the dogs at every opportunity, climbing hills and wandering through the badlands, while Sarah tended camp or sat in a chair and read books. He was forced to navigate by sighting landmarks because the GPS satellites had not been launched yet. Being entirely dependent on himself to find his way back required him to turn around and periodically identify the route he had just taken, noting the location of the particular canyon he had followed to arrive at the point where he now stood, from this distance nothing more than a small, undistinguished cut in the complicated series of mounds and crevices that formed the panorama of the desert. His knowledge of the terrain would be tested when he sought to return to camp. In order to reach camp he would have to successfully identify this particular gap, and the next one and the next one, no mean task considering the wealth of gaps to choose from and the similarities of all the hills and valleys.

One time years ago, he and Sarah had set up camp north of Yuma, nothing special, just out in the desert well away from the highway. They had a few drinks after dark to take the edge off of a long drive. Being moderately inebriated, Jonathan wanted to go for a walk. He headed out from camp with the dogs in the dark of night and came upon a magnificent saguaro. He had an idea. He'd memorize the shape of this particular saguaro and use it as a signpost, a waypoint, to guide him back to camp. He walked another hundred yards or so and came across another saguaro. He memorized all the details of this one. He couldn't miss it. He kept doing this over and over again. Finally he started to sober up. Time to head back.

He quickly realized that he didn't recognize any of the saguaros that he came across. He wandered around for a while, but it was no use. There were lots of saguaros and none of them looked familiar. He was probably a mile and a half from the truck and hopelessly lost. The sky was cloudy and the night was too dark to navigate by any other means. He could be walking in any direction. He had really done it this time. At least the night was mild. There was no way to find the truck until daybreak, and he'd probably be out there until then.

Still, Jonathan walked in the direction where he thought the truck would be. Like most people, he tended to walk in circles when he had nothing else to guide him. From his past experiences, he knew that his circles went counterclockwise. If after a certain distance he stopped, rotated his body 90 degrees to the right, then took this new direction as his proper course, he'd reach his intended goal in due time. So he executed this maneuver after he'd walked about three quarters of a mile. The act was simple in itself, but unnatural and hard to actually perform. He was always convinced that he'd been walking in a straight line and that he was already walking in the right direction, and therefore the destination was up ahead—not ninety degrees to the right.

Suddenly, about a quarter mile directly in front of him, he saw the truck lights flash on and off. Sarah had gotten worried about him and figured that he'd gotten lost. The truck was right where he had expected it to be and he would have reached it soon enough, but he was glad for the reassurance.

Three years after they were married, Sarah was diagnosed with cancer. She'd had five major surgeries since then and could no longer accompany him on his adventures. Jonathan now split his time between being on the road with Emma and hanging out with Sarah and Henry at the house, along with Sarah's cats. Sarah had a coterie of friends and family living nearby and didn't require Jonathan's constant companionship. She told him that she was glad he was doing the things he wanted to do in life.

Each day Sarah played the piano for the animals, classical pieces that she performed with the same flowing tempo. The dogs and cats all swooned and closed their eyes, imagining themselves to be in some sort of heaven. Sarah's mother had longed for her

319

daughter to become a concert pianist and provided her with lessons at a very young age, then sent her off to a prestigious music school. In a similar way, Jonathan's father had hoped that Jonathan would someday become a professional golfer. His father loved golf and envisioned himself traveling around the country while his son competed in major tournaments, events that were typically held at the most prestigious golf courses.

But Jonathan and Sarah were alien children and they did not belong to their parents. They came from god-knows-where, the true decedents of people who had lived long ago and left no historical record. The personalities and traits of these distant forbears had been carried silently from generation to generation with no hint of their lurking presence, yet these nameless individuals were somehow still alive, hiding in the molecules of gametes and in the recesses and dark corners of living cells, waiting for the day when they would rise from their biological graves and be resurrected. Jonathan occasionally had visions of an unfamiliar world he believed once belonged to some forgotten soul in his ancestry, glimpses of the life this person had lived, vague feelings of what this person had experienced. The images were either pictures and emotions that his subconscious generated out of the mishmash of his childhood experiences, or else memories of events that had actually taken place and were preserved over many decades, or perhaps even centuries, in living bodies. People were free to choose between these conflicting interpretations, but Jonathan didn't believe that the idea of these visions coming from another person who was long dead was entirely groundless, because the images didn't appear to come from anything he had ever experienced as a child.

When Sarah played the piano, she did not make up the dissonance and consonance of musical notes in her head. People generally felt that a musician perceived an independent reality that existed in the world and then wrote it down, much in the way that the physicist wrote down the laws of physics. The relationships between chords and notes referred to the structures and activities of material bodies and Sarah could not arbitrarily decide these things herself. A musician could not say "I want harmony to be something other than what it was," although the creative spirit surely pined for such freedom. The musical notes were what they had to be and the relationships between them were determined by something other than whim, and we could call this "something" the physical reality of the world.

Sarah read sheet music with the same ease that a normal person read the printed pages of a book. She translated the meanings of the symbols into the geometrical relationships of the keyboard, then manipulated the ivories with her fingers, causing hammers to strike taut wires of precise length and tension. The rigidly defined mechanical actions had all been calculated to reproduce the original stream of carefully conceived sounds with its delightful harmonies. The precise configurations of the parts of the piano showed that the concept of harmony was really a mathematical one.

Music was based on numerical ratios, however, these ratios told Jonathan nothing about the natural world. They did not come from the nature, but were instead imposed on nature. Nature didn't play music because nature couldn't base its actions on mathematical ideas or use mathematical ideas to make things, but if each rock had played a tune when Jonathan picked it up that would be cool in a way. He was glad that they didn't because the effect would have ruined what he cherished most about the desert—the silence. He could hear himself think out in the desert because he didn't have the chatter and noise of modern society to distract him and continually capture his attention. The clouds didn't sing, the winds didn't hum catchy jingles, the sunrises didn't play complex melodies and syncopations, the hills did not burst into four-part harmonies. If Jonathan wanted stuff like that he put a CD into the player or turned on the radio. Music was a purely human invention because only the human mind could understand the ratios that were involved and then produce just the specific ones that resonated together. Only the musician could consciously select the harmonics that matched, the ones that vibrated in unison, and the ones that didn't, then play the dissonance and consonance against one another in a balanced way.

The birds were often said to sing enchanting melodies but the calls they made did not entice Jonathan to tap his foot or strum along with his guitar. He didn't have the same feeling he got when a musician hit the blue note in a nightclub, or a rock band hit the power chords on a big stage, or when a classical piece resolved to the tonic in a concert hall. Using the term 'music' to describe bird calls was only metaphorical. The birds were not really playing music. Likewise the crickets only made a rhythmic noise in the late summer, a ritual to chronicle the end of their brief existence. While the synchronized chirping of thousands of crickets was an amazing phenomena to behold, rising and falling with the hypnotic regularity of swells on the ocean, the sound was not a genre of music that could be placed alongside contemporary bluegrass or death metal, nor was it comparable to an operatic chorus. Nature made all kinds of sounds, but none of it qualified as music.

The natural world did not give us music. It did not perform and compose musical pieces. People had to make the music themselves. Just as with geometry, music didn't exist until someone created it. If the world had made music on its own, then Jonathan would have been forced to admit that the world was in fact mathematical, because the creation of a cohesive set of harmonies and rhythms required an understanding of ratios and these would have to be built into phenomena in the same way that they had been built into musical instruments. The ratios between diameters, lengths and tensions were what allowed musicians to play stringed instruments, and the ratios between pressures, lengths and velocities allowed them to play wind instruments. Music depended on the ratios between lengths and intervals, frequencies and vibrations, rhythms and tempos. Music only came into being through the employment of mathematical ideas and this was why nature couldn't make music. Someone might argue that the mere potential for music to exist showed that the world had the necessary mathematical components already built it, but the world could be many things, including silent, noisy, and discordant. The world was not structured by non-existent possibilities. If music required the manipulation of physical reality according to mathematical ideas, then music would be impossible only if mathematical thoughts were impossible to think or physical reality was impossible to manipulate.

Geometry and music were human inventions based on human imaginations. Nature did not possess the cognitive faculties to envision relationships between mathematical forms and then deliberately fabricate objects that embodied these relationships. The ability of some people to make music didn't mean that everyone lived in a musical universe, just as the ability of physicists to create the laws of physics didn't mean that everyone lived in a mathematical universe. Just as with scientific experiments, music required instruments that were built according to mathematical principles, devices that had predetermined lengths and ratios built into them. Music existed only after someone had constructed these instruments, just as physical laws existed only after someone had designed and built the apparatuses of a laboratory, and that was why there was no music and no physical laws in nature. Both depended on human ingenuity and intellect.

The continuously variable vibrations of the auditory spectrum formed a medium of artistic expression much as the potter's clay. The potter similarly considered proportion and symmetry in creating a harmony of shapes. A skilled craftsman in either medium could make anything out of their respective raw materials. A musician molded the notes into his preconceived patterns much in the way that a potter molded the clay into his visionary forms and the scientist assembled mirrors and coils of wires into a similar work of modern art that he dreamed up while sitting at a desk writing equations. The sky was the limit here. Sure, the scientist was bound to put the pieces of the instrumentation together in certain ways in order to get the results he wanted, just as the violin maker was bound to meet certain requirements in order to get the sound he desired out of his instrument, and although these were primarily mathematical restraints, they also included the limitations of the materials involved. The bottom line was that all of this resulted from ideas and not from nature. Nature was just a pawn in each of these schemes, something to be manipulated and coerced into performing selected chores, an actor given a script to read. The scientist created the perception of mathematical relationships between events in the same way that a musician created the perception of harmony between notes, and each told nature what to do in order to achieve that end.

Whenever Jonathan watched one of the great guitarists performing on stage, he always wanted to go home and start playing the instrument himself because it looked so easy, but as everyone knew, the simplicity was just an illusion. The master musician executed the most outrageous maneuvers on the fretboard while coordinating these digital acrobatics with the rapid motions of his picking fingers, and he did this so smoothly and effortlessly that the whole presentation seemed as if it were the most natural thing in the whole world, and of course, to the musician, it was. The mechanics of playing the guitar appeared to represent a natural harmony, but if Jonathan went back to the beginning and retraced the steps that had been necessary to establish this state of affairs, he saw that the components of good musicianship were acquired only through great strains and stresses, colossal efforts fraught with innumerable difficulties and failings, trying to establish patterns that were anything but natural and obvious. The accomplished musician could take any song title, any riff or motif, any chord progression or musical style, and turn it into a beautiful song, and in a similar manner, the physicist could take any situation and turn it into an harmonious array of mathematical concepts. Seeing the world in mathematical terms became entrenched in the physicist's mind and the well-rehearsed habits gradually developed into a way a life, but the task was not easy to learn because it required an amazing set of skills that could only be obtained after years of practice and dedication. The transformation of the world's events into mathematical concepts was anything but natural and straightforward, and the resulting schemes in no way embodied a natural order of things.

The day was still young but Jonathan had already walked a long distance through open country. He stopped for a moment and looked up at the line of rugged hills ahead, imposing stone formations that blocked any hope of further progress—except for one opportunity, a steep and narrow canyon climbing to a saddle wedged in between the vertical spires of rocks and towering piles of interlocking boulders. As he approached the pass from the valley floor his optimism faded because the abrupt channel was strewn with large boulders and clogged with vicious-looking cactuses and sharp-pointed agaves, but a shoulder of land elevated somewhat from the bottom of the arroyo showed signs of an indistinct and slender game trail—not well used, but passable.

Emma chose to follow the proper course of the arroyo, stepping from boulder to boulder and peeking into the dark recesses beneath each one, cautiously sniffing the bases of the yuccas and prickly pears that were jammed into the openings, suspicious of the clogs of sticks and dead grass mingling with the rocks. Strong smells emanated from every corner of this medley of plants and rubble and woody debris, and some of them were familiar to her, but others were not. Emma wisely refrained from poking her nose recklessly into those holes.

Jonathan found his footing along the game trail to be uncertain and at times treacherous so he focused all his attention on each carefully crafted step and did not spend too much time watching Emma explore the arroyo below. To his surprise, the arroyo did not lead to a saddle, but bent around and continued well beyond the opening afforded by the cliffs, making an abrupt change in direction that he could not have foreseen. He followed the course of the arroyo as it snaked between the hills, maintaining a path halfway between the backbone of the ridge and the river bottom below, both of them gradually leveling out and allowing him a quicker pace.

Once they were beyond the steep, rocky passage, the arroyo opened up. Emma raced down the sandy riverbed with long strides similar to those of a greyhound chasing the mechanical rabbit at a dog track, then disappeared into a crevice. Jonathan saw no further sign of her and wondered what she was up to. After a short distance he had an opportunity to stand on a rock ledge and survey the arroyo for a good stretch in either direction, but he didn't see Emma anywhere. He then turned around and looked up at the hills, admiring their interesting shapes, when he caught sight of her high atop the ridge looking down at him. She was obviously beckoning him to join her. He couldn't imagine how she had gotten up there so quickly and without him noticing her progress, but that kind of thing happened all the time. Jonathan frequently lost track of Emma and searched the terrain where he had last seen her only to find the land empty and barren as far as he could see. He usually lingered a few minutes, puzzled by her sudden disappearance, then broke down and whistled for her. She often came running toward him from the opposite direction, wondering what the hell was wrong with him.

Emma continued to watch Jonathan from her lofty perch, expecting him to follow her, while Jonathan contemplated the task. The ascent up to her position would be a grueling climb and at first he balked at the prospect of scaling such a height, but Emma was insistent so he figured that he could let her lead for once. He tautened his shoulder straps and began the trek up to the top. Emma was elated that Jonathan was going to follow her and she set off exploring the other side of the hill without him. He labored up the grade in short stretches of intense effort, working his legs with all the energy he could muster, then he stood in one place for 20 seconds or so and breathed heavily. He repeated the cycle, covering another 20 yards. This was the basis of interval training in a gym, but he didn't do it this way because it was healthy for him. He did it because this was the only way he knew to make it up the hill.

Emma came running over to greet Jonatan as he reached the crest. She had been right. The view from up here was fantastic. He stuck to the network of interconnecting hills and ridges maintaining his elevation as well as his superior perspective of the surrounding country until he reached a point where the ground fell away into a very steep canyon. His water supply was dwindling so this was a good place to turn around. He rested for a moment, taking in the magnificent sight, then summoned Emma with a brisk "C'mon Emma, let's go home!"

Only one path would take him back to the truck, the one he had followed on his way out, so he set a course for the arroyo that had led him here, roughly retracing his original steps. As he approached the opening in the initial rock wall, he surveyed the valley below. He paused and meditated on his journey today. His cross-country treks on foot were what people had done since the first appearance of humans on earth. Through the ages, native Americans and indigenous peoples around the world had lived in direct contact with nature. They studied the natural environment in great detail, lacking the diversions and distractions offered by modern technologies, the toys and games that were an integral part of our civilized lives, the endless hobbies and entertainments available to us to satisfy our curious and creative minds. Instead they focused on the world around them, examining every aspect of that world, noticing every nuance of the objects and events, every pattern of behavior. If mathematics had been a part of that world, they surely would have brought it to the forefront of their cultures long ago. If mathematics were a matter of observation, it would have been invented at the dawn of human existence. But that was not where mathematics came from.

Mathematics began with a turning away from the world in a highly unnatural act, something that was utterly unthinkable to primitive peoples, therefore the idea of mathematics never occurred to them. The technique was probably not discovered by Plato, but it was certainly championed by him. The philosopher withdrew from the world of the senses in order to contemplate a higher reality provided by his intellect. The philosopher didn't actually have to close his eyes—a blank stare would do—but his mind had to focus on his own thoughts and labor at putting these thoughts together, assembling them into systems of ideas. Mathematics was not based on the world that everybody lived in. It was not the world of the tribesman immersed in the vast wilderness. Rather it was the way that certain ideas fit together. To understand these relationships, the philosopher had to put aside the distractions of the material world, preferably locking himself in his study for long periods of time. The philosopher did not hide in the tall grass at the edge of a wetland hoping for a rare and beautiful mathematical idea to alight on the surface of the water, or skulk behind the rocks hoping for a glimpse of a herd of abstractions grazing on the plains. If a person wanted mathematical ideas, then that person didn't look at the world. Plato understood this and developed his "flight in order to approach" method for understanding the world. He would lose himself in thought and mingle with the Pure Forms, then return to the world triumphant. After his protracted journey into this ethereal realm, he now felt that he understood everything perfectly well. Well, he lived in a simpler time.

Plato failed as a philosopher because he never questioned the origin of these Pure Forms. He assumed that his thoughts were somehow the foundations of everything that existed, but he needed to prove that. In Jonathan's opinion, the best critique of Plato's philosophy was: "Don't believe everything you think." If ideas didn't come from the world, shouldn't that make everyone suspicious of them, particularly when they tried to make sense of the world around them? However, Plato was a sucker for the notion of harmony. He believed that the truth coincided with the good and ideas were the basis for material realities, but his age of optimism and belief later gave way to periods of doubt and skepticism. Perhaps nature hadn't given us a consciousness capable of formulating mathematical ideas so that we could understand the universe. Perhaps when nature provided us with the mental faculties necessary to create mathematics, it was just playing another cruel joke on us. The root of the human condition was that no one ever explained to us what this world was all about or why we were put on this earth. The scientific notion that we were given the tools to find this out for ourselves was equally naive. Perhaps we were given the ability to grasp mathematical concepts not so that we could understand the universe, but so that we could construct our own universe and set ourselves apart from the whole of creation. Mathematics seemed to function in this capacity quite well. The thunderheads and sunrises were still mysteries and Jonathan observed them with a sense of awe, but he had his silver palace, an assemblage of parts with precise relationships, and he was able to drive around on round tires over flat surfaces. He was grateful to mathematics for what it had done for him, even though that didn't include giving him an understanding of the world he lived in.

In physics, an electron was replaced by mathematical forms where it could be viewed as the product of intricate chains of logical deductions, but in fact the infinite particularity had been replaced by a set of mathematical constructions that were gratuitously added to the realities of the world. The physicists' electron was made up of concepts that were by no means necessary assumptions, a set of abstract parameters that had been turned into the properties of a now fictitious entity. The problem was that the quantities didn't reflect the labyrinthine details of whatever was out there, the infinity of particulars. Quantities were amalgams of ideas and techniques of measurement, characteristics of the instrumentation, the substitution of a reality that could not be grasped with a mathematical analogue that could easily be understood in terms of precise definitions and logical categories. The physicist then played with markers and formal relationships trying to get the numbers to match, but this wasn't reality anymore. It was just a silly game with little or no relevance to the actual events of the submicroscopic world.

Physicists claimed that quantum physics had brought about about the fall of determinism, yet they could have seen the failure of this doctrine simply by looking out the window. If only they had taken the real world seriously and stopped fussing over their experiments, they could have grasped the truth long ago. The macroscopic world refuted determinism so thoroughly and decisively that no one in their right mind would ever cling to such a ridiculous doctrine. Determinism referred to the rigorous forms of mathematical objects, the neat interplay of logical ideas, and the beautiful structures of mathematical theories, but none of this had anything to do with the existence of the world. Quantum mechanics told us nothing we didn't already know. The world was clearly not made up of concepts and we should have understood that long before the invention of this most recent version of mechanics.

So inevitably the day came when the electron betrayed the physicist's trust. He thought that he had known what he was talking about, that he knew his familiar friend the electron by its outward behavior, by its mathematical properties, but the flickering chimera on the horizon of his imagination was just a mirage, an illusion. The electron was not a mathematical entity as he had hoped. He had created something in his mind that didn't exist in the world, a spinning particle with mass and momentum, a bundle of wave forms superimposed over one another, a distribution of probabilities, and his dreams were shattered by the cruel intrusion of an inscrutable existence, an unfathomable material reality far beyond the scope of symbols scribbled on a blackboard. The quantum world wasn't weird at all—it was physics that was weird. It was the method that didn't make sense. The physicist scratched his head and said that he couldn't figure out the results he was getting, the behaviors he was observing. Well, he could have had the same experience if only he had stopped long enough and taken a good, honest look at a thunderhead, a sunrise, or a fallen branch.

When a particle plunged into the fray, when it flew off into the void, when it disappeared into the anonymity of the faceless masses, no one could know what happened next, no one could sort out the ensuing shambles. When a gazillion other particles did the same thing at the same time, no one could untangle the resulting mess, no idea could encompass the bedlam. So the physicist said "Let's not talk about this electron or that electron. Let's talk about no electron in particular, but instead let's talk about a hypothetical electron. We can start with the idea of an electron, an electron in general, just as we had talked about a thunderstorm in general. We can construct an electron out of ideas and say that the ideas came from real electrons. Well, to be fair, the ideas came from mathematics, but if electrons were really mathematical entities, then it was all one and the same thing anyway."

A perfectly linear stream of electrons striking the exact center of a tangential metal plate with two parallel slits cut into it was not something that Jonathan would ever encounter on one of his walks. Electrons normally traveled wherever they chose so the physicist was forced to take all of the electron's options away. The electron could go through either slit A or it could go through slit B, but there was no other way out. The physicist had trapped the electron inside the apparatus, just as the rats had been trapped by Jonathan's steel mesh and cement capped walls. But the clever little rats found ways around Jonathan's barriers, just as the clever little electrons found ways around the physicists' barriers.

When it could not be determined which slit the election had passed through, instead of turning to more mathematical concepts and attempting to explain the results through more abstract and convoluted reasonings, physicists needed to realize that it was the fallacy of mathematical thinking that had brought them to this point. Physicists were facing a reality that was vastly different from the trifling forms and superficial associations they had been imagining in their minds and sketching on paper. Particles and waves were not the realities of the universe, but merely convenient fictions. Physicists had assembled the electron out of the concepts of force and mass and energy, things that didn't exist in the world, mathematical forms that were in turn assembled out of the false concepts of time and space, abstractions continually leading them astray at every juncture, generating more and more misrepresentations and falsehoods. The truths of the real world had long been forgotten but they had not disappeared and suddenly they reared their heads to confound the physicists' neat little conceptual universe. Things were not what they had thought them to be because they had been making it all up along the way, piling mathematical concepts on top of one another until they had amassed an incredible heap of intellectually worthless junk. Mathematics had never captured the reality of what existed and by its very nature could never possibly do so, but this rather obvious fact had been ignored from the outset. The world was not a mathematical system and all the attempts to make it appear as if it were amounted to nothing.

The world was always more complicated than it seemed and concepts always fell short of reality. In previous experiments, electrons had behaved as if they were the empty forms of the physicists' dreams, the particles and waves they had envisioned to be the ultimate realities of the universe, but all along physicists were dealing with mere appearances. Electrons couldn't possibly be mathematical forms. The forms were only manifestations of some deeper underlying reality, a wealth of unobservable details that constituted the true universe, and it was the same type of reality that everyone saw all around them in their daily lives, a reality that could not be accounted for with blank forms and simple outlines.

Yet the general feeling in the scientific community was that the double-slit experiment could be explained successfully with the idea of probability waves. Congratulations. The physicists solved the puzzle and won the game and Jonathan saluted their ingenuity. Their skills as players were to be commended, but the question still remained: what was reality? Jonathan stopped and scanned the horizon, as he often did, searching for herds of antelope staring back at him with looks of curiosity on their faces, a lone coyote hurriedly running away and glancing backward to see if he was pursuing it, or a few cattle grazing lazily off in the distance and oblivious to his presence. These animals did not demand the existence of probability waves. Being alive in no way depended on the values of the numbers generated by these equations. The physicist's apparatuses were the only things that required probability waves. The physicist had constructed a world out of ideas and in this world he had given probability waves a place and a purpose, but only his ideas demanded their existence. The rest of the world was fine without them.

Quantum mechanics told Jonathan no more about the world than the seesaw in the playground. Quantum mechanics wasn't about the world, and neither was the seesaw. Quantum mechanics was about the mathematics of the apparatus, just as the law of the lever was about the mathematical configuration of the seesaw. Without the seesaw, the law of the lever was nothing. The law of the lever depended on the geometry of the seesaw just as quantum mechanics depended on the geometry of the laboratory. Quantum mechanics could never explain anything other than this narrow, little world of fabricated mathematical objects, but Jonathan didn't care about such things. Nothing that happened in the laboratory made any difference to him. As far as he was concerned, the results of experiments were absolutely meaningless. No one other than physicists cared about them.

When Jonathan studied photographs of the Large Hadron Collider, along with most people he was stupefied by the complexity of the device, the overwhelming number of mathematical objects that had all been tied together with precise arrangements that copied the relationships between ideas. Physicists believed that somewhere buried deep within this labyrinthine hardware a truth about the world could be found, a truth generated by the countless mathematical alignments all fitting together and forming a higher order of mathematics, the final result of all the mathematical forms that had been laboriously imposed upon each material object. How could such an enormous amount of mathematical shapes and configurations ever produce anything other than a mathematical truth? A substantial truth, on the other hand, could never be revealed by such a device. Rather the machine created its own truth, a truth that did not exist in the world or have a meaning apart from the mathematics residing in the minds of its makers. The truth of the machine was a direct consequence of the arbitrary definitions that underpinned all the sophisticated designs of harmonious components, and thus it did not have a life of its own. Out in the world the end result of all this industry was a worthless tidbit of information that in no way qualified as a genuine knowledge of anything that existed in the universe. Physicists had simply raised the art of gaming to astronomical heights, but the plays on the board still had no meaning outside of the rules of the game.

Probability waves could only explain the phenomenon they had been created to explain, a laboratory phenomenon that was found nowhere else in the universe—except in the unlikely event that an alien race on a planet half way across the galaxy had exactly the same ideas and had created an identical setup, matching the one erected here by terrestrial scientists in every detail. That still left a lot of phenomena out of the picture—essentially the whole universe. As with Newton's Law, physicists imagined that probability waves were everywhere, working behind the scenes, giving everyone the world they knew. No one needed to think about them. Everyone could rest assured that these entities were out there, completely out of sight and out of mind, making all of our lives possible. None of this could ever have existed without them.

While the physicist found the desert peaceful and relaxing, nothing here engaged his intellect. Jonathan found that strange, because to him the existence of the desert was fascinating, full of mysterious processes and exotic communities. He longed to take his place among the circle of stones, the brotherhood of rocks, the league of leafy plants, the alliance of stars and planets, the fraternity of owls. The physicist only wanted to rush back to his laboratory to feed his addiction of playing mathematical games with intricate devices —a compulsion worse than smoking or gambling. The laboratory would tell him nothing about the desert, but the desert was not his concern. He wanted no part in any of this— and for good reason. Mathematics was hard if not impossible to come by in the desert. The machines he had built on the other hand, gave him all the mathematics he could ever want. He'd rather play with them.

The physicist had only to deal with his apparatuses, but how tiny and insignificant they were when stood alongside the grandeur and spectacle of the whole of creation. Experiments were trivial in comparison, the products of petty minds full of insipid imaginations and dry theoretical speculations. The physicist didn't actually have to construct the entire universe out of these elements, only believe in the possibility of doing so. Once again, the program was nothing more than a visualization. He'd have to leave the friendly confines of his laboratory in order to carry it out and he was not going to do that for obvious reasons, so he'd complete the remainder of his project with a few words and a little hand waving, gesturing toward the great outdoors with a flick of the wrist as if he'd subjugated the whole of creation with a couple of slick moves on a game board. Perhaps he needed to take a walk, breathe in some fresh air, and confront his adversary full in the face. This simple exercise might put everything back into perspective for him— assuming of course that he opened his eyes and had more than a blank stare on his face.

Jonathan wondered where the idea that mathematics was a depiction of reality ever came from? When Jonathan took a good honest look at what existed, mathematics was the most puzzling invention ever created by human beings and the one thing that most needed justification simply because it did not receive any support from the world. The physicist had to explain to everyone what made him think that he could use mathematics to formulate an account of the world when the program was anything but obvious. The world seemed so thoroughly non-mathematical that any attempt to turn it into a mathematical system would invariably fail and the very prospect itself was utterly insane. The cleverness of fabricating a laboratory built entirely out of mathematical objects and then taking this artificiality to be the true reality of the world could not be overstated.

Jonathan felt that mathematics should not only be questioned, but dismissed on the grounds that it was not a genuine part of the world. He understood that mathematics could not be taken for granted the way physicists always assumed it could, that the interplay of ideas did not have the relevance and value they ascribed to it, but if mathematics lost its legitimacy, then physics was left completely unfounded. The truth was that physicists added mathematics to something that didn't require it. Physicists imposed their will in the laboratory, but out in the real world they could only influence the course of events in general ways. They could not make a particular storm appear in the fullness of its being, exactly as nature had originally created it, but they could seed the clouds and make the storm yield more precipitation, or perhaps ionize the atmosphere and make the electrical activity more intense. Physicists dealt with the storm as a system of ideas because physicists could not deal with the storm at the level of particularity that defined its existence. The real storm existed only in terms of its particulars because every detail was produced through other details. No ideas were involved, so ultimately the storm could only be understood as an infinite particularity. This was what made the storm what it was. As soon as the physicist started treating the storm as a system of ideas, he'd lost the very essence of the storm.

The physicist knew that the real storm was out here somewhere marching across the desert instilling fear and wreaking havoc, but he was not going to talk about any of that in his scholarly lectures. He was not trapped in the storm as Jonathan was. When he expounded his theories in the luxury and comfort of a classroom, the storm was conveniently very far away, granting it an air of unreality. What did the physicist actually know about the storm? If only the physicist would admit his ignorance Jonathan would have been satisfied, but the physicist insisted that he knew the storm better than Jonathan did.

The physicist swore that the storm could not have possibly attacked Jonathan up on the ridge, because those kinds of things just didn't happen. The physicist was reasoning from ideology here. He didn't need to look at what happened. If Jonathan had said to him: "Well, it seemed to me that the storm attacked us," the physicist would have simply told Jonathan that he was wrong. The physicist would give Jonathan his interpretation of events even though he wasn't even there and had no idea of what Jonathan was talking about. He could refute Jonathan's observations on principle. This exposed physics for what it was. Physics wasn't based on observations. Physics was based on premises. Physicists imagined mathematics in their minds, then built mathematical objects that copied these ideas, then observed the mathematical behaviors of these objects, but how could that possibly be observing the world? Jonathan was observing the world when he saw the thunderclouds attacking him. He was observing the world when he saw the winds stealing his hat. He was observing the world when he saw the dust devils dancing across the desert. He was observing the world when he watched the sunrises unfolding in the sky. Physicists weren't observing the world at all—in fact, they no longer even knew that it still existed. They were perpetually glued to their "televisions sets"—the computer screens, dial faces and digital readouts of their instrumentation—binge watching the latest series of data displays, thoroughly entertained by the multiple channels showing the clever and sophisticated performances of computer models, sitting in chairs transfixed by the hypnotic glows of the devices, eagerly anticipating the next episode. This was what they called "observation."

Even as a young child, when Jonathan watched Gilligan, the Professor, Ginger, the Howell's, Mary Ann and the Skipper come out of their neat and tidy grass huts in freshly laundered and pressed clothes, he posed the question: "Exactly what kind of island were these people stranded on?" The scenes and dialogues did not reflect the plights of actual shipwrecked people abandoned in desolate places and cut off from civilization where they were forced to struggle for survival and fend for themselves. The data displays that physicists watched every day were similarly unrealistic representations of a world that was utterly different. Physicists clothed objects in neat and tidy mathematical forms, then had them play roles and act out parts against a backdrop of improbable order and regularity. People who sat in front of their television sets all day were typically regarded as being out of touch with reality. Everyone had their favorite programs, but what did it really matter what anyone watched? If they were staring at the screen, then they were watching television.

When the physicist confronted Jonathan with the notion that what Jonathan had envisioned wasn't real, Jonathan replied "touché!" Jonathan pointed out that physicists also saw things that weren't real, fictitious lines and sets of mathematical entities, numerical scales and graphs of magnitudes, forces and energies. In the laboratory, the physicist forced the world to behave according to his ideas, thoughts he had entertained in his head, regardless of what normally happened in the world. The physicist turned a blind eye, refused to listen, denied what appeared to be true, in order to institute his program. He didn't care what Jonathan saw. He would make the world behave according to his principles and if anyone told him otherwise he would show him the door.

If an actual physicist had been out there with Jonathan that day, he might have experienced events in the same way that Jonathan had. He might have said to Jonathan later: "You know, you were right, and it really did appear as if the clouds attacked us up there on the ridge. How do you explain that?" Another physicist might butt in with: "I can explain it. There is no explanation, so by saying that there is no explanation, I've explained it." Jonathan took this point to heart and continued the thread: "Just because there was no scientific explanation didn't mean that there wasn't some other kind of explanation. Scientific explanations required reproducibility. No thunderstorm had ever attacked him in the past making this one unique, and such a thing would probably never occur again. The mechanics of the event could not be explored and for that reason the scientific method could not be applied. Thus there was no possibility of a scientific explanation. Since the scientist felt that every other way of thinking was automatically invalid, Jonathan had nowhere else to turn. The scientific method failed, but there was nothing wrong with the scientific method, so therefore it was the phenomenon that failed." An argument of this type was what was commonly known as pretzel logic. If Jonathan bought into the physicist's way of thinking, then he was trapped, just as the physicist was trapped, but if Jonathan insisted that the world didn't normally act the way that it did in a laboratory, then he was free, just as the world was free.

The clouds gathered in the afternoon yet again, but this time they were serious about bringing rain. An intimidating squall approached Jonathan, originating from beyond the hills, portions of the clouds torn away and draping over the land. After observing lightning storms come and go over the years, he'd noticed something interesting. An invisible boundary surrounded each storm cell. When a storm overtook him and the leading edge passed overhead, a momentary rush of wind swept over the ground. This was the most dangerous time to be out in the open because this was the most likely time for lightning to strike. He could make up a story to account for this. The impulse of wind created friction between the air molecules which caused a buildup of static electricity in the atmosphere. As the potential difference increased, the separation of charges eventually overcame the resistance of the air and a lighting bolt was produced. After many years of conditioning, the telltale gust of wind caused him to tense up and nervously peer over his shoulder at the hostile clouds overrunning his position.

A similar danger zone followed the storm as it departed, and this point was also prominently marked by a corresponding pulse of wind. The storm often sent a lightning bolt to the earth at this moment, perhaps as a final farewell, or maybe more as an afterthought, taking the opportunity for one last chance to nail Jonathan as it sailed away and said goodbye. Lightning appeared at times other than these, but here was where he'd most expect to see it, although he was often disappointed.

In his daily conversations with other people, Jonathan availed himself of every opportunity to tell his story. Everybody listened carefully and nodded in assent. His words made sense. Then he would let them in on the joke. He had just made it all up. Surface winds bore no resemblance to the winds aloft. The air movements up in the clouds would be the only thing that mattered, but the air movements on the ground told Jonathan nothing about the patterns occurring up in the sky. Friction was created by rubbing two

330

material objects against one another causing the charges to accumulate on the surfaces of the objects. Air molecules collided and didn't actually rub up against each other, but even if they did, there were no surfaces to collect and store the charges. In fact the very idea of air friction causing lightning was crazy. If abrupt wind bursts typically resulted in the buildup of static electricity, lightning bolts would be zinging in every direction whenever the wind blew. When Jonathan thought about it, the association of pulses of wind with lightning strikes was a very puzzling idea. Any connection between these two events would at the very least be extremely complicated, incorporating a host of other factors.

His story was just a bunch of ideas and lightning had nothing to do with ideas. On a number of occasions he had seen lightning travel in a complete circle up in the sky. The potential difference between a point and itself could never be anything other than zero, so he would have to assume that the potential of the initial point changed rapidly as the lighting bolt traveled around the circuit, starting out at one value only to leap in a brief instant to a completely different value and thus be able to attract the same bolt that it had just issued. Certain people had reported seeing lightning bolts travel upward from the ground, so instead of seeking ground, the bolts were trying to flee from ground, apparently searching for something else. The idea that lightning was a static buildup of charges seeking ground was not only implausible, but apparently an impossibility. Everyone needed to get busy and start making up new stories so they could tell them to each other and through repeated conversations dupe themselves into believing them—at least until someone let them in on the joke.

Lightning was caused by a million different things all coming together at once and no one really understood it, why it happened this way and not that way in a given situation, why it happened in this case and not in another case. No idea could be the equivalent of this infinity of particulars and no idea could replace it. The ideas might make sense, but lightning didn't have to make sense. Lightning was not bound by logic and rationality—but the problem was more than that. Lightning didn't make sense because it couldn't possibly know what that was. How could lightning know what made sense to someone?

As a result many people would go to their graves thinking that lightning was simply the discharge of static electricity seeking a ground, just as certain other people would go to their graves thinking that Jesus Christ was going to save them. It didn't really matter what anyone thought. People believed whatever made sense to them, whatever fit their personal dispositions. Making sense was a sensation in the brain. It was the relationship of one idea to all of the other ideas that one had accumulated over time. What made sense to one person was incomprehensible to the next person because they had different ways of thinking about things. Making sense was all in the mind and it had nothing to do with the physical processes that occurred in nature.

Was it possible, Jonathan thought, that the truth made absolutely no sense? Could the truth directly contradict itself? Only logic said that it couldn't. Should Jonathan therefore place logic above the truth? Did anyone honestly believe that logical propositions were more powerful than the entire universe, that the authority of these blank symbols were greater than the authority of the whole kingdom of creation? The universe could do whatever it wanted and that included saying to hell with the logicians. Perhaps we all needed to stop making sense. Perhaps we needed to put an end to this logical gibberish and start thinking in ways that were more responsible, more appropriate to the universe.

When Jonathan thought back to the tree next to his truck, the one that had been hit repeatedly by lightning, he realized that the idea of a mineral pocket was just as ridiculous as the idea of a subterranean spaceship. Most people would dispute this, but only because the type of explanation represented by the mineral pocket was more acceptable and therefore more plausible, and not because it was the actual explanation. Lightning strikes were fantastically complex and mineral pockets probably had little or nothing to do with them, yet people generally believed that the actual explanation would be something like that, something along the lines of a mineral pocket. They considered a mineral pocket to be a rational explanation, while a spaceship was too far-fetched to be taken seriously, even though such an occurrence was not completely outside the realm of possibility. He could send geologists out to the spot and have them use sonar equipment to determine if there was a mineral pocket or a spaceship underneath the tree. He guessed that there was neither, but even if they had found a mineral pocket, what did that prove? The presence of the mineral pocket might just be a coincidence. So now he had to go out and look for other trees that had been repeatedly hit by lightning and see if they also had mineral pockets underneath them. A certain percentage of them would have mineral pockets. But then he had to look for mineral pockets under trees that had never been hit by lightning. A certain percentage of these trees would have mineral pockets underneath them. Jonathan could see that this was going nowhere. Mineral pockets in certain instances might be associated with other factors and he couldn't know if lightning was attracted by the pockets themselves or by one of these other factors. He ended up gathering numbers and playing games with them. Tracing connections through statistics was utterly impossible, yet that didn't seem to deter anyone. He could calculate as many probabilities as he wanted, but this would never lead him to the truth about the tree.

Could it be, Jonathan asked himself, that the battered tree did not have a cause, that no idea could possibly encompass its infinite presence in an infinite world. Perhaps Jonathan had merely witnessed the various ways that storms developed over the mountains, the ways that the valley shaped and layered the atmosphere causing clouds to gather in certain places, the ways that the winds spread over the hills and compressed between the opposing ridges causing the movements of ions and water molecules to develop peculiar patterns, the ways that trees affected and influenced each other, sometimes shielding and sometimes exposing their neighbors, the ways that lightning directed itself in the sky, single-mindedly pursuing its own goals, the ways that barometric variations and electromagnetic radiations reflected off the rocks and cliffs—along with many other unknown and intangible factors. The tree was not a nodal point in a web of causalities because that notion referred to a map of something and here he had something that could not be mapped, a diversity of incompatible factors that could not be put on single scale. Perhaps he needed to abandon the causality of ideas and stop thinking that way, stop searching for the one idea that would explain everything perfectly— or even proximally and for the most part. The existence of the tree depended on the whole of creation, a term that did not seek to include the farthest reaches of the galaxy but only the deepest depths of material existence, the unaccountable aspects and boundless associations that surrounded the tree and spread out from it in all directions to the horizon and beyond. Other trees could not share these details because they did not stand in the same spot as this tree, thus Jonathan could not understand this tree through other trees. This fact made the use of statistics utterly meaningless.

If only there had been something as mundane as a spaceship underneath the tree, it would have provided a perfectly believable explanation and the cause would have been an obvious conclusion reached by nearly everyone. He could understand intelligent beings using the tree for their own purposes and forcing the tree to get hit by lightning, but he couldn't understand the tree as having no cause at all. The idea that the existence of the tree was inscrutable left him feeling hollow inside. All he could do was move on and think about something else, walk away empty-handed. The actual truth of the tree was far more outrageous than any alien spaceship from another galaxy. As part of an alien experiment, the tree would have had meaning and purpose, but as it stood, the tree was lost in an infinity of infinities that could not be mapped or navigated by anyone.

Physicists were satisfied with their explanations only because they had agreed ahead of time on the forms that the explanations would take. They had devised methods for generating these types of explanations and had proceeded to crank them out using the stock concepts of physics and mathematics. If Jonathan asked questions, the physicist would invoke the laboratory as if summoning a genie. The genie would explain everything to him, the physicist said, but the genie talked about things that were unfamiliar to Jonathan and he could not tie the genie's words to anything around him. Did this mean that nature was fluent in multiple languages, saying one thing to the physicist and another thing to Jonathan? Or had the physicist created his own language and made a dummy out of nature, moving the jaws of a wooden doll sitting on his lap while surreptitiously enunciating the words of his own theories, desperately trying to make it appear as if the words were coming from the doll and not him, the expressionless figure turning its head and spinning sensational stories to a curious and highly attentive audience.

The physicist still believed that nature was supporting him in his endeavor. The physicist asked nature: "What do you think of my new coordinate system?" Nature replied: "I think $(x,y,z)=(12.1, 6.9, 1.0)$" Jonathan had walked for thousands of miles and he had listened to nature his whole life, and this didn't sound like nature talking. It was the physicist who had invented the 'x,y,z' language and then gave nature a voice, but nature was actually mute and communicated only via signs and gestures. Physicists played the roles of interpreters. They observed the signs and gestures and translated them into words for everyone to hear. Yes, they told everyone, the 'x,y,z' language was definitely the language of nature. As the physicist talked in the terms of coordinate systems, the focus of the audience's attention now shifted away from him and he seemed to disappear. His role in all of this was forgotten and by all appearances the world was delivering a soliloquy directly to the audience, telling everyone its own personal story, when in fact they were witnessing a clear case of ventriloquism.

But speaking for someone who could not speak was easy to do. Jonathan asked himself: would a person who was mute otherwise speak Persian or Arabic, Hindi or Urdu, Mandarin or Cantonese, Welsh or Gaelic? In what language did this person formulate his or her thoughts? If they couldn't write anything down, then how could Jonathan answer the question? Emma clearly had ideas of her own without having the benefit of language, so perhaps the person spoke no language at all. Did nature use the language of mathematics as the physicist had supposed, or was nature like Emma, going about its business perfectly well without ever employing a single word in any dialect? Nature by its silence submitted to the physicist's interpretations and put up with his verbosity. The thought that nature was speaking to him via mathematics was absurd. Jonathan often looked at the phenomena of the desert and asked himself: what was nature trying to tell him? What was the message encrypted in the crow's caws as it swooped down and flew over his head? The crow seemed to be trying to tell him something, but he couldn't decipher what it was. The rocks also spoke to him. When he was near to them, they seemed to put words into his head. When he heard their voices, could he really trust what they were saying? He didn't know who they were, so he couldn't make a judgement. He shrugged his shoulders and continued walking, much as he did after another human being had spoken to him and he departed pondering the weight of the person's words. How could he possibly trust what anyone said?

Everyone understood the need to have purposes so they projected them onto everything and inserted them into every situation. Jonathan walked around with all these thoughts in his head and he was satisfied that he had explained something and that he knew all about it. He convinced himself that he had thoroughly thought it through when in reality he hadn't a clue about the infinite complexity of something that was utterly beyond his ken.

As Jonathan followed the road over a small pass and gained elevation, he had a grand view of the valley ahead. The openness made him feel wonderful. Ever since he was a child he'd been a bit claustrophobic. Tight spaces didn't actually cause him to panic, but he was unhappy with them. For that reason, the forest had never been his favorite place. He had stumbled upon many fire rings in the course of his wanderings, often tucked away in groups of trees or wedged into the narrowest parts of canyons, and he always wondered why the people had chosen those sites. He would have been uncomfortable staying in such places for any length of time, and his enjoyment of the outdoors would have been grossly impaired by his sense of confinement. He preferred grand vistas where the mountains were set back a great distance, ideally strung out along the horizon. Open spaces made him feel free and unencumbered. They made his spirit soar.

Jonathan was constantly studying the sky, but under a canopy of leaves and branches he couldn't even see the sky and he dearly missed its companionship. In the forest he felt as if he were locked in a room, catching a glimpse of a mountain peak through a small opening as if he were looking out at the world through a window. Trees were fine. He adored trees. He just didn't like being surrounded by them, or being hemmed in by great numbers of them, or being suffocated by a dense layer of leaves and branches hanging over his head. He also liked rocks, and if he had to choose between rocks and trees, he'd choose rocks. Rocks were less embracing than trees, and much more compatible with his viewing of the earth and sky.

Jonathan slowly descended from the pass. A sheet of gray hung in the sky spread out before him, a curtain of haze and fog draped behind a bank of dark clouds. Secret activities were taking place in the hollow between the curtain and the bank of clouds. Flashes of light illuminated the vertical backdrop as if a welder were hard at work in his shop engaged in the metal fabrication of some strange object whose purpose Jonathan could not fathom, the intense light from the arc reflecting off the walls of his celestial room. The persistent atmospheric toil emitted a constant rumble. Jonathan surmised that lightning bolts must be streaking inside the clouds, hidden from view yet darting about somewhere deep within the cottony folds of water vapor.

He stopped and stared at the clouds in fascination as the flickering continued and the reverberations jostled the heavy air, wondering exactly what was going on in there, when he suddenly caught a glimpse of something unbelievable. Through a small rent in the clouds he saw a series of golden threads emerging from one cloud and swarming toward another cloud, the glowing strands woven into a network of fine capillary structure accompanied by floating flakes of gold, incandescent sparks washing over some invisible surface as if across a pane of glass. Electricity was flowing from one place to another, but not in the traditionally violent means of a lightning bolt. Instead the golden sparks wafted gracefully across the empty space between the clouds, drawn in harmony with the spreading web of golden filaments. Obviously the event had been deliberately hidden from view, so he suspected that he wasn't supposed to see it, but for a brief second the mysterious proceedings were revealed to him, a mere mortal standing on a hilltop.

He continued to peer at the spot when suddenly it happened again, the electricity spreading out and being received by the companion cloud, the golden network flowing from one region to another, transferring unknown assets and information. The glowing embers and threads were being drawn into an invisible receptacle offered up by the second cloud, then unexpectedly a curtain of silvery lightning rained down upon the ground sweeping from right to left directly in front of him. And that was it. The storm showed no further signs of life and gradually the two clouds moved away in silence, retreating into the distance and inconspicuously mingling with the other clouds.

As Jonathan walked onward, the nature of the event suddenly dawned on him. The series of ejaculations of cosmic energy had been succeeded by an intense orgasm. He wondered what kind of love child had been conceived in the sky and what kind of life it would lead in the fleeting society of clouds. He had no doubt in his mind that right there before his eyes, a supernatural entity of some sort had been created by the successful union of two clouds. He felt as if he were a small boy who had inadvertently walked into a secret chamber and caught his parents in a passionate embrace. He witnessed the act and saw everything for himself, the details of this exotic form of heavenly copulation, and he was no longer innocent. From now on, when he looked up and saw two clouds coming together and shooting out lighting bolts, he would know exactly what was going on. Of course, he would not avert his eyes in embarrassment, but stare intently at the clouds in order to glean more information about this strange and extraordinary process. The structures within the clouds were so complex and the processes so involved and intricate as to defy our meager intellects.

The show was obviously over and the intensity of the clouds weakened. As he continued to descend into the valley, Emma's barking jarred the silence and instilled a panic in him. He rushed forward in the general direction of the raucous outburst, but the location of the sound was still far away, apparently down in the arroyo.

He finally came upon the scene. Emma was frantic but had not yet nabbed the creature as he approached, her rear end sticking up out of the bushes. There was still time to grab her by the tail and pull her backward, but just as he zeroed in on this convenient handle, Emma leapt forward and seized the prey in her mouth. Jonathan feared that it might be a porcupine. She continued forward and emerged with it on the other side of the bushes. Unshakably amicable, Emma loved people and other dogs—and even cats. Whenever another dog snarled at her, she stared back with a look of incomprehension, seemingly unable to understand the meaning of violence and aggression, yet the smell of a porcupine turned her into a monster, unleashing a fury so startling, a meanness so incongruous with her nature, that Jonathan began to wonder about this dog that he thought he knew. Emma held the hapless rodent down with her paws and viciously bit into its neck, ripping its head clean off, then ran away with the gruesome booty in her mouth, bloody entrails dangling from her jaw. He didn't chase her or attempt to take the head away from her because he knew that she'd swallow it before she'd give it to him and he desperately wanted to avoid that consequence.

Emma's behavior was obviously inherited from her forebears, the result of a long history of confrontations between canines and porcupines spanning countless generations. Emma immediately recognized her long-established enemy whenever she encountered it and instinctively recalled the injustices of the past, the pain and agony that had been inflicted on her ancestors by these devilish creatures, and now she sought revenge. This was payback time. How did Emma know all this? Her entrenched knee-jerk reaction was part of the mysteries of genetics. She would perform her duty today and spare an unfortunate coyote the miserable death, the long hours of suffering, the torture of being unable to eat or drink anything until its body slowly withered away. She would nobly sacrifice herself and play her small role in ridding the earth of this terrible scourge.

The decapitated body was still laying where she had left it, so Jonathan stepped up to examine it. At first glance the animal didn't look like a porcupine at all. He was not really sure what it was, so he knelt down before the carcass trying to identify it. He could see the victim's heart still beating through the gaping hole between its shoulders. He grabbed a stick and turned the lifeless body over onto its stomach. He'd never seen anything quite like it before. He guessed that it was a very young porcupine, because its quills were undeveloped and they were completely black. Actually they looked like fur, but he was not willing to touch the seemingly soft coat with his fingers.

Emma came toward him from behind the bushes. She no longer had the head in her mouth so he didn't know if she had abandoned it or swallowed it. There was nothing he could do at this point other than avail himself of the opportunity to get Emma out of there and prevent further mishap. She was still excited by the unexpected event but followed Jonathan as he departed, stopping frequently to use her front paws in vain attempts to remove the painful spines, desperately trying to scrape the carpet of pins from her face. He saw that hundreds of quills were embedded in her mouth, covering her tongue, gums and cheeks, sticking out in all directions. Now that they were separated from the porcupine's coat, the bases of the quills were grayish white in color and short in length, sticking out from Emma's jowls much as the stubble of an old man's beard.

He finally reached the truck, deposited his pack onto the truck bed, and loaded Emma into the cab. Emma would not be getting her dinner tonight. As he approached the highway, he speed dialed the local vet's cell phone number and asked her to meet him at the clinic in an hour. Yes, he'd pay the extra hundred dollars. He was just grateful that she would do this for him on a Sunday night. Emma laid down on the seat next to him, gagging from all the foreign objects in her mouth. She set about working on the problem by pawing at her face, but Jonathan got her to sit up and he started massaging her throat. This took her mind off the discomfort and forced her to stare out the windshield and hold her mouth open. He found that he had to keep this up the entire way back to town, a distance of about 40 miles, driving with his other hand on the steering wheel. He didn't want her breaking off the spines because then they'd never get them out.

As he cruised down the largely deserted interstate, a small island of lights appeared in the distance. Storms had gathered just south of town and formed a united front in the darkness. The sky was black and the clouds themselves could not be discerned, but the electrical activity was quite visible and rather astounding. Not having to concern himself with other traffic, Jonathan focused all of his attention on the sky. The display started out with ordinary lightning bolts, the jagged lines flashing into view complete and intact, the entire bolt appearing from beginning to end all at once. But then something strange happened. A leading edge appeared at one point and steadily moved across the cloud bank trailing a streak of luminescence behind it. The streamer quickly lengthened, wavering as it progressed, reminding Jonathan of a snake slithering on the ground. After a certain distance, the bolt ran out of energy and dissipated.

As he leaned forward and peered upward at the unseen clouds through the windshield, the display got even more bizarre. A trailing edge now followed the leading edge forming a blazing segment of light that traveled across the sky as a unit, lacking the swiftness and urgency of an ordinary lightning bolt. Suddenly a circle of these segments radiated outward in all directions from a single point. The glowing pieces of light really looked like snakes wiggling and zigzagging as they escaped into the night, appearing to have all simultaneously crawled out of a hole in the sky, each one choosing a different path to follow. Jonathan had seen all kinds of strange lightning storms over the years, but in all of these examples, he'd never seen anything like this. He desperately wanted to keep driving down the interstate in order to see what happened next, but his exit was approaching and the doctor was waiting for him.

The lights were on inside the clinic as he pulled into the empty parking lot. An aide came to open the locked door. He led Emma into the back room where a table was waiting for her. The vet took one look at Emma and her first words were: "Where in the world did she find a porcupine around here?" Her assistant clarified the remark by volunteering the information that porcupines were much more prevalent at higher elevations. They were generally nocturnal creatures, but sometimes they were out during the day, a fact that Jonathan and Emma had just discovered.

—16—

Summer was gradually fading away and the predawn hours were now quite chilly, forcing Jonathan to don his long-sleeved shirt and fleece-lined windbreaker. As he struck out into the desert and left the truck behind, with Emma somewhere off in the distance, he heard a heavy, metallic thud on the ground. The sound was not familiar to him and he couldn't imagine what had caused it. The clip on his GPS wasn't very secure and sometimes the device fell off his belt and hit the ground with a light thud, but the sound it made, the sound that Jonathan had heard a number of times before, was very different from this. This object had been heavier and hit the ground with more force. He reached over and noticed that the GPS was not on his belt, but then he put his hand inside his jacket pocket and found it there, along with a leash. He felt around on the other side of his belt and discovered that his can of pepper spray was still in its holster. He had nothing else that could have fallen to the ground and made such a sound. He slipped his pack off his shoulders to see if he'd left a pocket unzipped, but every pouch was sealed tight. He turned around and backtracked, searching the ground he had just walked, but he had been right, he hadn't dropping anything. He paused to reflect on the situation. Jonathan saw no other explanation. Something must have fallen from the sky.

The item might have been a bolt or small bracket—or much worse—that had been ejected from a plane flying overhead, but this part of the country had few if any air corridors and he rarely saw a contrail in the sky. The object might have been a small meteorite. These chunks of rock rained down on everyone day and night, pelting the surface of the earth, possibly making thuds similar to the one he had just heard. He lived in a bizarre world where untold numbers of metallic rocks hurtled through space and fell through the atmosphere and crashed down upon the ground at every hour—even on quiet and serene mornings such as this.

Jonathan didn't understand the purpose of this phenomenon. He looked around at the universe and he couldn't fathom the point of this ever present danger. Why would anyone create such large spaces and then fill them with rocks flying through the emptiness at very high speeds? How could anyone look at this and say with a straight face that we lived in a rational universe? How could anyone think that they could make the universe rational by imparting their own rationality to it?

When Jonathan looked at what was happening in nature, he saw that the whole arrangement hadn't been well thought out. Obviously, the way things were set up, animal populations were going to get out of hand and there wouldn't be enough food to support them. Terrible suffering and starvation would result. Then there were these microorganisms lurking around in the environment and they were going to attack the animals whenever they got the chance, and all of the energy that had been used to create these stupendously complicated bodies of flesh and bone would have been wasted. This really showed a lack of forethought. It seemed that nature could have worked this out more and come up with a better plan. Rocks falling from the sky—who's idea was that? Wild climate swings, ice ages—that wasn't gonna work. Life was essentially based on death. Why didn't anyone see the flaw in this logic?

When people finally took control of their destinies through technology, few people realized that they'd also have to control population or else there would be global disaster. Ego clashing, greed, megalomania took over like cancers, dictators assumed power as the world spiraled out of control, conflicts emerged, and then the physicists gave everybody nuclear weapons. Great thinking. The civilized world was planned out no better than the natural world. Why couldn't people grasp the ideas necessary to build a good life for everyone, including the plants and animals that lived around them? Because everyone thought about things differently and they were driven by separate impulses that conflicted with one another. Was that, Jonathan wondered, also nature's problem?

If we didn't live in a rational universe, then why must there be a rational explanation for everything? When a dark spirit approached Jonathan in the middle of the night, the physicist said that Jonathan must believe that there was some rational explanation for what he experienced. Why was the existence of spirits prima facie irrational? And why was mathematics the only rationality—without equal or alternative? Did nothing else make sense? Why couldn't Jonathan develop his own logic, one that refuted and contradicted the crazy thinking and unreasonable assumptions that had given rise to mathematics in the first place?

The world didn't make sense, but long ago physicists had forgotten about the world and decided they would rather live in their conceptual funhouse, the highly artificial world of their imaginations. Physicists no longer came out to the desert because they much preferred their ideal world of straight lines and perfect circles. Jonathan saw that physics was really a form of escapism—a lot like drinking alcohol. As some of his buddies in college used to say: "Just stay drunk all the time." But eventually everyone had to sober up or die from cirrhosis of the liver. So Jonathan wondered: when did physicists plan to come back to the real world, to deal with the indisputable facts of their existence, to come to terms with the certainties of their lives? How long would it be before they kicked the habit and climbed onto the wagon? Could they ever find a way to eliminate their dependency on geometrical constructions and stop using them as crutches? When were they going to free themselves from the hypnotic influence of mathematical thinking?

Jonathan knew that physicists weren't going to appreciate being compared to alcoholics. Everyone had rosy images of themselves and thought of themselves in glowing terms, but that was not always the way others saw them. Being clever people, physicists would concoct arguments to refute everything Jonathan had said, but that was not the point. The point was could they find a way to break the spell of their addiction and escape their dependency on something that distorted their perceptions and impaired their judgements? Could they expand their horizons to include this, the desert, a place without mathematics?

Physicists would certainly respond to Jonathan's novel perspectives by arranging these thoughts into logical contradictions, however, devising a rational argument to counter his challenges wasn't going to save physics. If these allegations were at all true, then physics would collapse under its own weight. Debating issues took place in the artificial arena of logical statements and this ultimately had nothing to do with reality. The funny thing was, contradictions didn't seem to bother the world at all.

In order for physicists to have any chance of understanding the world they lived in they must (1) stop playing games. Physicists could not understand nature by playing games with it. A game created a new reality, a reality of definitions that did not copy the reality of the physical world. Physicists maintained that the games still "corresponded" to the real world because aspects of the world had been incorporated into the games, but the games corresponded only to the ideas that had created them. Physicists must (2) realize that the world of logical propositions was a fake world. They could have seen this clearly by simply looking out the window. Logical propositions did not—and could not—capture the essence of the physical world, but moreover, such propositions could not even be formulated without the introduction of a structured framework to provide the clearcut alternatives. Logic had nothing to do with the unbounded realities of physical existence, and in all cases, logic profoundly and maliciously misrepresented the true state of affairs.

Physicists must (3) understand that abstractions weren't real. Mental constructions could not possibly be the basis for the universe. The full weight of these concepts rested on mathematics, something that had been added to the phenomena in a series of independent and unwarranted steps. Mathematical concepts were imaginary structures built on top of what existed and not necessary assumptions or logical deductions from what existed. Physicists tried to ground their model of the universe in mathematical comparisons, but these comparisons presupposed a way of thinking that could not be justified, something that had to be assumed. Not everyone thought about things in the same way and mathematical thinking did not have a special status. Physicists must (4) see that mathematics was all about empty forms and these forms could never be more than superficial representations of a world that was utterly different, a world that was filled to the brim with substance and content. Mere patterns could not reveal what anything was and that was what they needed to know. Appearances were deceiving, and any account that relied heavily on them would eventually lead to gross errors in judgement and patent falsehoods. Physicists must (5) stop saying that physics was based on observation. A mathematical world would have been observable, but an infinite particularity presented only endless facades, illusions and appearances. Observing the behaviors of fabricated mathematical objects wasn't observing the world, because mathematics wasn't about the world. Physics was based entirely on the imagination. Physicists must (6) abandon the criterion of reproducibility. Reproducibility meant nothing in a world where every event of any complexity was utterly unique. Reproducibility was a characteristic of mathematics and it was valued only for that reason. Reproducibility could not be used to define reality or become the standard for what existed. Physicists must (7) concede that the indeterminate character of an infinite particularity meant that it could not be measured or quantified. These concepts required mathematical objects to define them and out in the natural world they quickly became nebulous and enigmatic. Physicists must (8) forget about calculating probabilities. These numbers were generated by simplistic games and arrangements with well-defined outcomes. In the laboratory this was done with instrumentation, by operating within a definite set of rules and possibilities, but out in the real world nothing even remotely resembled independent, identical trials. Probability took physicists further away from the world and further into mathematical ideas, representing yet a greater leap into the abyss of nothingness. Physicists had gone in this direction too far already and needed to turn around. Physicists must (9) come up with something to account for the creativity of nature. Forces could not serve in this capacity. Creativity in the human world was brought about by ideas and visions but it was hard to see how these aspects of consciousness might function in the natural world. Physicists must (10) stop focusing on the mechanical aspects of objects. Mechanics reduced objects to their basest levels, treating them essentially as nonentities. This was not the way to understand objects or any of their interactions. In physics, an object was primarily a nondescript material body that was also, by the way, an object of immense complexity. In the real world, an object was primarily an infinity of particulars that inadvertently became a nondescript material body in situations where its essential characteristics played no role in the outcome. This reversal was crucial to the formulation of all physical theories.

After subtracting all of these aspects from physics, nothing remained. The falsifications of the natural world that had been introduced at the very beginning of the enterprise, going all the way back to drawing lines, counting objects, and putting objects into sets, could not be corrected by any means other than by simply starting over with a completely different approach, one that paid close attention to the essential features of the world and cultivated these qualities instead of trying to override and negate them. Physicists would continue to construct their alternate realities in the laboratory, but at some point this would become impossible. The glaring discrepancy between the laboratory and the real world would remain in any case. No explanations of ordinary experiences were forthcoming and Jonathan was left to speculate about the causes of what he saw around him. Physicists could vehemently maintain that Jonathan's explanations were all wrong, but they could not offer any explanations of their own, so in each case the correct explanation remained a mystery, leaving Jonathan alone in the dark. Physics was supposed to shed light on the world, but after studying physics for many years, Jonathan was still unable to understand why strange things happened in the desert.

No one was going to listen to Jonathan. They would listen instead to those in positions of power, people who held prestigious posts at highly respected universities and institutions, people who had been granted certificates and who had won awards, written celebrated books and published important papers, given lectures and perhaps even appeared on television. Who was Jonathan to go around making statements about physics? He was a nobody, a drop-out, a loner. He had no credentials. He had no recognized status, no official title, no designated place in the academic world. He had spent most of his life living off the grid, wandering around in the wilderness with his dogs. His perspective didn't matter because he was an outsider, a person who couldn't possibly understand physics, at least not nearly as well as physicists did, but as a true inhabitant of the desert, Jonathan was a representative of a much larger world. He was not just spouting uninformed opinions. He was voicing an outlook that took precedent over all the others, one that had a very special status. Why was the desert so special? Because the desert was the real world. In truth Jonathan had a set of credentials that far outweighed all the accolades bestowed on others by renowned personages in so-called positions of authority.

But according to physicists, everything in the desert was basically happening by chance. The winds blew the sands in every direction, branches fell and the pieces were haphazardly thrown across the ground, clouds gathered at unscheduled times and places, storms suddenly appeared out of thin air, seeds fell on the ground and some germinated while others shriveled up and died, plants withered in the hot sun while others thrived nearby, rocks broke free from the cliffs and tumbled recklessly down the slopes following crazy trajectories. Ascribing all of this activity to randomness and chaos was a neat way to get rid of it, to dismiss it and characterize it as meaningless. The details were lost in the incessant flux of unknowable particulars, making nature's actions appear blind and undirected. The exact courses that natural events followed all existed for no reason, so no one need bother with them. Nobody could explain the antics of nature because these occurrences had no explanations, so let the scholars and rationalists forget about them and cast them aside. Let nature be the subject matter of poets and romantics, people with strong emotions and inferior intellects, and let the scientists concern themselves with the more important business of fabricating an artificial mathematical world in a laboratory, a world that could be controlled, calculated, predicted, and rationally explained, a world that consisted of thoughts and ideas. All the physicists' concepts about probability and statistics, chaos and randomness, coincidence and happenstance, were simply means of downplaying the events of nature, clever ways of denigrating them by discounting their significance and diminishing their value.

Since physicists could not even begin to complicate their ideas enough to match the complexity of the phenomena of nature, where all kinds of things were happening at once and at every level of organization, where influences both large and small were radiating in every direction and layering on top of each other forming a dense morass of causality that could not be unraveled and untangled, they could simplify phenomena to the point where the processes and events occurring within the phenomena could be matched to ideas, thereby creating the illusion that the ideas had been behind the phenomena all along. Ideas became the reasons why things happened, the causes for what was observed, the connections between objects and events, the relationships between convenient fictions. Everything could now be understood in terms of the interplay of ideas, proving that everyone did in fact live in a conceptual universe. If ideas could not rise up to the level of complexity required to match the infinite particularity of nature, then the phenomena of nature could be pared down to match the level of conceptualization that could be managed in everyone's minds.

An experiment existed for one reason and one reason only: to insert ideas into the world. The physicist put ideas together in his head, then built the entire experiment out of ideas, explained the experiment with more ideas, so that conceptualization lay behind all the arrangements and observations—only that was not the way a sunrise had come into existence. An experiment could not replace a sunrise and a sunrise could not be understood in terms of an experiment because they had radically different origins. When a physicist tried to understand a natural phenomenon by performing an experiment, he was seeking to replace something that had been never thought out in the first place with something had been built out of ideas from the very beginning. He did this in order to put those ideas into nature, to set nature on a firm conceptual foundation. Now all of his precious ideas were right where he wanted them to be, deeply implanted at the very roots of existence, and the way was open for the concepts to become the ultimate realities of the universe.

If a sunrise was not rationally assembled, then it existed for no reason and it was not grounded in any system of thought. No one could think about the sunrise by supplying an account of every atomic interaction, deducing the personal relationships of every individual pair of molecules in the entire sky, so they couldn't even begin to visualize what was really going on in the clouds, yet there it was nonetheless, the sunrise spreading out across the sky, glowing in glorious colors and taking on fantastic shapes, moving, transforming itself, becoming more, becoming less, finally fading away, becoming nothing more than a memory in the mind of the beholder.

The physicist chimed: "We have proven in the laboratory that events are based on ideas. We can use the laboratory as a model for the world and conclude that the world must also be based on ideas. The only difference is the world—unlike the experiment—is not based on simple ideas, but on very complex and intricate ideas. Well, ok, infinitely complex ideas. That's alright. We'll just multiply Newton's Law a gazillion times and then tie all the relationships together with invisible strings." Physicists were never going to admit that the world was not based on ideas. They would use the laboratory as the basis for their conclusions and the inspiration behind their visions rather than the world itself.

The physicist asserted that the molecules didn't actually have to think about following mathematical rules, because they could just follow them blindly for no reason. Well, the physicist added, there was a reason why they followed these rules. They were obeying the laws of physics. They were bound to geometrical constructions that didn't exist in the world. Well, the physicist continued, if their outward behaviors described mathematical forms, then these forms did exist, embodied in the behaviors of the molecules, and if these forms existed, then they could be woven into both chains of causality and lines of reasoning.

Jonathan pointed out that mathematics could only be brought into existence by thinking these ideas and then fabricating realities based on them, the molding of mathematical objects. If the molecules weren't thinking about mathematics, then what physicists saw wasn't mathematics. But for physicists, mathematics was also something else, something that had never been planned, never thought out, a mathematics that was no longer made up of ideas. Of course, this was just another idea added on top of the ideas of mathematics. Mathematics led a double life and it no longer required a mathematician to make it come alive. Mathematics existed long before mathematicians were born, long before anyone knew anything about these abstract relationships and the chains of deductions that were built around them. Mathematics was not the creation of the analytical mind, but rather the ultimate reality of the universe. Mathematics became something that no one needed to contemplate and the ideas were not really ideas anymore. They were imaginations that did not require anyone to imagine them, concepts that were beyond the realm of consciousness, an intellectual universe that existed entirely outside of the human mind.

If on the other hand, Jonathan abandoned mathematics and stuck with the fullness of being, he immediately found himself at an impasse. Brute existence offered no way to proceed. The whole enterprise was over before it had begun and the path led nowhere. All Jonathan could do was stare blankly at the events unfolding before him. Physicists had no choice but to throw reality out the window because reality was inscrutable and no one had ever found a way to deal with it. But if physicists tied reality to mathematics, something infinitely malleable, endlessly fascinating, supremely workable, then suddenly the way was open. Mathematicians could formulate new definitions and invent new rules freely. The intricacies of mathematics could engage their intellects in almost limitless ways. Physicists could jump onto the vehicle of mathematics and the network of conceptual highways and backroads stretching across the intellectual landscape would take them wherever they wanted to go—as long as they didn't look in the rearview mirror because as they sped away in their exuberance, they had left reality standing on the curb.

The idea that they could tie reality to the back bumper and drag it along with them was absurd. Once they embarked on their journey into the hinterlands of the mathematical countryside, all they could do was maintain a symbolic or pictorial channel of communication with the world they had left behind. They could send conceptual postcards at every opportunity: "Having a great time without you! Everyday you are in our hearts and minds!" And the world could send back coded messages explaining what they were missing. The relationship between theories and realities was now nothing more than a formal correspondence, a series of handwritten notes and scenic snapshots. No one could leave home without surrendering the life they had left behind.

Religious thinkers typically conceived of intelligent design as issuing from a central figure who conferred his own intelligence to the world, a world where everything made sense according to the particular way of thinking of this one individual. Although some people might not see the rationale behind certain events occurring in the world and take issue with the designer on these points, questioning the wisdom of his actions, others would point to the possibility of a hidden logic lying behind the apparent contradictions. Thus a given world could be intelligent to some people and not intelligent to others depending on their perspectives. Actually, the fact that the world was an infinite particularity made it possible that it was intelligent and non-intelligent at the same time. The world that Jonathan knew made sense to him in certain ways, but not in others, as if the creator had worked out certain details but failed to incorporate everything. These shortcomings were understandable. The universe was a vast place and managing all the details was a tall order. Stupidities were bound to creep in here and there. Oversights and omissions could not be totally avoided. Every nuance had multiple meanings and every element played diverse roles in the overall plan. Some things would be left out of the picture out of necessity.

Intelligent design—if there really was such a thing—traditionally proceeded from the general to the particular, but in Jonathan's view, intelligence moved in the opposite direction and went from the particulars to the general forms, the individuals to the societies. There was no intelligence in the abstract, but only multitudes of tiny bits of intelligence, each with its measly share, untold numbers of insignificant entities grappling with one another—much as human society itself. Instead of the supreme authority of mathematical laws reaching out across the great distances and dispensing its order upon objects and events, this intelligence gathered and coalesced, transcended the primitive tribalism of its baser forms, welled up in forums and consortiums that were challenged by bands of renegades, then triumphantly rose up out of the din to build empires and civilizations.

Since his student days, Jonathan had enthusiastically adopted the philosopher's upward gaze, his eyes always studying the clouds and sky over his head searching for answers, his mind always reaching for the unattainable truth of this world, and like Thales he was well aware of the danger of falling into a well and drowning while in this distracted state. Perhaps the real danger was that the truth was not up in the sky or located in the stratosphere of abstract thought, but rather below his feet, buried in the ground that he walked on. Perhaps in order to find the truth he needed to have the downward gaze instead. Maybe truth was not contained in theoretical principles and universal laws, sublime metaphysical notions and detailed academic analyses, but in the tiniest and most insignificant details of material existence, the minutest aspects of each and every phenomenon, the infinite and mind-numbing particulars that confounded the most brilliant minds and overwhelmed the astutest intellects.

Jonathan saw the hazy suggestions of intelligent life forms lurking beneath the surface of everything around him, myriads of sentient corpuscles rallying and surging, fighting and striving to elevate themselves, bubbling up from some subterranean vault, their triumphs and failures intermingling and producing only a uniform inconclusiveness. Sometimes they slept for ages or drifted in limbo and one could easily assume they were dead, yet after a time something new would always appear. Maybe such entities existed and maybe they didn't—who could tell? Jonathan perceived only the faint outlines of something, plans that had at some point been abandoned, statements that had never been fully articulated, artistic impulses that had been thwarted by the constraints of materiality, the birth of something unique. Could he honestly say that it was alive? Feeble and limited cognitive forces had attempted to create a world for themselves—but had they seen the end result beforehand? Did they know what they were doing? They tried their best to accomplish goals and they did what they could given the meager means available to them, but the final product was ambivalent, an intelligence of sorts perhaps, a sketchy and incomplete design, a patchwork of incongruous forms competing with one another striving for some sort of dominance and supremacy. Diversity and conflict were to be expected. From Jonathan's experience with human beings, it was damn near impossible to get everyone to agree on anything.

Biologists were currently attempting to construct processes such as photosynthesis out of the concepts of quantum superposition and quantum tunneling in the same way a physicist might construct the flight of the frisbee out of Newton's Law—and why not? Electrons found the most efficient route by being everywhere at once, exploring every option simultaneously, just as forces were everywhere at once and the frisbee was propelled by these tiny entities all acting together as one. However, the electron was as much an illusion as the component forces acting on the frisbee. They were all mathematical entities that did not exist in the world, but once the concept of quantum effects was in hand, the scientist could put it anywhere he chose, pull it out of his tool bag at will in order to fabricate explanations. If a concept became believable in certain situations, those imagined successes could be used as references to grant the concept credibility in other situations. The concept now had a proven track record so it became the plausible candidate whenever someone was looking for an explanation of something else, even in contexts where the concept might not be appropriate. Once ideas gained traction they became the ideas of choice in any situation that could benefit from such ideas, and because ideas were abstract, they could be fitted into many frameworks, plugged into many different chains of reasoning and systems of thought. The possibilities for playing with ideas were endless. They could be molded, modified, transposed, rearranged, and reinterpreted as necessary, because they weren't real. Brand new associations presented themselves, schemes and ideas appeared in the mind, visualizations materialized in the imagination, and everyone could then see the new reality quite clearly, but what was actually happening in the complex world of atoms and molecules? Had quantum mechanics really shed light on anything, or was quantum mechanics just another fantastic flight of fancy? Suddenly quantum effects were everywhere, not because they were widely distributed throughout the universe, but because these ideas, generated out of certain laboratory experiments, could be adapted to fit many situations. The ideas, once in hand, proved useful in fabricating all sorts of highly original explanations.

Biochemical reactions proceeded more quickly and easily than they should have given the mechanics of ordinary chemical reactions, and this observation could be explained by resorting to the idea of quantum effects, but Jonathan could also account for the uncanny efficiency of biological processes by saying that complex organic molecules somehow knew what they were doing. They had a rudimentary intelligence and their actions were guided by an awareness not only of their surroundings, but a basic understanding of what they must do in order to make everything work. They had a sense of purpose and a knowledge of how to attain the goals of molecular biology that had been set by their peers. This explanation was not taken seriously by scientists because consciousness did not lend itself to either mathematical formulation or experimental test. The scientific method was primarily the application of mathematical ideas to generate mathematics in the world, and this could only be done by focusing on the mechanical aspects of the phenomena, but what exactly were the mechanical aspects of consciousness? No distances were involved, so there were no positions, no movements, no velocities. The program of mathematization was stymied and therefore the hypothesis wasn't taken seriously. Molecules might in fact be conscious, but this would be the death of the scientific method and the end of the illusion of a mathematical universe and physicists couldn't have that. The assumption of a mathematical universe was the very beginning of the scientific method and in order for the mathematical universe to become a reality, everything in the world had to be reduced to simple mechanics.

The idea that consciousness was nowhere to be found in the universe until suddenly it appeared out of nothingness and became a central part of higher organisms did not seem to be a rational supposition. Where consciousness came from was anybody's guess. Consciousness was certainly not the consequence of the physicists' empty abstractions. Physicists had not yet found a way to assemble consciousness out of the measurements of distances or the vibrations of particles, but perhaps consciousness was tied to the mysterious interplay of electricity and magnetism, arising out of a strange cosmic dance of forces. Associating the blue spark of an electrical discharge with the flickering flame of consciousness was a natural impulse. Physicists had no idea what electricity was and they could easily suppose that it had some hidden powers, properties that had not yet been discovered. Neurons in the brain clearly had electrical characteristics, but it was also true that atoms and molecules had electrical characteristics. One thing was certain— electrical activity was not unique to the brain. Electricity could be traced to the very basics of matter, so if electricity did play a role in consciousness, then perhaps consciousness originated in the submicroscopic world and consciousness was an elemental force that arose out of electricity and magnetism. And if consciousness had nothing at all to do with electricity and magnetism, then what else was there to fall back on? Nothing suggested that consciousness could stand alone, completely divorced from all material existence, but how then was it tied to the world?

If molecules were in some sense conscious, then perhaps their lives weren't as completely blind and undirected as physicists had supposed. They might not actually possess ideas of their own, but even if they did, these ideas would certainly not be the ideas of mathematics and logic. Direction and purpose could exist through simple urges and innate compulsions without the necessity of having a preconceived plan or any clear ideas about accomplishing specific goals. Investigators had only to find out what the molecules wanted to do and determine the roots of their desires, the things that motivated them to act. And if they actually knew what they were doing, then what were their hopes and dreams and aspirations? Jonathan assumed that modern society and the human race were the intended products of their labors since that was where all of their actions had led them. Long ago they had set out to construct a world, not only the one that everyone now found all around them, but the one everyone had created for themselves.

Imparting motives to living things was natural and easy to do. One author claimed that in the course of evolving, organisms had actually been solving physics problems, finding solutions to the constraints posed by their environments. However, solving math problems was only what physicists did, yet they typically envisioned their personal behaviors being repeated in nature. Everything in the universe was just like them and all life forms analyzed their situations in terms of the mathematical relationships between quantities. The world was populated by little calculators all trying to figure out what the right answer was.

In his later years, Einstein dedicated himself to the problem of developing a unified field theory, discovering a common mathematical framework that could be used to tie the different forces together: nuclear, electromagnetic and gravitational. There was really no point in doing this since it would not lead to new technologies or advance anyone's understanding of the universe, because tying these forces together mathematically wouldn't mean that they were actually related to one another, but Einstein had gotten so obsessed with playing games and solving puzzles that he couldn't pull himself away from this senseless drive to come up with the right answer. He never succeeded, and in this regard at least he died a frustrated man. He had taken the game-playing to new levels, but he needed to step back and realize what he was doing, only he was too involved to ever see any of this clearly. The idea of fitting all the pieces together and finally completing—if not the world's largest jigsaw puzzle, then certainly one of physics' largest jigsaw puzzles —was something that utterly consumed him and he couldn't let the project go, simply because it was so intriguing and tantalizing, pushing his analytical skills to the limit. He was convinced that there must be some way to arrange the symbols into the correct patterns, and the more his efforts were stymied the more the problem preyed on his mind. The challenge itself was the true allure. Everyone wanted to be a winner at something.

Obviously the author of the book had speculated that organisms were working out problems in applied physics rather than engaging in purely theoretical speculations as Einstein had done, but in that case where were all their failures and miscalculations? Survival and competition meant one arrangement working better than another in large populations over the long run, but all designs had to be successful to a certain degree simply in order to appear in the first place, in contrast to the fact that calculations didn't necessarily produce meaningful results and often produced nothing but pure nonsense. Early human attempts at flight tried to mimic the flapping wings of birds and bats, solutions that worked quite well for these creatures, yet the contraptions failed miserably for human pilots. Organisms must have also come up with some harebrained schemes that never worked. Where in nature was the evidence for this host of truly stupid ideas that never made it into the panoply of established species?

The natural world seemed to be alive, so the idea that this perception was just an illusion and that the natural world was actually dead and lifeless seemed strange to Jonathan, a proposition that required some sort of justification beyond the fact that mathematics was also dry and lifeless and physicists had deliberately set out to cast the entire world into such a bleak and spiritless format. If consciousness was the ultimate reality of the universe and not mechanics, that would certainly be an ironic turn of events for physicists.

Most people equated consciousness with free will, meaning erratic behavior and the unpredictable impulses and desires of someone acting irrationally. Most people did not strive for total creativity at every moment of their lives because that required maniacal dedication, a sustained and very substantial conscious effort, a relentless determination to be absolutely original at every possible moment and in every possible way. Generally peoples' lives had a more stable structure where actions were directed toward set, specific goals, where daily routines were firmly entrenched, producing an existence full of repetitive behaviors that were rather predictable. People did things for reasons that were relatively fixed and definite, reasons that persisted over time, often long periods of time. Even though options were always available and a person's behavior could change radically at any moment, overall, people traditionally were bound to the external constraints of their society as well as the limitations of their natural environment and have acted accordingly, yielding to the demands of their surroundings without surrendering their faculties of perception and understanding, yet letting the world mold their characters to a large extent. Insect societies such as ants and bees were even more rigid. Individual roles were well-defined and there were few if any dissenters.

Although insects had an awareness of their environment indicating some level of consciousness, behaviorists tended to say that insects acted not out of pure volition but were in the grip of instincts—invisible, innate forces that bound their consciousnesses to predetermined patterns. The insects acted without thinking about what they were doing. Although they were not considered to be free to think for themselves, they still knew what a flower was. They knew where the flowers were located and how to get to them. They knew what time of day it was and remembered where the hive was located so they could find their way back. In fact, bees seemed to be aware of a lot of things. They were at liberty to roam the countryside, apparently making decisions about which way to go next and what to do now. Did they listen to the birds chirping in the background as they worked diligently collecting pollen and enjoy the melodious songs of these creatures? Did they love the fragrances of the ferns and grasses as well as those of the flowers? Did they like the feel of sunshine on their backs? Did they enjoy a beautiful summer day as much as Jonathan did? Or did they despise their lives and hate their jobs, resentful that they were trapped in a rigid society of drones and workers? Sensations, feelings, and volitions were the hallmarks of consciousness, and these faculties did not necessarily imply erratic and unpredictable behaviors.

The arroyo narrowed and sank into a basin largely occupied by mesquite trees, sage, apache plumes and bushy junipers. The area appeared to be completely deserted but unbeknown to Jonathan about twenty antelope were hiding in the dense cover. They had been peacefully relaxing in the cool shade prior to his arrival, basking in the fine morning air after spending a night of sleeping out under the stars. They finally panicked when Jonathan got too close and bounded away, expending more energy going up and down than moving forward. Emma was well off to one side, but she finally saw them. Instead of pursuing the antelope as they fled, she ran up to the area where they had been standing and busily started inhaling their scents, weaving a pattern across the bedding ground, enjoying all the wonderful smells they had left behind. Emma was not stupid enough to make a fool out of herself. She had tried to chase antelope when she was younger, but she was older and wiser now and didn't fall for that trick anymore. The antelope stopped a short distance away, noticing that neither Jonathan nor Emma were showing any interest in them. The antelope watched the newcomers with great fascination, peering over the crowns of bushes and peeking through the tangles of branches, curious about the behaviors of these strange creatures. The antelope suspected that these were animals who could be easily outrun and therefore they posed no danger to them.

A large jackrabbit burst from its hiding place and sprinted up the rocky slope. Emma paid no attention to it and continued sniffing the scents of the departed antelope. This summer she had learned to distinguish between black-tailed jackrabbits and desert cottontails. She had now caught 19 rabbits and 16 of them had been young cottontails. She knew that adult jackrabbits had longer legs and they ran faster, but she could gain on a juvenile cottontail and close the gap between them even when it had gotten a bit of a head start, so now adolescent cottontails were the only rabbits she'd chase. The fact that she had figured this out showed a remarkable degree of intelligence. Jonathan turned away from the scene and continued following the arroyo as it drove headlong into a region of rather steep, high ridges.

Many years ago Jonathan had canoed a section of one of the big rivers that flowed into James Bay, located at the southern tip of Hudson Bay. His college roommate had enlisted a couple of his friends in the expedition, and of course Jonathan was all in. The four of them were only on the river for two weeks, but in that time they forgot about their old lives and became accustomed to the Ontario wilderness: the precambrian granite outcrops, the towering pines, the surging whitewater rapids, the reflections of sunlight on the rippling waters, the splash of otters swimming alongside the boats, the sudden appearance of deer wading in the shallows at dusk. They bonded with the wilderness and became one with it.

Then one day they rounded a bend in the river and the highway bridge came into view, two hump-backed, steel-girder arches supported by three rectangular concrete abutments. The spans were a regular pattern of supports and trusses, all tightly riveted together. The canoeists were stunned by the precise configuration of the lines, so unlike anything they had seen for the past two weeks. No one spoke a word. They just drifted along with the current, staring in disbelief at the outrageous monstrosity coming toward them. None of them were marveling at the amazing engineering feat that had been involved in its construction. They were shocked by the sudden appearance of mathematics in a world that had no mathematics. They realized for the first time in their lives just how strange geometry really was.

They docked at the landing and pulled the canoes onto the shore. Their '52 chevy pickup was parked nearby waiting for them. They threw their gear into the back, hoisted the canoes onto the racks, and started driving down the lonely, deserted highway. Not a single word was uttered by any of them. They sat in complete silence, the four of them compressed together on the front seat, staring blankly out the windshield with the same dumbfounded looks on their faces, the same thoughts running through their heads. They were leaving the incredible beauty of the wilderness behind, trading it in for a mathematical world of flat surfaces and straight lines. They all wanted to turn around and plunge themselves back into the irregularities of nature. They wanted to shun this abrasive, angular world of walls and sidewalks and flee the striking mathematical shapes of highways and bridges.

The four of them arrived at a roadside cafe for some food. Jonathan had stupidly begun the journey with already well-worn blue jeans. The acidity of the river water ate them away during the course of his daily soakings and by the end of the trip they were rags. He had continually sewn them together with fishing line as they fell apart, but it had been a losing battle. With tattered clothes, disheveled hair, and a wild look in his eyes, Jonathan was quite a sight.

They all sat down at a table, still not speaking. A waitress came running over with menus and in a cheery voice exclaimed: "Oh, you boys just come in from the bush?" No one answered her. The waitress directed her gaze from one to the other with a smile on her face. She understood. "Can I get you boys some coffee?" Jonathan's canoe partner finally mustered a response: "Sure." The waitress took that to mean for everyone and promptly went to get four mugs and a coffee pot. The menu offered a hearty breakfast designed specifically for people like them who had been out in the backcountry for an extended period of time, a generous sampling of everything else on the menu. They all recognized this selection as the obvious choice. The shock started to wear off and their voices slowly returned. They remembered their former lives, really not that long ago, when they had coexisted with mathematical forms and thought nothing of it.

The trip was over and none of them were continuing on down the river, although Jonathan and his canoe partner had expressed an interest in doing so in front of the rest of the group. The others replied that they had responsibilities at home. The plan wasn't practical. They were right. Jonathan and his partner just figured that they'd come back at a later date and resume their adventure sometime in the future, but Jonathan left town and that was the end of it for him. He never heard from any of those guys again.

The physicist claimed that mathematical forms had existed in primitive cultures prior to the formal development of mathematics. The equally spaced teepees of an Indian village spread out on a grassy field were examples of such forms, the cones that gave rise to the conic sections of Euclidean geometry. Jonathan pointed out that they were only approximately cones, rather hazy and blurred images of the precise, mathematical shapes, as if seen indistinctly through a fog. The roughness of their construction caused the teepees to blend in somewhat with the more erratic forms of nature that surrounded them. The physicist could bring the rather nebulous mathematics of the teepees into focus by constructing more precise cones out of plastic or steel, structures that more closely followed the exact designs of the ideas themselves, but this would only accentuate the starkness of the mathematics and expose its total foreignness. The absolute uniformity of perfectly smooth cones would be entirely out of place in this world. Such objects didn't fit in at all with the abundant works of nature. The fact that they didn't belong here was obvious to everyone, regardless of their backgrounds and personal perspectives, because mathematics was incongruous with the physical world, a jolting experience, the sudden appearance of the unexpected, a puzzling enigma. Jonathan's first reaction to perfect metal cones standing in a field would be: "What in the hell are those things? Where did they come from and what are they doing there?"

Mathematical thinking had a peculiar logic that embraced certain perspectives and orientations. It employed specific conventions and procedures. Far from being a natural way of looking at things, mathematical thinking routinely engaged in mental operations that were far-fetched, and given what Jonathan knew to be true about the world, they were moves that could not be justified. When the world was left out of the picture, mathematics became its own justification, a body of thought that was self-contained and self-determined. The physicist exclaimed "Let's just think about everything in mathematical terms," yet in doing so, he was not only shaping a world view, but creating an entirely new world in his mind. This spectral world of vacant ideas had no connection to the physical world because the only ties it had were mathematical in nature and mathematics connected things that weren't connected, that is, the supposed ties didn't actually exist in the world, but were artificially placed there using fanciful mathematical imagery. The physicist was trapped in his own mathematical universe yet he didn't realize it because he had constructed mathematical bridges to material realities and felt free to use them whenever the urge struck him, confident that they were not simply mirages. In truth he could reach the world only by walking on air, by putting all of his weight on airy abstractions and visualizations—the very same ones he had begun employing at a young age when he was sitting in a classroom staring at the abyss of nothingness commonly known as mathematics.

Physicists just started reasoning mathematically without ever questioning this approach. They began looking at the world through the concepts of mathematics as if that were the most natural thing in the world, what any normal person would automatically do in this situation, when in fact it was the most outrageous step, something that no one in their right mind would ever seriously contemplate. Mathematics changed reality and transformed it into something very different. Just because physicists saw these new forms as being highly desirable didn't justify the move. Sure, they got what they wanted—but what happened to the world in the process?

Physicists created precision in the laboratory for no other reason than to mimic the relationships of mathematics, to make the world of mathematical ideas come to life, to cause this invisible kingdom to become incarnate on earth—even though it had no place among the phenomena of nature. The laboratory stood out as a glaring discrepancy, similar to the arrival of an extraterrestrial spaceship in a science fiction movie, an alien environment controlled by otherworldly beings who had come to live among us and transform society, as well as the desert, into something it was not. Just as the fearsome race of beings from outer space depicted in the movie, physicists had come to institute a new world order of their own, a world of mathematical forms and relationships. Such designs were utterly foreign to nature. Nature had created itself to be what it was, and not to be something else.

Jonathan wondered what a truly mathematical world would look like. The atmosphere, rather than a chaotic hodgepodge of molecules, would be a crystal latticework of precise angles and distances. The ground, rather than a tangle of rocks and plants, would be a precise mathematical surface with geometrical forms fitting together perfectly in symmetrical constellations. A thoroughly mathematical world was hard to imagine and perhaps even a nonsensical idea. Exact algebraic ratios would have to exist between every two aspects of every two objects in every conceivable relationship. An infinite particularity could not be corralled in this way, made to appear mathematical in terms of every possible interconnection and in every detail, so at best physicists were only talking about a partially mathematical world, a hybrid of mathematical and non-mathematical elements fused together, or more realistically, a slightly mathematical world with a few precise, numerical relationships scattered here and there.

Modern societies had made great strides in constructing cities with geometries incorporated into them, streets and highways with parallel lanes laid out on a rectangular grids, buildings with planar walls and the floors stacked vertically with a uniform spacing—well, the list went on and on. The cities were a long way from being mathematical utopias, however, and there was still much work left to be done. Laboratories had also done their share in creating a mathematical world, but here too the project had only just begun.

The thing Jonathan found most shocking and disturbing in his college studies was that all the ratios and formulas in physics textbooks were merely approximations, and in a truly mathematical world that wouldn't be the case. In a mathematical world, every relationship would be exactly true, just as within mathematics itself. Physicists were working constantly to minimize the discrepancies by complicating their theories, and reducing the errors to insignificant levels at least in the controlled environments of their laboratories, but as soon as they moved out into the world-at-large the task became utterly hopeless. As soon as the physicist stepped out the door he was looking at the opposite extreme: a world with no mathematics whatsoever, a world that could not even be mathematized to any significant extent, a mishmash of objects and intractable events all behaving badly. He could make the objects and events behave properly in the laboratory, just as the classroom teacher could force the students to sit quietly at their desks and listen attentively to their lessons, but once the students were released into the schoolyard, pandemonium broke loose with a cacophony of screeching and yelling and a furious three-ring circus of running, jumping and fighting. Tell me, Jonathan said, which of these scenarios fits the real world? The answer was not hard to find.

Physicists were still holding out on the possibility that they could straighten everything out in the laboratory, but they were pining their hopes on a falsehood. They were waiting for the second coming of Einstein, or rather, optimistically anticipating "The Ultimate Einstein" in all his glory, coming to save the equations and usher in the Kingdom of Mathematics to the weary, frustrated calculators. Jonathan was willing to wager that Jesus Christ would return long before that and usher in his Kingdom of Heaven to the poor and the righteous. Physics was based on faith, an unquestioned belief in the triumph of mathematics over chaos. Jonathan was placing his bet on the world. He saw the world everyday and he could see that the world was way too much for mathematics. He also felt that evil would win out over the feeble forces of good. Jesus didn't stand a chance in this vile, corrupt world. Jesus had no more hope of reigning over the earthly empire than had the physicist. Jonathan wasn't rooting for one side or the other—the world against these purveyors of ideologies. The contest wasn't fair, even given the inspirational story of David and Goliath. The opposition was utterly overwhelming. Surely the physicist could see that, as could the preacher standing in his pulpit. Everyone wanted to believe in rationality and goodness so they kept stringing us along. We told ourselves that the day would soon come when the world would be righted, even though deep down inside we knew that this would never happen.

So did this mean that physics was yet another religion? The answer depended on how physicists wanted to think about physics. If they believed that mathematics was somehow built into the universe and that they were working in their laboratories to uncover a hidden realm of deliberate mathematics, then yes, physics had many of the qualities of a typical religion. Physicists imagined that they were surrounded by an invisible world of entities that controlled the outcome of events, much as a shaman saw a world populated by spiritual allies and evil forces, ideas that acted in the world and existed on their own. But if physicists realized that mathematics had nothing to do with nature, then no, because in that case they must surely know they were just playing games with numbers.

Physicists often talked about doing great things, regardless of their philosophical views on the nature of physics. The prolific science fiction writer Arthur C. Clarke laid out the mythology of physics quite nicely. He wrote stories about the future of mankind showing that someday physicists would be able to transfer consciousness onto the copper wires and silicon wafers of computer circuit boards, and then eventually onto the fabric of space itself. With this act, physicists would grant everyone immortality by turning them into disembodied spirits. Physicists currently had no idea what consciousness was or how this feat could possibly be accomplished, yet the greatest stumbling block appeared to be the fact that the fabric of space was just a mathematical concept and that it didn't exist. But if Clarke was saying that physicists were going to ultimately turn everyone into mathematical entities so they could share in the eternal nature of mathematical truths and mingle with the laws of physics directly, and thereby become one with the numerical godhead, then we had a clear case of typical religious folklore. Although these goals were not the officially sanctioned objectives of physics, many physicists had similarly rosy ideas about the future of their program.

Other people were skeptical of the promises made by these advocates of the new science, yet true believers gloated about the accomplishments that had been made so far, pointing out that the skeptics had been wrong all along, and this optimism engendered a believe in the attainment of a promised land, a heavenly realm of perfection where everyone could live in complete bliss. C'mon, don't say this wasn't in the back of the minds of a majority of physicists, and also deeply ingrained in the popular consciousness. Life will be great once the physicists discover all of the mathematical laws of the universe and then create an artificial world based entirely on these principles. But could people look to physicists to be their saviors? Would the streets of heaven be paved with mathematical entities instead of gold, or were these just empty promises, as in the case of traditional religions? Was physics anything other than the worship of mathematics and a reverence for the mathematical constructions that transformed the world into an ideal place of pure rationality, thereby imparting a surrealistic and otherworldly character to an ordinarily mundane physical existence?

The physicist imagined that he would just keep tracing outlines until eventually he had a complete picture of reality. The omitted parts, infinite in character and composition, would gradually be penciled over and blotted out, and at long last he would have cornered reality and tied it down with imaginary strings—or at least made it go away underneath all the scratch marks. He'd had a certain amount of success with this approach in the laboratory, and if he had no intention of extending his method beyond the narrow confines of his workshop, then maybe he wouldn't have a problem with it, other than the customary struggles and strains of trying to fit theory to experiment. But outside his window the world exploded into a vast landscape where the intricacies of the designs went on and on all the way to the horizon and beyond and where the hopelessness of the project was quite clear. "That's alright," he said, "I'll just draw the curtains. I'll get to all of that stuff later."

He was convinced that sometime in the future a day would come when physicists would put their pencils down, the glorious day when they would ride on magnificent steeds through the center of their new empire in a triumphant victory parade, brandishing the flags and banners of the fundamental constants and the essential ratios, leading the grand celebration marking the long-awaited arrival of the wonderful mathematical universe. The self-imposed rulers of this new world order would gratefully soak up the adoration of the general public and strut their stuff before the uproarious cheers of the assembled masses who really had no idea what the physicists were talking about. Perhaps the physicists should stop dreaming and take another look out the window.

Physicists thought that someday they would conquer the world, but empires rose and fell and many things were not what they seemed. The world wasn't under anyone's control. Jonathan did not fear the day when machines would take over the world and then as our masters treat us badly. The real problem was that the machines we had built would not help us when the world suddenly turned on us and destroyed everything we had done. The dominion of our machines over the natural world was an illusion, a temporary sway in the ebb and flow of human beings and their environment, the tides of war. The balance of power would inevitably shift and the hard-fought territory we had gained would be reclaimed by its original owner. Machines were not gods—just look at Jonathan's poor old truck standing in the tall weeds all dented and scratched, a pathetic assembly of dirty, worn-out pieces of metal and plastic.

But physicists thought that in building their complex machines they had acquired some understanding of nature's mysteries, however that too was an illusion. They had merely forced nature to behave in mathematical ways by molding objects and events into mathematical forms. Nature didn't ordinarily behave that way, as anyone out here in the desert could plainly see. The mathematics they had obtained was all just fakery. Sooner or later the enigma of existence would return in full force and become the final victor, because an infinite particularity could never be understood with mathematical ideas—or any other ideas for that matter.

The day came when Jonathan also put his pencil down, not because he had completed the monumental task of mathematizing the entire universe, but because he had suddenly realized the utter futility of such an undertaking. Mathematics was trivial and could never be more than mere appearance, meaning that physics wasn't anything other than a game played with the sums and products of numbers and the truth values of propositions. Mathematics was in no way the essence of existence or the ultimate reality of the physical universe. He saw no point in going further with any of this.

But the physicist was so in love with his beautiful mathematics that he not only put her on a pedestal, but on a throne, bestowing her with the royal title of "Queen of the Universe." As he laid out the derivations to his equations, he followed her down paths of logical reasoning, worshipping the very ground she walked on, blind to the many serious shortcomings in her character, and the betrayal that would surely one day come. She was frivolous and superficial. She was demanding and took control of everything. She was all looks and no substance, and these were not the qualities that people generally held in high esteem, or the traits that led to a stable, permanent marriage. The physicist wasn't going to come to his senses anytime soon. He hadn't realized yet that his love affair with mathematics was doomed.

Jonathan was enthralled with another damsel altogether, one who shared no ancestry with the sweetheart of the mathematicians, a corporeal woman of many moods who was often difficult to deal with, prone to outrages and tantrums, frequently fickle and disobedient, a whirlwind of conflicting impulses. She was dangerous and unpredictable, and one day she might turn on him and murder him in cold blood without the slightest provocation or cause. No one could tame that shrew, yet he loved her just the same. He accepted her for who she was and didn't want to change her. He sought only to embrace his wonderful desert as often as he could. He took every opportunity to be with her and walk alongside her in order to admire her infinite loveliness. He wanted to discover all of her secrets for himself, that one day he might understand what she was all about.

Perhaps Jonathan had been blinded by his love of the desert as much as the physicist had been blinded by his love of mathematics. Each one had deluded himself into thinking generous and flattering thoughts about his beloved, each had become captivated with his own little darling, failing to see clearly that in doing so he had lost touch with reality. But the physicist was trapped inside a mathematical world of his own making while Jonathan was free to roam, traveling unchecked through both the philosophical and terrestrial landscapes, not locked into any way of thinking or tied down to any particular locale. After all these years, Jonathan was still a free spirit who had never planted his flag and never settled down. He had no stake in the outcome of his investigations and the world could be whatever it was. This granted him an objectivity that the physicist couldn't even imagine.

The physicist was obliged to turn everything into mathematical forms because that was the only thing he knew how to do. It was all he had. The physicist was stuck in his laboratory, his dreary little existence, while Jonathan had all of the great outdoors at his disposal.

Out in the badlands Jonathan could live free, safe from the dictates of society and the oppressive laws of physics. He shared with nature the ability to do as he pleased whenever the urge struck him. He could create his life as he saw fit, not bound by strict rules and regulations, not tied to the predictable outcomes of controlled experiments, just as nature created unusual storms and calms, strange winds and rains, one-of-a-kind sunrises and sunsets. The world was wide and anything was possible. Physicists would never find him here, happily lost in the wilderness, resting in the shadow of a juniper, sauntering across the bottomland of a narrow arroyo cut beneath the rolling hills, or climbing a steep trail under a dense canopy of tall pines, wandering aimlessly for many miles in the cool mountain air. Emma could run with him as long as she was able to move her legs and hold her nose to the wind. Her company would always be welcome, until that day when he laid her to rest in a shallow grave on some lonesome hillside, a place he had chosen especially for her, so that her everlasting spirit could always have a spectacular view of the desert she had so dearly loved, something Jonathan had done several times in the past for his previous companions, now the ghost dogs who once shared with him the joy of a glorious sunrise.

Emma galloped up a particularly tall ridge, then wheeled about to look at Jonathan from the crest. She reminded him of the Lone Ranger when he raised a hand in salute to the crowd assembled on the streets below, a farewell to the townspeople he had just helped, before riding off into the desert on his mission of doing good for those less fortunate than himself. Jonathan waved back at Emma. She saw the gesture and disappeared over the ridge. He wondered if he would ever see her again.

He'd probably catch up with her soon enough in some grassy arroyo or on some rocky ridge. Following the hero on wild adventures and being the sidekick wasn't so bad, and he didn't mind playing the part. The sidekick was the one who saved the Lone Ranger when he got into trouble, the one who backed him up when he was outnumbered by the outlaws and bandits. Jonathan was here in a supporting role only because Emma was the star of the show, the magnetic personality that drew others toward her. He was grateful for his privileged position as her trusted cohort, the one she turned to when times got tough, the one who was always there for her when she needed him. He spoke softly to himself as he bore down on the steep grade, "Hold on, Emma, I'm coming," and with measured steps he walked upward into the unknown dangers that lay ahead.

www.ingramcontent.com/pod-product-compliance
Lightning Source LLC
Chambersburg PA
CBHW070525220526
45467CB00003B/843